# Smart Innovation, Systems and Technologies

## Volume 40

**Series editors**

Robert J. Howlett, KES International, Shoreham-by-Sea, UK
e-mail: rjhowlett@kesinternational.org

Lakhmi C. Jain, University of Canberra, Canberra, Australia, and
University of South Australia, Adelaide, Australia
e-mail: Lakhmi.jain@unisa.edu.au

## About this Series

The Smart Innovation, Systems and Technologies book series encompasses the topics of knowledge, intelligence, innovation and sustainability. The aim of the series is to make available a platform for the publication of books on all aspects of single and multi-disciplinary research on these themes in order to make the latest results available in a readily-accessible form. Volumes on interdisciplinary research combining two or more of these areas is particularly sought.

The series covers systems and paradigms that employ knowledge and intelligence in a broad sense. Its scope is systems having embedded knowledge and intelligence, which may be applied to the solution of world problems in industry, the environment and the community. It also focusses on the knowledge-transfer methodologies and innovation strategies employed to make this happen effectively. The combination of intelligent systems tools and a broad range of applications introduces a need for a synergy of disciplines from science, technology, business and the humanities. The series will include conference proceedings, edited collections, monographs, handbooks, reference books, and other relevant types of book in areas of science and technology where smart systems and technologies can offer innovative solutions.

High quality content is an essential feature for all book proposals accepted for the series. It is expected that editors of all accepted volumes will ensure that contributions are subjected to an appropriate level of reviewing process and adhere to KES quality principles.

More information about this series at http://www.springer.com/series/8767

Ernesto Damiani · Robert J. Howlett
Lakhmi C. Jain · Luigi Gallo
Giuseppe De Pietro
Editors

# Intelligent Interactive Multimedia Systems and Services

 Springer

*Editors*
Ernesto Damiani
Department of Information Technology
University of Milan
Crema
Italy

Robert J. Howlett
KES International
Shoreham-by-Sea
UK

Lakhmi C. Jain
Faculty of Education, Science,
    Technology and Mathematics
University of Canberra
Canberra
Australia

Luigi Gallo
Institute for High-Performance Computing
    and Networking (CNR-ICAR)
National Research Council of Italy
Naples
Italy

Giuseppe De Pietro
Institute for High-Performance Computing
    and Networking (CNR-ICAR)
National Research Council of Italy
Naples
Italy

ISSN 2190-3018          ISSN 2190-3026  (electronic)
Smart Innovation, Systems and Technologies
ISBN 978-3-319-38642-3      ISBN 978-3-319-19830-9  (eBook)
DOI 10.1007/978-3-319-19830-9

# Preface

Dear Readers,

We introduce to you a series of carefully selected papers presented during the 8th KES International Conference on Intelligent Interactive Multimedia Systems and Services (IIMSS-15).

At a time when computers are more widespread than ever and computer users range from highly qualified scientists to non-computer expert professionals, Intelligent Interactive Systems are becoming a necessity in modern computer systems. The solution of "one-fits-all" is no longer applicable to wide ranges of users of various backgrounds and needs. Therefore, one important goal of many intelligent interactive systems is dynamic personalization and adaptivity to users. Multimedia Systems refer to the coordinated storage, processing, transmission, and retrieval of multiple forms of information, such as audio, image, video, animation, graphics, and text. The growth rate of multimedia services has become explosive, as technological progress matches consumer needs for content.

The IIMSS-15 conference took place as part of the Smart Digital Futures 2015 multi-theme conference, which groups AMSTA-15, IDT-15 and SEEL-15 with IIMSS-15 in one venue. It was a forum for researchers and scientists to share work and experiences on intelligent interactive systems and on multimedia systems and services. It included a general track and three invited sessions.

The general track (pp. 1–140) focuses on issues ranging from intelligent image or video storage, retrieval, transmission, and analysis, to knowledge technologies and smart, knowledge-based applications. The invited session "Intelligent Video Processing and Transmission Systems" (pp. 141–250) specifically focuses on functionalities and architectures of systems for video processing and transmission. The invited session "Autonomous System" (pp. 251–312) considers issues such as ethical, legislation, and regulatory matters for what concerns mobile robots and autonomous systems. Finally, the invited session "Innovative Information Services for Advanced Knowledge Activity" (pp. 313–378) focuses on advanced functionalities for information and knowledge-based services.

Our gratitude goes to many people who have greatly contributed to putting together a fine scientific program and exciting social events for IIMSS 2015. We acknowledge the commitment and hard work of the program chairs and of the invited session organizers. They have kept the scientific program in focus and made the discussions interesting and valuable. We recognize the excellent job done by the program committee members and the extra reviewers. They evaluated all the papers on a very tight time schedule. We are grateful for their dedication and contributions. We could not have done it without them. More importantly, we thank the authors for submitting and trusting their work to the IIMSS conference.

We hope that readers will find in this book an interesting source of knowledge in fundamental and applied facets of intelligent interactive multimedia and, maybe, even some motivation for further research.

Crema, Italy                                                        Ernesto Damiani
Shoreham-by-Sea, UK                                              Robert J. Howlett
Canberra, Australia                                                  Lakhmi C. Jain
Naples, Italy                                                           Luigi Gallo
Naples, Italy                                                   Giuseppe De Pietro

# Organization

## Honorary Chairs

T. Watanabe, *Nagoya University, Japan*
Lakhmi C. Jain, *University of South Australia, Australia*

## Co-general Chairs

Ernesto Damiani, *University of Milan, Italy*
H. Kosch, *University of Passau, Germany*

## Executive Chair

Robert J. Howlett, *University of Bournemouth, UK*

## Program Chairs

Giuseppe De Pietro, *CNR-ICAR, Italy*
Gwanggil Jeon, *Incheon National University, Korea*

# Publicity Chairs

Luigi Gallo, *CNR-ICAR, Italy*
Fulvio Frati, *University of Milan, Italy*

# Invited Session Chairs

## Intelligent Video Processing and Transmission Systems

Margarita N. Favorskaya, *Siberian State Aerospace University, Russia*
Lakhmi C. Jain, *University of South Australia, Australia*
Mikhail Sergeev, *SUAI, Russia*

## Autonomous System

Milan Simic, *RMIT University, Australia*
Reza Nakhaie Jazar, *RMIT University, Australia*

## Innovative Information Services for Advanced Knowledge Activity

Koichi Asakura, *Daido University, Japan*
Toyohide Watanabe, *Nagoya Industrial Research Institute, Japan*

# International Program Committee

Marco Anisetti, *University of Milan, Italy*
Koichi Asakura, *Daido University, Japan*
Vivek Bannore, *Brisbane, Australia*
Monica Bianchini, *University of Siena, Italy*
Dumitru Dan Burdescu, *University of Craiova, Romania*
Mario Döller, *University of Applied Sciences Kufstein Tirol, Germany*
Dinu Dragan, *University of Novi Sad, Serbia*
Gianluca Elia, *University of Salento, Italy*
Colette Faucher, *LSIS, UMR CNRS 7296, Marseille, France*
Margarita N. Favorskaya, *Siberian State Aerospace University, Russia*
Abdelaziz El Fazziki, *Cadi Ayyad University of Marrakech, Morocco*
Christos Grecos, *Independent Imaging Consultant, USA*
Hsiang-Cheh Huang, *National University of Kaohsiung, Taiwan*
Lakhmi C. Jain, *University of South Australia, Australia*
Dimitris Kanellopoulos, *Department of Mathematics, University of Patras, Greece*
Mustafa Asim Kazancigil, *Yeditepe University, Istanbul, Turkey*
Roumen Kountchev, *Technical University of Sofia, Bulgaria*
Marcello Leida, *Taiger, Spain*

# Contents

# Touchless Target Selection Techniques for Wearable Augmented Reality Systems

**Nadia Brancati, Giuseppe Caggianese, Maria Frucci, Luigi Gallo and Pietro Neroni**

**Abstract** The paper deals with target selection techniques for wearable augmented reality systems. In particular, we focus on the three techniques most commonly used in distant freehand pointing and clicking on large displays: wait to click, air tap and thumb trigger. The paper details the design of the techniques for a touchless augmented reality interface and provides the results of a preliminary usability evaluation carried out in out-of-lab settings.

**Keywords** Touchless interface · Freehand pointing and clicking · Wearable augmented reality · Wait to click · Air tap · Thumb trigger

## 1 Introduction and Background

In recent years, we have witnessed a rapid evolution of wearable augmented reality (AR) technologies. In particular, smart glasses, and monocular or binocular AR displays now allow users to visualize computer-generated information directly in their field of view (FOV), without having to wear cumbersome head mounted displays. Accordingly, these systems can be profitably used in many different application

N. Brancati · G. Caggianese · M. Frucci · L. Gallo (✉) · P. Neroni
Institute for High Performance Computing and Networking,
National Research Council of Italy (ICAR-CNR), Naples, Italy
e-mail: luigi.gallo@na.icar.cnr.it

N. Brancati
e-mail: nadia.brancati@na.icar.cnr.it

G. Caggianese
e-mail: giuseppe.caggianese@na.icar.cnr.it

M. Frucci
e-mail: maria.frucci@na.icar.cnr.it

P. Neroni
e-mail: pietro.neroni@na.icar.cnr.it

© Springer International Publishing Switzerland 2015
E. Damiani et al. (eds.), *Intelligent Interactive Multimedia Systems and Services*,
Smart Innovation, Systems and Technologies 40,
DOI 10.1007/978-3-319-19830-9_1

areas, such as entertainment, marketing, education, training, navigation and tourism, mainly because the device becomes invisible to the user.

The possibility of mixing real and virtual objects in a single space also demands more natural interaction approaches. Nowadays, interaction techniques for wearable AR systems have been mainly tailored to the specific application. The most commonly used interaction modalities exploit vocal control, touch control or a combination of the two (e.g., the Google Glass project [15], in which the user interacts by means of a touchpad embedded in the rim of the glasses and/or by using vocal commands to execute specific actions). Nevertheless, achieving a direct manipulation through a point-and-click paradigm could be an interesting alternative, especially if the user does not need to wear any specific device to control the system. In fact, in wearable AR systems the computer-generated information visualized in the user's FOV can easily become dense, and in this case selecting a target via pointing and clicking is more practical than by using vocal control or a touchpad.

**Fig. 1** The user's FOV is augmented with a set of POIs, which can be selected via a touchless point-and-click interface

Different clicking approaches have been developed to interact with virtual content. Most techniques require the user to wear instrumented gloves [7, 12, 14], although some exploit depth or color cameras to track the hands [1, 13]. However, virtual object manipulation when using head mounted displays (HMD) presents some similarities with distant interaction on large displays; in fact, the augmented environment is visualized as if it were a large screen projected at a certain distance from the user.

In this paper we analyse, by tracking the hand with a near-range depth camera [11], three free-hand pointing and clicking techniques frequently used to interact with large displays, in order to evaluate their suitability for wearable AR systems. In more detail, we present an implementation of a touchless point-and-click interface for a cultural heritage AR system [4], whose objective is to allow users to navigate within the tangible cultural heritage. As already achieved for a cultural exhibition [5], the AR system exploits the user's position to show only the relevant information about the surrounding buildings and monuments that is manipulable by using hand gestures (see Fig. 1). Three target selection techniques are described and evaluated in an AR context: Wait to Click, Air Tap and Thumb Trigger [17]. In the paper, we detail the design of theese techniques, adapted to the specific application in which the depth sensor is worn by the user, describe the problems encountered and discuss the results of a preliminary qualitative evaluation.

**Fig. 2** The wearable AR system

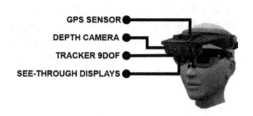

The rest of the paper is structured as follows. In Sect. 2, we describe the wearable AR system. In Sect. 3, we describe the three target selection techniques and the main problems encountered in implementing them in the AR system. After that, Sect. 4 focuses on the qualitative evaluation tests performed, providing a discussion of the results. Finally, in Sect. 5 we present our conclusions.

## 2 The Wearable AR System

The wearable AR system integrates a wearable AR display and low consumption inertial and depth sensors. In more detail, the home-made wearable smart glasses (see Fig. 2) have been realized by integrating the Vuzix STAR 1200XLD, an augmented reality eyewear equipped with a binocular see-through display with a FOV of 35° diagonal, together with the time-of-flight camera Softkinetic Depth Sense 325, which is light enough to be worn on top of the glasses. To track the position and gaze direction of the user, we have further used a GPS sensor, a magnetometer and a gyroscope.

The AR application allows a citizen/tourist to visualize georegistered points of interest (POI) relative to cultural, historical and tourist information, overlaid on the real world view. These POIs are retrieved by using web services that exploit both the user's position and her/his gaze direction. In this way, only the POIs nearest to the user are visualized. Moreover, the POIs can be further filtered by type and distance from the user. The system allows the user to interact with the POIs by using touchless point-and-click interface. The user can select each POI to visualize additional information about it or navigation aids to reach it, which are superimposed on her real world view.

When the user executes the pointing gesture (all the fingers closed except for the index finger), the system visualizes a virtual pointer over the fingertip position. The virtual pointer is visualized close to the tip but not always exactly on it since a velocity-based pointing enhancing technique has been used [6] to smooth the pointer movements and to simplify the selection of small (i.e. far) POIs in the scene. To use the pointing metaphor in a virtual space which is bigger than the working space of the user, defined as the area reachable with an extended arm, we have used the ray casting technique [10]. In the ray casting approach the pointer controlled by the user is connected to a virtual ray that works like a picking ray, since each time it intersects a virtual object this object is returned to the system as the selected one. When the user selects a POI, it is highlighted so as to provide her/him with a visual feedback. To allow the user to select a POI in the augmented scene, three different target selection techniques have been developed, namely Wait to Click, Air Tap and Thumb Trigger, which are described in the next Section.

**Fig. 3** The Wait to Click clicking procedure: in positioning the pointer is moved on the POI that is highlighted; in press the pointer is still on the POI for more than $t_{press}$ seconds; in release the pointer is still on the POI for more than $t_{release}$ seconds (Color figure online)

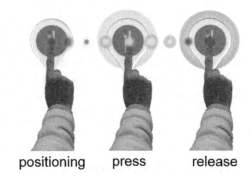

positioning      press      release

## 3 Clicking Techniques

### 3.1 Wait to Click

In the Wait to Click technique, the user has to keep the pointer (controlled with her/his index finger) over the object she/he would like to click on, for a predefined time. The main advantage of this technique is that it is not affected by the Heisenberg effect [2], since the user does not need to execute any movement of the hand to confirm the selection. However, it is affected by the Midas Touch effect [8], since the confirmation task can result in an unwanted selection. In order to avoid this problem, we provide the user with a visual feedback to better inform her/him about the virtual object currently selected (see Fig. 3).

The choice of the amount of time required to trigger the confirmation of the selection becomes an important issue. In fact, with an excessively short time window, the user does not have enough time to realize her/his selection; conversely, an excessively long time window results in more fatigue for the user. We set two time thresholds: if the user keeps her/his index finger over an object for $t_{press}$ seconds, a selection event is triggered; then to confirm the selection, her/his index finger must remain still on the object for $t_{release}$ seconds, when a release event is generated. In addition, the time required to trigger the click action is also associated with a visual feedback, consisting in a change of the pointer colour gradually from green to red.

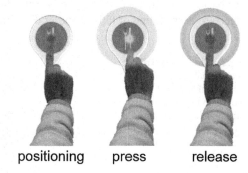

**Fig. 4** The Air Tap clicking procedure: in positioning the pointer is moved on the POI that is highlighted; in press the index finger is moved quickly down; in release the index finger is moved quickly back to its start position (Color figure online)

positioning      press      release

## 3.2 Air Tap

The Air Tap clicking scheme is similar to the one used when clicking a mouse button. Usually, in this technique, the cursor is connected to the user's palm position instead of the index finger position in order to avoid the Heisenberg effect. In fact, moving the fingertip down to trigger a selection affects the position of the fingertip, and so also that of the pointer.

However, such a solution cannot be used in wearable AR applications. Placing the pointer on the centre of the hand would result in occlusion problems, since the targeted object would be covered by the user's hand. For these reasons in our implementation we lock the X,Y position of the fingertip by analysing when the fingertip's velocity along the Z axis becomes higher than a threshold (see Fig. 4). We use two velocity thresholds, $v_{press}$ and $v_{release}$: once the fingertip velocity $v$ measured along the Z axis is higher then $v_{press}$, the press event is triggered; when $v$ is lower than $v_{release}$, a release event is activated. If the index fingertip velocity generates a press followed by a release in a short period of time, the target selection action is performed.

## 3.3 *Thumb Trigger*

In the thumb trigger technique, the click event is triggered when the thumb is moved in and out towards the index finger side of the hand (Fig. 5). In our implementation, we fix two distance thresholds: bringing the thumb close to index finger generates a press event and moving it away generates a release event. Similarly to Air Tap, if the thumb movement generates a press and release event in a short period of time, the click action is performed.

Theoretically, this click style has a distinct advantage over the above clicking techniques since the thumb by touching the side of the hand provides a kinaesthetic feedback. As a drawback, the need to use also the thumb, may cause occlusion problems besides a feeling of fatigue and/or discomfort.

**Fig. 5** The Thumb Trigger clicking procedure: in selection the pointer is moved on the POI; in press the thumb finger is moved quickly to touch the side of the hand; in release the thumb finger is moved quickly back to its start position (Color figure online)

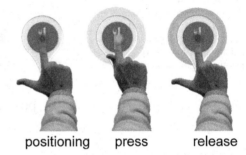

positioning    press    release

## 4 User Study

### 4.1 *Experimental Procedure*

We recruited 10 unpaid volunteers to perform the study. The age of the participants ranged from 28 to 41 years old, with an average age of 31.8. Four of the participants were female and all of them were right-handed.

The experiment was performed by using a single pair of smart glasses with the same simulated scenario for all the users, guaranteeing the same set-up and conditions for each tester. The simulated scenario was designed for an open space and populated by POIs with different sized circular areas in order to simulate different distances from the user. Moreover, since the chosen domain of use dense of information, we arranged for the POIs of the simulation to be organized similarly to a normal scenario so that each participant was required to deal with the selection of both isolated and groups of POIs.

Before starting, a facilitator showed each volunteer a brief video to introduce the system and the main elements of the user interface. Specifically, the video was used to explain how to use the three proposed target selection techniques paying attention to the exact sequence of movements to confirm the selection of a POI. In this phase, the facilitator was allowed to answer participants' questions.

After the demonstration video, in a practice session designed to make the participants feel more comfortable and relaxed, the volunteers were left free to familiarize themselves with the interface without any limitations, being able to switch between the three different target selection techniques. By undertaking such practice, the participants increased their confidence in wearing the smart glasses, in dealing with AR environments, and with the three target selection techniques. In this phase the facilitator was only allowed to switch between the approaches if requested by the testers.

Next, the test session started with the facilitator giving precise instructions. The test session consisted in the repetition of two different and consecutive stages for each technique. In fact, each tester was first allowed to use a technique and then, systematically was asked to evaluate it by using a post-hoc questionnaire. No time limit was imposed on the participant during the use of the technique. In order to better gather information, each user's interaction was observed to collect all the possible impressions of her/his experience and, additionally she/he was asked to think aloud, describing her/his intentions and possible difficulties. In the same way, no time limit was imposed on the filling in of the proposed questionnaire. However, the users were asked to record the answer to each item as quickly as possible, rather than thinking about the items for a long time.

The questionnaire, used to measure both the usability of the technique and the user's experience, is a five point Likert-scale questionnaire [9] structured to fulfill all of the criteria listed by Uebersax [16]. The participants answered the questions in a scale from 1 (very low) to 5 (very high). In detail, the users were asked to complete the System Usability Scale (SUS) [3], which allows a rapid evaluation of the system interaction expressed as a single number which ranges from 0 to 100.

## 4.2 Results and Discussion

Our intention was to investigate the possibility of using target selection techniques, developed for interacting with large and high resolution displays, in wearable AR systems.

The majority of the participants complained about fatigue when using the Air Tap and the Thumb Trigger techniques, mainly due to their normal attitude of tilting down the hand to reproduce the normal gesture of pointing to an object. Especially for Air Tap this attitude caused a hight rate of wrong selections, forcing the tester to continuously correct the hand position. Most of the users expressed satisfaction with the kinaesthetic feedback provided by the Thumb Trigger technique (e.g., the thumb touching the side of hand). However, after a few minutes, some users reported a stiffening of the hand. Moreover, sometimes the selection mechanisms resulted in an unwanted variation of the cursor position.

The Wait to Click technique was the one most appreciated by the participants, mainly because of its ease of use. Although Wait to Click was supposed to require more time to trigger a selection action compared to the other two techniques, during the test we observed that this was not the case. In fact, when using the other two techniques, most of the users spent more time in correcting the pointer position, which was affected by the selection mechanisms.

These observations were also confirmed by the results of the SUS questionnaire. The average SUS score, whose values range from 0 to 100, was 82.45 for Wait to Click, 60.35 for Thumb Trigger and 43.5 for Air Tap.

## 5 Conclusion

Wearable AR technology is now rapidly evolving. Many start-up companies are releasing low-cost, wearable AR technologies that could have the potential to enter into everyday life. In this context, a challenge to face is the design of a truly natural way of interacting with augmented worlds.

In this paper, we have detailed the design of three different target selection techniques for wearable AR systems, specifically designed for the enjoyment of cultural heritage sites. We have also described the results of a preliminary usability evaluation performed with ten volunteers.

The results show that fatigue is the main challenge in touchless, mid-air interaction. Future work will focus on enhancing the interface by following the user's comments collected during the tests. Moreover, we intend to perform a quantitative evaluation with a larger group of users.

**Acknowledgments** The proposed techniques has been developed within the project OR.C.HE.S.T.R.A. - ORganization of Cultural HEritage for Smart Tour-ism and Real-time Accessibility, funded by the European Community, Regione Campania, the Ministry of Education Universities and Research (MIUR), and the Ministry of Economic Development, under the Call for Smart Cities and Communities and Social Innovation.

## References

1. Bader, T., Rpple, R., Beyerer, J.: Fast invariant contour-based classification of hand symbols for hci. In: Jiang, X., Petkov, N. (eds.) Computer Analysis of Images and Patterns. Lecture Notes in Computer Science, vol. 5702, pp. 689–696. Springer, Berlin (2009)
2. Bowman, D.A., Wingrave, C.A., Campbell, J.M., Ly, V.Q., Rhoton, C.J.: Novel uses of pinch gloves for virtual environment interaction techniques. Virtual Reality 6(3), 122–129 (2002)
3. Brooke, J.: Sus: a quick and dirty usability scale. In: Jordan, P.W., Weerdmeester, B., Thomas, A., Mclelland, I.L. (eds.) Usability Evaluation in Industry. Taylor and Francis, London (1996)

4. Caggianese, G., Neroni, P., Gallo, L.: Natural interaction and wearable augmented reality for the enjoyment of the cultural heritage in outdoor conditions. In: De Paolis, L.T., Mongelli, A. (eds.) Augmented and Virtual Reality. Lecture Notes in Computer Science, pp. 267–282. Springer International Publishing, Cham (2014)

5. Chianese, A., Moscato, V., Piccialli, F., Valente, I.: A location-based smart application applied to cultural heritage environments. In: 22nd Italian Symposium on Advanced Database Systems, SEBD 2014, Sorrento Coast, Italy, 16–18 June 2014, pp. 335–344 (2014)

6. Gallo, L., Minutolo, A.: Design and comparative evaluation of smoothed pointing: a velocity-oriented remote pointing enhancement technique. Int. J. Hum. Comput. Stud. **70**(4), 287–300 (2012)

7. Gallo, L.: A glove-based interface for 3d medical image visualization. In: Intelligent Interactive Multimedia Systems and Services, pp. 221–230. Springer, Berlin (2010)

8. Jacob, R.J.K., Leggett, J.J., Myers, B.A., Pausch, R.: An agenda for human-computer interaction research: interaction styles and input/output devices (1993)

9. Likert, R.: A technique for the measurement of attitudes. Arch. Psychol. **22**, 1–5 (1932)

10. Mine, M.R.: Virtual environment interaction techniques. University of North Carolina at Chapel Hill, Chapel Hill, NC, USA, Technical report (1995)

11. Park, G., Ha, T., Woo, W.: Hand tracking with a near-range depth camera for virtual object manipulation in an wearable augmented reality. In: Shumaker, R., Lackey, S. (eds.) Virtual, Augmented and Mixed Reality. Designing and Developing Virtual and Augmented Environments. Lecture Notes in Computer Science, vol. 8525, pp. 396–405. Springer International Publishing (2014)

12. Poupyrev, I., Billinghurst, M., Weghorst, S., Ichikawa, T.: The go-go interaction technique: non-linear mapping for direct manipulation in vr. In: Proceedings of the 9th Annual ACM Symposium on User Interface Software and Technology, pp. 79–80. UIST '96, ACM, New York, NY, USA (1996)

13. Schick, A., van de Camp, F., Ijsselmuiden, J., Stiefelhagen, R.: Extending touch: towards interaction with large-scale surfaces. In: Proceedings of the ACM International Conference on Interactive Tabletops and Surfaces, pp. 117–124. ITS '09, ACM, New York, NY, USA (2009)

14. Simon, A.: First-person experience and usability of co-located interaction in a projection-based virtual environment. In: Proceedings of the ACM Symposium on Virtual Reality Software and Technology, pp. 23–30. VRST '05, ACM, New York, NY, USA (2005)

15. Starner, T.: Project glass: an extension of the self. IEEE Pervasive Comput. **12**(2), 14–16 (2013)

16. Uebersax, J.S.: Likert scales: dispelling the confusion. http://www.john-uebersax.com/stat/likert.htm

17. Vogel, D., Balakrishnan, R.: Distant freehand pointing and clicking on very large, high resolution displays. In: Proceedings of the 18th Annual ACM Symposium on User Interface Software and Technology, pp. 33–42. UIST '05, ACM, New York, NY, USA (2005)

# Improving User Experience of Cultural Environment Through IoT: The *Beauty or the Truth* Case Study

Angelo Chianese and Francesco Piccialli

**Abstract** Internet of Things (IoT) computing applied to the Cultural Heritage domain is an emerging discipline which consists of the application of intelligent sensors and technologies within cultural sites; it is strongly related to the development of systems able to be pervasive and ubiquitous with the definitive goal of rethinking such spaces. IoT paradigm can constitute a powerful tool to enhance people fruition and enjoyment of such spaces; thanks to ICT technologies, a cultural object can be effectively "dressed" of its context and juxtaposed into it. In this paper, an intelligent IoT system, designed with the aim of improving user experience and knowledge diffusion within a cultural space, is presented. The paper describes the hardware/-software system components, and presents a case study of a sculptures exhibition named *the Beauty or the Truth* (http://www.ilbellooilvero.it) in Naples where the system was deployed. Furthermore, the paper provides the results of an users behaviour analysis which revealed up a significant increase in user satisfaction and cultural knowledge diffusion.

**Keywords** Internet of things · Cultural heritage · Mobile systems

## 1 Introduction

Cultural Heritage represents a world wide resource of inestimable value, attracting millions of visitors every year to monuments, museums and art exhibitions. It has been playing an increasingly important role in the cultural fabric of society; in the current rapidly changing and globalization world, museum collections, ancient ruins, and artefact exhibitions represent at the same time sources and instruments of education that should to be available to a wide range of people. Indeed, to achieve

A. Chianese · F. Piccialli (✉)
Department of Electrical Engineering and Information Technologies,
University of Naples "Federico II", Naples, Italy
e-mail: francesco.piccialli@unina.it

A. Chianese
e-mail: angchian@unina.it

© Springer International Publishing Switzerland 2015      11
E. Damiani et al. (eds.), *Intelligent Interactive Multimedia Systems and Services*,
Smart Innovation, Systems and Technologies 40,
DOI 10.1007/978-3-319-19830-9_2

a wide fruition of a cultural space and its objects that is effective and sustainable, it is necessary to realize smart solutions for visitors interaction to improve their experiences. In this context, new emerging technologies offer new opportunities and environments to create, exchange, dis- cuss and disseminate cultural content. The adoption of Future Internet (FI) technology, and in particular of its most challenging components like the Internet of Things (IoT) and Internet of Services (IoS), can constitute the basic building blocks to progress towards unified ICT platforms for a variety of applications within the large framework of smart cities. IoT paradigm supports the transition from a closed world, in which an object is characterized by a descriptor, to an open world, in which objects interact with the surrounding environment, because they have become intelligent [5, 6]. Accordingly, not only people will be connected to the internet, but objects such as cars, fridges, televisions, water management systems, buildings, monuments and so on will be connected as well. Indeed, thanks to recent advances in miniaturization and lower cost of RFID, Bluetooth Low Energy, sensor networks, NFC, wireless communications, technologies and applications, IoT is gradually acquiring an important role in several research fields [7]. In this paper, we focus specifically to design an IoT system that is able to enhance the users' experience during a cultural visit within a smart environment, improving the related knowledge diffusion. Moreover, we present and discuss an useful case study of the proposed architecture, immersed in an art exhibition of sculptures, *the Beauty or the Truth*, located in Naples. The paper is organized as follows: Sect. 2 explains the background, Sect. 3 describes how IoT can be applied to the Cultural Heritage domain, Sect. 4 describes the case study. Finally, Sect. 5 concludes the paper with some considerations and future works.

## 2 Background and Related Work

In order to better understand motivations behind the design of the proposed IoT architecture supporting the development of a smart cultural space, it is important to deeply analyse the kind of relations that exists between such spaces and people. Accordingly, the behaviour of a person/visitor, when immersed inside a space and consequently among several objects, has to be analysed in order to design the most appropriate architecture and to establish the relationship between people and technological tools that have to be non-invasive. For this reason, it should be preferable to provide cultural objects with the capability to interact with people, environments, other objects and transmitting the related knowledge to users through multimedia facilities. In an *intelligent* cultural space, technologies must be able to connect the physical world with the world of information in order to amplify the knowledge but also and especially the fruition, involving the visitors as active players which offer the pleasure of perception and the charm of the discovery of a new knowledge. In the last months, the authors of this paper have experienced the design and the application of location-based services and technological sensors applied to Cultural Heritage environments (especially indoor), in [1–4]. These presented prototypes aimed to transfer a *smartness* to cultural sites, applying different communication technologies and sensors. In addition, several papers and projects have been proposed, by using technological and multimedia facilities to enhance cultural items; since the promotion and the fruition of cultural heritage are probably the most interesting and useful applications of modern technologies. Accordingly, the authors in [8] stated that technology can play a crucial role in supporting museum visitors and enhancing their overall museum visit experiences; content and delivery must provide relevant information and at the same time allow visitors to get the level of detail and the perspectives in which they are interested. The authors in [10] propose a mobile recommender system for the Web of Data, and its application to information needs of tourists in context-aware on-site access to cultural heritage. In [9] the initial steps of a project aimed at creating mobile apps to facilitate the usability of museum visits for differently-abled and special-needs users are discussed. *DALICA* [11] is another agent-based Ambient Intelligence for outdoor cultural-heritage scenarios that it sends information about nearby points of interest from sensors, while in [13] the authors propose a general architecture of a SNOPS (Social Network of Object and PersonS) Platform and present a specific smart environment related to the archaeological site of Herculaneum. The authors in [12] present a first prototype of a wearable, interactive augmented reality (AR) system for the enjoyment of the cultural heritage in outdoor environments by using a binocular see-through display and a time-of-flight (ToF) depth sensor. In [14] a system, called SMART VILLA, based on a set of mobile applets, each interfaced with a NFC based subsystem, related to particular sites (SMART BIBLIO for ancient books, SMART ROOM for particular rooms and SMART GARDEN for surrounding historical gardens) is presented. The diversity of the mentioned methods and applications, highlights that in most cases, they remain isolated "exercises" and do not arouse effective interest due to the lack applicability and difficulty of reuse in different environments and scenarios. It is evident that, for improving users experience and knowledge diffusion, in all its forms and needs, there is the necessity of designing an integrated system following the IoT paradigm, that can be exploited and adapted to the different scenarios.

## 3 IoT and Cultural Heritage: Designing a Smart Exhibition

In this section, the architecture of an IoT system, the technological sensors immersed in the environment and the communication framework are presented. The sensors aimed to transform cultural items in smart objects, that now are able to communicate with each other, the visitors and the network; this acquired identity plays a crucial role for the smartness of a cultural space. Indeed, as stated in [15, 16], smart objects represent an important step on an evolutionary process that is affecting modern communication devices and has been triggered by the advent of IoT. Accordingly, in order that this system can perform its role and improve end-users cultural experience transferring knowledge and supporting them, a mobile application has been designed; in this way people have the opportunity to enjoy the cultural visit and be more at ease simply using their own mobile device.

## 3.1 The IoT Architecture

To describe the proposed system we resort on the three-layer architectural model for IoT presented in [17]. It consists of: (i) the sensing layer, which is devoted to the data transfer and acquisition, and nodes collaboration in short-range and local networks; (ii) the network layer, which is aimed at transferring data across different networks and applications; and (iii) the application layer, where the IoT applications are deployed together with the middleware functionalities. Figure 1 shows the resulting three-layer architecture. The three basic elements of the proposed system are: the CHIS (Cultural Heritage Information System) server, the gateway, and the sensor layers.

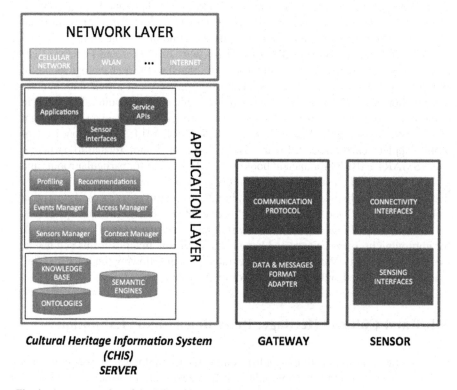

**Fig. 1**   A representation of the IoT architecture for a cultural space

**CHIS Server** As depicted in Fig. 1, the Cultural Heritage Information System server is composed by the Network and the Application layers. The Application Layer is modelled by three sub-layers. The first sub-layer includes (i) the knowledge base for the storage and management of the content, (ii) the ontologies used to represent a semantic view of the cultural heritage domain and (iii) the semantic engines used to provide a framework for representing functional and non-functional attributes and operations of the IoT objects. The second sub-layer includes the instruments that implement the core functionalities of the CHIS system (Fig. 1).

**Gateway and Sensor** The *Gateway* module is enabled to manage the overall communication between the sensors and the CHIS server; moreover it is responsible to adapt and deliver the environmental data captured by the sensors. The *Sensor* module is aimed to provide the connectivity interfaces that enable the sensors functionalities according to the different types of sensor nodes.

According to IoT requirements, two types of sensor nodes are designed in order to make *smart* the cultural items inside a museum or an art exhibition.

– SERVER node: this type of node stores on board the content related to the items where it is placed and creates a Wi-Fi coverage area; the App automatically connects to this network and retrieve the content. Moreover, in absence of connectivity, this node can be equipped with a UMTS/GSM module, in order to manage the status of this node and communicate with the CHIS server.
– SLAVE node: this type of node can placed (i) near or on a single artwork, (ii) near a subset of artworks very close together; the user mobile device can be sensed by this node, since it creates a Bluetooth Low Energy surrounding area, and requests to the SERVER node transferring to the App the related multimedia content.

From a point of view of communication and interaction, the sensor nodes are equipped with the following features:

– Discovering the neighbours: A sensor node, thanks to a proximity technique using the Bluetooth Low Energy (BLE) protocol, is able to sense the neighbours SLAVE nodes; this feature allows any node to contextualize itself inside the space and enable mechanisms of content recommendations or visiting paths inside the cultural space.
– Discovering the visitors: A sensor node, thanks to a proximity technique using the Bluetooth Low Energy (BLE) protocol, is able to sense the visitors inside the surrounding area; this feature allows any node to present itself to an user and deliver to him multimedia content.

## 4 The Case Study: The Beauty or the Truth Sculptures Exhibition

In this section we present the case study, it consists of an art exhibition consisting of 271 sculptures, divided into 7 thematic sections and named *the Beauty or the Truth*. This exhibition shows, for the first time in Italy, the Neapolitan sculpture of the late nineteenth century and early twentieth century, through the major sculptors of the time. The sculptures are exhibited in the monumental complex of San Domenico Maggiore, in the historical centre of Naples.

**Fig. 2** The main screens of the mobile application

## 4.1 Implementation Details

The proposed IoT system was entirely deployed inside the exhibition, as illustrated in Fig. 3. Each sculpture of subset of them were equipped with a sensor SLAVE node, while in each room was deployed one or more sensor SERVER node. In detail we deploy over than 70 SLAVE nodes and 10 SERVER ones. The mobile application, named *OPS Opere Parlanti Show* (the Talking Artwork Show) is currently available on the main smarphone app stores (see Fig. 2). Visitors can download and install it on their mobile device in order to start a novel visit experience. The multimedia collection is constituted by about 1500 images, 500 audio files (Italian and English languages), 300 video files and over than 1000 text files, all about the exposed sculptures. The graphic elements placed in the rooms are represented by captions (one for each sculpture), indicating the name, the author, the historical period and the material, and information panels in each section. Currently, the exhibition records about 45,000 visitors from the date of opening (30 October 2014).

## 4.2 Analysis of User Behaviour

In order to analyse user behaviour during the visit and consequently (i) the user satisfaction related the proposed system and (ii) a real improvement in knowledge diffusion, we perform a number of trials recruiting 297 people. They have been divided in two groups, the first (151) that used the IoT system, the second (146) that represents a control group visiting the exhibition without any technologies. Three indicators related to the users behaviour and satisfaction have been analysed in both groups. The three indicators are: (i) the average time of total duration of the visit, (ii) the average number of artworks on which a visitor focused his attention, (iii) the average rating about the overall appreciation of the system.

Table 1 provides a comparison that emphasizes the increasing of the average duration of the users visits (by using the system) and the increased dwell on the presented artworks, thus allowing a more in-depth cultural and consequent diffusion of knowledge. Although the increase of visits total duration can be attributed to a playing time of the visitor with the new technology, this can be considered a positive factor since the role played by technology is to put in contact users to the art through the game, the novelty, curiosity, etc. Finally, users that used the IoT system have a greater appreciation respect to the entire exhibition

**Table 1**  A comparison between using the proposed IoT system and without using it

| Indicator | With IoT system | Without system |
| --- | --- | --- |
| Average total duration of the visit (minutes) | 72.7 | 48.5 |
| Average number of artworks on which a visitor focalized his attention | 70.3 | 40.2 |
| Average rating about the overall appreciation of the exhibition (between 0 and 5) | 4.1 | 3.5 |

To deeply analyse user behaviour during the use of the proposed system, the mobile app builds and sends to the SERVER nodes a LOG file (one for each visitor) structured as follows

**Listing 1.1.** The user behaviour structured LOG file.

```
1   <?xml version="1.0" encoding="UTF-8"?>
2     <USER ID='UI001'>
3         <START_SESSION></START_SESSION>
4         <END_SESSION></END_SESSION>
5     <TRANSACTION>
6       <REQUEST>
7         <HTTP_METHOD>GET</HTTP_METHOD>
8         <PATH_INFO>/opera</PATH_INFO>
9         <REQUEST_PARAMETERS>
10           <CODEARTWORK>ART0224VICTA</CODEARTWORK>
11           <DATE>20/11/2014</DATE>
12           </REQUEST_PARAMETERS>
13           <REMOTE_ADDRESS>192.168.1.6</REMOTE_ADDRESS>
14       </REQUEST>
15         <PARAMETERS_LOG>
16           <HOUR_LISTEN_START>20/11/2014 13:58:12</HOUR_LISTEN_START>
17           <HOUR_LISTEN_END>20/11/2014 14:00:42</HOUR_LISTEN_END>
18           <AUDIOS>
19         <TOT_NUMBER>3</TOT_NUMBER>
20         <AUDIO ID='AU1111'>
21             <HOUR_END>20/11/2014 14:00:42</HOUR_END>
22           <LENGTH>180</LENGTH>
23           <RATE>4.5</RATE>
24       </AUDIO>
25         </AUDIOS>
26           <IMAGES>
27         <TOT_NUMBER>11</TOT_NUMBER>
28         <IMAGE ID='IM1122' />
29         <IMAGE ID='IM1134' />
30         <RATE>5</RATE>
31           </IMAGES>
32             <VIDEOS>
33         <TOT_NUMBER>2</TOT_NUMBER>
34         <VIDEO ID='VI3333'>
35             <HOUR_END>20/11/2014 14:20:12</HOUR_END>
36           <LENGTH>180</LENGTH>
37             <RATE>4</RATE>
38       </VIDEO>
39         </VIDEOS>
40           <TEXTS>
41         <TOT_NUMBER>4</TOT_NUMBER>
42         <TEXT ID='TX4455' />
43             <TEXT ID='TX4456' />
44           <RATE>4</RATE>
45     </TEXTS>
46         </PARAMETERS_LOG>
47     </TRANSACTION>
48         <TOTALRATE>4.5</TOTALRATE>
49   </USER>
```

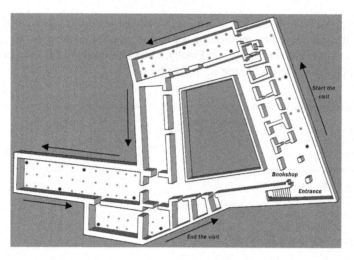

**Fig. 3** The exhibition layout: The blue circles represent the SLAVE nodes whereas the red circles represent the MASTER nodes. The arrows indicate the exhibit itinerary

This LOG file stores the following visitor behaviour: (i) the beginning and the end of the visit (ii) all the mobile App requests to the retrieval of multimedia content (e.g audio, video, text, image) (iii) the start time and the end of listening/watching a audio/video file. Moreover, it collects a rate (between 0 and 5 stars) that the user can assign to each multimedia content. An ad-hoc algorithm parses the LOG, extracting some implicit information about the user, such as:

– User ID
– Average time observation of an artwork
– Percentage of observed artwork
– Visit duration
– Percentage of audio heard
– Percentage of videos viewed
– Average number of images displayed
– Average number of texts read
– Average of audio/text/video/images rating

An analysis of the collected LOG files of 151 people shows interesting results. The percentage of observed sculptures exceeds 50 % of the total (135 sculptures) in 82 % of the selected log files. The average rating about the audio/text/video/images is respectively 2.8, 4.2, 3.9, 4.5; this results suggest to improve the overall quality of the audio files. The percentage of viewed video file, for each log file, is extremely low; an average of 5.8 video files played for each visitor. This result suggests that usually users are not interested in the observation of supplementary video content during their visit. For what concern supplementary images related to the sculptures, we observe that the average of displayed images for each visitor is 70.1; 2.4 is the average for each sculpture.

At the end of the visit, each person of the two groups answered to three questions about the exhibition, in order to assess the knowledge diffusion improvement using our IoT system. The questions are: (1) *What are the names of the three main sculptures authors of the exhibition ?*, (2) *What is the material mainly used by the sculptures of this historical period?*, (3) *What is the theme that links the different sections of the exhibition?* (multiple choice question with three possibilities).

**Table 2** Questionnaire results

| Question | Correct answers with IoT system | Correct answers without IoT system |
|---|---|---|
| 1 | 92 | 61 |
| 2 | 101 | 89 |
| 3 | 84 | 49 |

Table 2 shows the results of the questionnaire; the observed values indicates that the technology can significantly improve the user cultural experience during and art exhibition, increasing the knowledge diffusion related to the observed cultural items.

## 5 Conclusion

IoT constitute a powerful tool to address the design of the complex connection between new technologies, knowledge to be transmitted and visitors experiences of Cultural Heritage environments. As an effort in this direction, this paper define an architecture to represent and manage the smartness inside cultural spaces, adopting the IoT paradigm and supporting this direction with the design of a set of sensor nodes. The technologies cover the fundamental role of connector between the physical world and the world of information, in order to amplify the knowledge but also and especially the enjoyment. For these reasons, our research has been primarily focused on the design of IoT architecture for Cultural Heritage spaces. These sensor nodes have the capability to observe the environment and support the people enjoyment process, establishing multiple connections among the end-users through which convey information, stories and multimedia content. A case study, the *Beauty or the Truth* art exhibition in Naples, has been devised as a feasibility test of our system, the related sensor nodes and the users' satisfaction through an implicit (LOG files) and explicit analysis (questionnaires) of user behaviour.

## References

1. Chianese, A., Marulli, F., Moscato, V., Piccialli, F.: Smartweet: a location-based smart application for exhibits and museums. In: International Conference on Signal Image Technology and Internet Based Systems, pp. 408–415 (2013)
2. Amato, F., Chianese, A., Mazzeo, A., Moscato, V., Picariello, A., Piccialli, F.: The talking museum project. Procedia Comput. Sci. **21**, 114–121 (2013)

3. Chianese, A., Piccialli, F., Riccio, G.: Designing a smart multisensor framework based on Beaglebone Black Board, computer science and its applications, pp. 391–397 (2015)
4. Chianese, A., Piccialli, F.: Designing a smart museum: when cultural heritage joins IoT. In: Proceedings—2014 8th International Conference on Next Generation Mobile Applications, Services and Technologies, pp. 300–306 (2014)
5. Atzori, L., Iera, A., Morabito, G.: From smart objects to social objects : the next evolutionary step of the IoT. IEEE Comm. Mag. **52**, 97–105 (2014)
6. Sanchez Lpez, T., Ranasinghe, D., Harrison, M., McFarlane, D.: Adding sense to the internet of things: an architecture framework for smart object systems. Pers. Ubiquit. Comput. **16**(3), 291–308 (2012)
7. Zheng, L.: Technologies, applications, and governance in the Internet of things. IoT Global Technological and Societal Trends (2011)
8. Kuflik, T., Stock, O., Zancanaro, M., Gorfinkel, A., Jbara, S., Kats, S., Sheidin, J.: A visitor's guide in an active museum: presentations, communications, and reflection. JOCCH, vol. 3 (2011)
9. Buzzi, M.C., Buzzi, M., Leporini, B., Marchesini, G.: Improving user experience in the museum. In: Proceedings of the IADIS—Interfaces and Human Computer Interaction, pp. 327–331 (2013)
10. Ruotsalo, T., Haav, K., Stoyanov, A., Roche, S., Fani, E., Deliai, R., Makela, E.: SMART-MUSEUM: a mobile recommender system for the Web of Data. J. Web Semantics **20**, 50–76 (2013)
11. Costantini, S., Mostarda, L., Tocchio, A., Tsintza, P.: DALICA: agent-based ambient intelligence for cultural-heritage scenarios. IEEE Intell. Syst. **23**, 34–41 (2008)
12. Caggianese G., Neroni P., Gallo L.: Natural interaction and wearable augmented reality for the enjoyment of the cultural heritage in outdoor conditions. LNCS, pp. 267–282 (2014)
13. Amato, F., Chianese, A., Moscato, V., Picariello, A., Sperli, G.: SNOPS: a smart environment for cultural heritage applications. In: Proceedings of International Conference on Information and Knowledge Management, pp. 49–56 (2012)
14. Angelaccio, M., Basili, A., Buttarazzi, B., Liguori, W.: Smart and mobile access to cultural heritage resources: a case study on ancient Italian Renaissance Villas. In: Proceedings of the Workshop on Enabling Technologies: Infrastructure for Collaborative Enterprises, pp. 310–314 (2012)
15. Atzori, L., Lera, A., Morabito, G.: From "smart objects" to "social objects": the next evolutionary step of the internet of things. IEEE Commun. Mag. **52**, 97–105 (2014)
16. Sanchez Lpez, T. and Ranasinghe, D.C. and Harrison, M. and McFarlane, D.: Adding sense to the Internet of hings: an architecture framework for Smart Object systems. Pers. Ubiquit. Comput. **16**, 291–308 (2012)
17. Zheng, L.: Technologies, applications, and governance in the internet of things. IoT Global Technological and Societal Trends (2011)

# A Multimedia Summarizer Integrating Text and Images

Antonio d'Acierno, Francesco Gargiulo, Vincenzo Moscato, Antonio Penta, Fabio Persia, Antonio Picariello, Carlo Sansone and Giancarlo Sperlì

**Abstract** We present a multimedia summarizer system for retrieving relevant information from some web repositories based on the extraction of semantic descriptors of documents. In particular, semantics attached to each document textual sentences is expressed as a set of assertions in the $\langle subject, verb, object \rangle$ shape as in the RDF data model. While, images' semantics is captured using a set of *keywords* derived from high level information such as the related title, description and tags. We leverage an unsupervised clustering algorithm exploiting the notion of semantic similarity and use the centroids of clusters to determine the most significant summary sentences. At the same time, several images are attached to each cluster on the

F. Gargiulo · V. Moscato · F. Persia (✉) · A. Picariello · C. Sansone · G. Sperlì
Dip. di Ingegneria Elettrica e Tecnologie dell'Informazione,
University of Naples "Federico II", Naples, Italy
e-mail: fabio.persia@unina.it

F. Gargiulo
e-mail: francesco.grg@unina.it

V. Moscato
e-mail: vmoscato@unina.it

A. Picariello
e-mail: picus@unina.it

C. Sansone
e-mail: carlo.sansone@unina.it

G. Sperlì
e-mail: giancarlo.sperli@unina.it

A. d'Acierno
ISA - CNR, Avellino, Italy
e-mail: dacierno.a@isa.cnr.it

A. Penta
Dipartimento di Informatica, University of Turin, Turin, Italy
e-mail: penta@di.unito.it

© Springer International Publishing Switzerland 2015

21

E. Damiani et al. (eds.), *Intelligent Interactive Multimedia Systems and Services*,
Smart Innovation, Systems and Technologies 40,
DOI 10.1007/978-3-319-19830-9_3

base of keywords' term frequency. Finally, several experiments are presented and discussed.

**Keywords**  Web summarization · Information extraction · Multimedia

# 1 Introduction

Multimedia data allow fast and effective communication and sharing of information about peoples' lives, their behaviors, works, interests, but they are also the digital testimony of facts, objects, and locations. The widespread availability of cheap media technologies (e.g., digital cameras and video cameras, MP3 players, smart phones, and tablet computers) dramatically increased the availability of multimedia information, their production and utilization. As an example, images and videos are mostly used by both media companies and the public at large, to record daily events, to report local, national, and international news, to enrich and emphasize Web content, as well as to promote cultural heritage.

At the same time, the exponential growth of the Web has made the search and track of such kind of information apparently easier and faster, but the described huge multimedia data overload requires algorithms and tools for a fast and easy access to the specific *desired* information, discriminating between "useful" and "useless" information, especially in the era of *Big Data*.

In particular, this new situation requires to investigate new ways - e.g. *summarization* techniques - to handle and process information, that has to be delivered in a rather small space, retrieved in a short time, and represented as accurately as possible.

In such a scenario, *multimedia summarization* techniques can be profitably used to provide a short, concise and meaningful version of multimedia information spread out in various and heterogeneous documents. Such techniques typically exploit low level features of multimedia data together with high level information as *metadata*, knowledge of the context or user-generated content, etc. to produce a summary that can be expressed in different shapes (e.g. only natural language sentences, text with images and videos, etc.).

The first summarization techniques have been applied to texts and the automatic textual document summarization has been actively researched since the original work by Luhn [14] from both Computational Linguistics and Information Retrieval points of view [5, 11, 12, 16, 20]. Within such research area, *extractive* techniques are more feasible and have become the *de facto* standard in document summarization [23]. To extract the most significant part of a text, the majority of the proposed techniques follows the *bag of words* approach, which is based on the observation that words occurring frequently or rarely in a set of documents have to be considered with a higher probability for human summaries. Other approaches leverage the semantics attached to each sentence to produce non repetitive summaries with the most salient information [6, 13, 23].

Concerning the other kind of multimedia data [2, 3], the majority of literature mainly focuses on the *video summarization* problem aimed at deriving a synthetic representation of the related contents characterized by a limited loss of meaningful information [1]. A large part of the work is based on the idea that the summarized video should be a subsequence of the original one containing clips with most important visual information (e.g. key frames) [18]. In turn, several systems [7, 8] exploit natural language engines to create textual summaries based on video/audio features. In their vision, a good textual summary will help the user obtain maximal information from the video, without having to watch the video itself or parts of it.

In this paper, we propose a novel approach for multimedia summarization integrating *images* and *texts* from web repositories (i.e. news portals and online social networks) and based on the extraction of semantic descriptors of documents. In particular, semantics attached to each document textual sentences is expressed as a set of assertions in the ⟨*subject, verb, object*⟩ shape as in the RDF data model. While, images' semantics is captured using a set of *keywords* derived from high level information such as the related title, description and tags.

We describe a framework designed and developed to facilitate the different phases of a summarization process, i.e.: (i) search on the Web of "relevant" documents (texts and images); (ii) collection of the most important "pieces of information"; (iii) ranking of the relevance related to the information; (iv) composition of the pieces of information in a summary; (v) presentation of the summary in a readable format, also suitable for visually impaired users.

The paper is organized as follows. Section 2 outlines a motivating example for our work. Section 3 presents the summarization framework describing the adopted data model, the summarization process and some implementation details on the realized summarizer. Section 4 is dedicated to the experiments we performed to test the effectiveness of the system. Finally, Sect. 5 discusses some conclusions and future work.

## 2 Motivating Example

In order to better explain our idea, in the following we briefly describe an example of a typical workflow related to a news reporter that has just been informed of terrorist attempt happened on January 7th, 2015 in Paris and that would like to gather more details about this event.

We suppose that the following textual information[1] on the terrorist attack, extracted from some web sites already reporting the news, is available together with several images (coming from news portals or social networks):

*"Kouachi brothers forced their way into the offices of the French satirical weekly newspaper Charlie Hebdo in Paris. They killed 10 employees, 2 French National*

---

[1]The analyzed texts come from news articles, thus they are generally quite simple as concerning grammar and syntax.

*Police officers and injured 11 others. Charlie Hebdo had attracted attention for its controversial depictions of Muhammad. Police detained several people during the manhunt for the two main suspects. Police described the assailants as armed and dangerous. France raised its terror alert to its highest level and deployed soldiers".*

*"Masked gunmen attacked the French Charlie Hebdo satirical newspaper in Paris, killing at least 10 people. They were identified as Kouachi brothers, known to intelligence services, had been born in Paris, raising the prospect that homegrown Muslim extremists were responsible of the attack. The police organized an enormous manhunt across the Paris region for the suspects".*

Our basic idea consists from one hand in finding, from each textual sentence, a sequence of relevant information in the shape of a list of triples formed by ⟨*subject, verb, object*⟩ (using for instance some NLP facilities). The obtained triples are then clustered into subsets with similar information content; thus, we would like to discard repeated sentences and to consider only the relevant ones for each cluster.

For the previous terrorist attack example, the extracted triples are[2]: ⟨brothers, force, newspaper⟩, ⟨ brothers, kill, employees ⟩, ⟨Charlie Hebdo, attract, attention⟩, ⟨Police, detain, people⟩, ⟨Police, describe, assailants⟩, ⟨France, raise, alert⟩, ⟨France, deploy, soldiers⟩, ⟨gunmen, attack, newspaper⟩, ⟨gunmen, kill, people⟩, ⟨gunmen, identify, brothers⟩*, ⟨gunmen, be born, Paris⟩, ⟨ gunmen, be responsible, attack⟩, ⟨Police, organize, manhunt⟩.

**Table 1**  The clustering results on a set of triples (centroids are underlined)

| Cluster 1 | ⟨brothers, force, newspaper⟩ |
|---|---|
| | ⟨ brothers, kill, employees ⟩ |
| | ⟨gunmen, attack ,newspaper⟩ |
| | ⟨gunmen, kill, people⟩ |
| | ⟨gunmen, identify, brothers⟩ |
| | ⟨gunmen, be born, Paris⟩ |
| | ⟨ gunmen, be responsible, attack⟩ |
| Cluster 2 | ⟨Charlie Hebdo, attract, attention⟩ |
| Cluster 3 | ⟨Police, detain, people⟩ |
| | ⟨Police, organize, manhunt⟩ |
| | ⟨Police, describe, assailants⟩ |
| Cluster 4 | ⟨France, raise, alert⟩ |
| | ⟨France, deploy, soldiers⟩ |

Successively, a clustering algorithm creates the clusters reported in Table 1. The clusters are obtained by computing the *semantic similarity* between each pair of triples. Finally, we assume that it is possible to obtain a useful summary by just

---

[2]Particular cases in which verb is a modal verb or is in a passive or negation form - see triples marked with '*' - have to be opportunely managed.

considering the sequence of centroids and eventually re-loading the associated original sentence, as in the following:

*"Masked gunmen attacked the French Charlie Hebdo satirical newspaper in Paris, killing at least 10 people. Charlie Hebdo had attracted attention for its controversial depictions of Muhammad. The police organized an enormous manhunt across the Paris region for the suspects. France raised its terror alert to its highest level and deployed soldiers".*

From the other hand, we can associate to each cluster a set of images with a similar semantics by considering the *term frequency* of the related keywords (derived from title, description, tags, etc.) with respect to elements hosted by the different triples in the cluster (i.e. subjects, verbs and objects). Thus a set of images can be attached to each sentence in the summary, making it more appealing.

As an example, images of Kouachi brothers can be linked to the first summary sentence, images with the slogan "Je suis Charlie" or of newspaper to the second sentence, images of manhunt by French police to the third sentence and so on.

# 3 The Multimedia Summarization Framework

## 3.1 The Model for Automatic Summarization

The summarization problem may be stated as follows: given a set of multimedia documents, let us produce an accurate and all-sided summary that is able to reflect the *main concepts* expressed by the original documents in different shapes (e.g. text and images), matching some *length restrictions* and without introducing *additional* and *redundant* information.

The adopted model is inspired by the text summarization models based on *Maximum Coverage Problem* [10, 22] and represents an extension of that which some of the authors have recently proposed in [6].

In our vision, multimedia summarization is initially driven by the textual sentences within a document. In particular, in the proposed model, the documents are segmented into several linguistic units - named as *summarizable sentences* ($\sigma$) - in a preprocessing stage, and each linguistic unit is then characterized by a set of conceptual units (named as *semantic atoms*) containing the meaning of a sentence in the shape of $\langle subject, verb, object \rangle$ triples ($t$) as shown in Sect. 2. Our main goal is to *cover* as many conceptual units as possible using only a small number of sentences. Given a set of documents, a *summary* is a set of summarizable sentences together with a set of multimedia data (e.g. one or more images for each sentence). Let us assume the existence of a similarity function $sim(t_i, t_j) \in [0, 1]$ able to compute the semantic similarity between two semantic atoms and another function able to *score* the semantic atoms based on their importance, we introduce the concept of "Summarization Algorithm" as follows.

**Definition 1** (**Multimedia Summarization Algorithm**) *Let* $\mathcal{D}$ *be a set of documents, a* Multimedia Summarization Algorithm *is formed by a sequence of two functions* $\phi$ *and* $\chi$. *The semantic partitioning function* ($\phi$) *partitions* $\mathcal{D}$ *in* K *sets* $\mathcal{P}_1, \ldots, \mathcal{P}_K$ *of summarizable sentences having similar semantics in terms of semantic atoms and returns for each set: (i) the related information score by opportunely combining the score of each semantic atom, (ii) a set of multimedia data (e.g. images, videos etc.)* $\mathcal{M}$ *with a similar semantics:*

$$\phi : \mathcal{D} \to S^* = \{\langle \mathcal{P}_1, \hat{w}_1, \mathcal{M}_1 \rangle, \ldots, \langle \mathcal{P}_K, \hat{w}_K, \mathcal{M}_K \rangle\} \tag{1}$$

s. t. $\mathcal{P}_i \cap \mathcal{P}_j = \emptyset, \forall i \neq j$.
   *The* Sequential Sentence Selection *function* ($\chi$):

$$\chi : S^* \to S, \hat{\mathcal{M}} \tag{2}$$

*selects a set of the sentences* S - *together with a set of multimedia data* $\hat{\mathcal{M}}$ - *from original documents containing the semantics of most important clustered information sets in such a way that:*

1. $|S| \leq L$,
2. $\forall s_i \in S, \exists M_i \in \hat{\mathcal{M}} : |M| \leq N$
3. $\forall \mathcal{P}_k, \hat{w}_k \geq \iota, \nexists t_j, t_j \in_\sigma S^* : sim(t_i, t_j) \geq \gamma, t_i \in_\sigma S$.

*$\iota$ and $\gamma$ being two apposite thresholds. With abuse of notation, we use the symbol* $\in_\sigma$ *to indicate that a semantic atom comes from a sentence belonging to the set of summarizable sentences* S.

   Once obtained a partition of the space in terms of clusters of semantic atoms, we select a number of sentences equipped with other kinds of multimedia information, trying to: (i) maximize the semantic coverage - the most representative sentences of each cluster should be considered starting from the most important clusters, i.e. those having the highest average information score; (ii) minimize the redundancy by selecting one sentence for each cluster that is most representative in terms of semantic content and not considering similar sentences.

   In our model, critical issues to be addressed are: (i) how to extract semantic atoms of a document, (ii) how to evaluate the similarity between two semantic atoms, (iii) how to calculate a score for each semantic atom, and finally, (iv) how to define suitable semantic partitioning and sentence selection functions.

**Extracting semantic atoms from a text** We adopted the principles behind the RDF framework used in the Semantic Web community to semantically describe web resources. The idea is based on representing data in terms of a triple $\langle$*subject, verb, object*$\rangle$. In the Semantic Web vision, subjects and objects are web resources while verbs are predicates/relations defined in schemata or ontologies, in our case instead we attach to the elements of triples the tokens extracted by processing documents using NLP facilities. Thus, the semantic content of a document can be modeled by a set of structured information $\mathcal{T} = \{t_1, \ldots, t_n\}$, where each element $t_i$ is a semantic atom described by the following couples $t_i = \{\langle sub, val \rangle, \langle verb, val \rangle, \langle obj, val \rangle\}$.

**Semantic similarity function** In our model, we decided to compare two semantic atoms based on the similarity measure obtained by the comparison of the elements hosted by a summarization triple, namely subject, predicate, and object. In particular, let us consider two sentences and assume to extract from them two triples $t_1$ and $t_2$; we define as similarity between two $t_1$ and $t_2$ the function:

$$sim(t_1, t_2) = F_{agr}(F^{sim}(sub_1, sub_2), F^{sim}(pred_1, pred_2), F^{sim}(obj_1, obj_2)) \qquad (3)$$

The function $F^{sim}$ is used to obtain the similarity among values of the semantic atoms, while $F_{agr}$ is an aggregation function. If the $F^{sim}$ takes into account the information stored in a knowledge base, the function is computed based on the *"semantic"* aspects of its input, otherwise we can apply any similarity among words like the one based on the well-known edit distance. In particular, we use the *Wu & Palmer* similarity [24] for computing the similarity among elements of our triples. This similarity is based on the *Wordnet Knowledge Base*, that lets us compare triples based on their semantic content.

**Semantic atom score** There is a no absolute and objective criterion to evaluate a relevance of an information [11, 20]. Here, we developed a method that combines the values of *frequency* and *position* with a value of *source preference* of semantic atoms.[3] Our hypothesis is that interesting facts to be included in the summaries are frequently reported and are described in the first part of the document and not every news portal is judged as useful from the user. Similar hypotheses are also adopted in summarization systems like the ones described in [9] and [17].

**The semantic partitioning function** For the purposes of semantic partitioning, we could select any unsupervised clustering algorithm able to partition the space of texts coming from multimedia documents into several clusters, where the number of clusters is not defined a-priori, on the base of the discussed semantic similarity among semantic atoms. In our context, the clusters have different shapes and they are characterized from density-connected points; thus, we decided to use in our experiments the *OPTICS* algorithm [4], that is based on a *density-based* cluster approach. In our context the clustering procedure is applied on semantic atoms, thus we can use as distance $d(t_i, t_j)$ the value of 1- $sim(t_i, t_j)$. In this way, we will have clusters where the sentences that are semantically more similar are grouped together.

The second step is to associate a set of other kinds of multimedia objects (e.g. images) to each cluster. To this goal, we leverage high level information attached to multimedia data usually in the shape of *metadata* (i.e. titles and descriptions) and *tags*.

In particular, a set of keywords is extracted from each multimedia object using the approach that some of the authors have described in [19] for image categorization. We then calculate the related average *term frequency* of such keywords within the list of subjects and objects of each cluster:

---

[3] A value given by the user through a qualitative feedback that is assumed to be the *value of reputation* of each sources.

$$tf_\mathcal{P}^m = \sum_{i=1}^{l_m} (\frac{n_{k_i}}{n})/l_m \qquad (4)$$

$l_m$ being the number of keywords for the multimedia object $m$, $n_{k_i}$ the number of times that the keyword $i$ matches with any triple's subject or object of cluster $\mathcal{P}$ and $n$ the number of different subjects and objects.

The multimedia object is finally linked to the cluster with the highest value of term frequency.

**Sentence selection function and summary presentation** In this section, we discuss how we choose the sentences from the clusters and the related multimedia objects by maximizing the information score and minimizing text redundancy. We exploit to this goal the sentence selection greedy algorithm proposed in [22].

We order the clusters according to their scores, then we use a given threshold to select the most important ones with respect to the length restrictions. After this stage, we select the sentence that has the most representative semantic atoms, minimizing, at the same time, the redundancy. As several sentences may have multiple semantic atoms, we need to eliminate the possibility that the same sentence can be reported in the summary several times, because its semantic atoms are spread all over the clusters. We penalize the average score of each cluster if we have already considered it during our summary building process. We also note that each partition can be considered several times in the summary building process until the length restriction condition has been verified. When the cluster has been already considered, we apply a penalty strategy that affects the initial order within the summarizable sentence set.

Once the optimal summary has been determined, the sentences are ordered maximizing the partial ordering of sentences within the single documents.

Finally the *Top-K* of multimedia object for each cluster is considered and multimedia data are presented together with the related sentences in the final summary.

### 3.2 The Summarization Process and Some Implementation Details

The summarization process consists of the following steps Fig. 1:

- *Web Search* - this activity has the task of retrieving a set of documents that satisfy some search criteria using a *Search Engine* external component. Using the available API of the most common search engines (e.g. Google, Bing, Yahoo, etc.), several web pages are retrieved and stored with the related multimedia items and their metadata and tags into a local repository.
- *Text and Image Extraction* - (i) the textual sentences are extracted from the several web sources by parsing the textual content of the related HTML pages and analyzing the HTML tags (using the *JSOUP* API), (ii) images' keywords are computed.

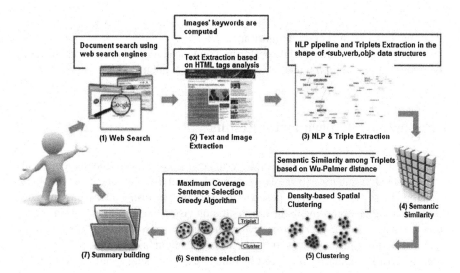

**Fig. 1** The summarization process

- *NLP and Triples Extraction* - NLP processing techniques are performed on the input sentences, in particular *Named Entities Recognition* (NER), *Part Of Speech* (POS) tagging, Parse Tree Generation, Anaphora and Co-Reference Resolutions; successively, in this stage, semantic triples are detected for each sentence, analyzing the related Parse Tree[4] and using appropriate heuristics [21]. NLP algorithms were implemented using the well-known *Stanford NLP Libraries*[5] combined with the *FreeLing*[6] NLP suite.
- *Similarity Matrix Builder* - a matrix containing the similarity values for each couple of triples is computed.
- *Clustering* - a proper clustering algorithm is applied on the input matrix and each image is associate to a given cluster using the related term frequency. In our implementation, we used, as already described, the *OPTIC* clustering algorithm. To accomplish the clustering task, we also need a *Semantic Similarity Performer* component that computes, by using the WordNet lexical database, the semantic distances for each couple of summarizable sentences, disambiguating the word senses using a context-aware and taxonomy-based approach as described in[15], if necessary.

---

[4]http://www.cis.upenn.edu/treebank.

[5]http://nlp.stanford.edu/software/.

[6]http://nlp.lsi.upc.edu/freeling/.

- *Sentence Selection* - this activity performs a sentence selection to generate the summary using our proposed algorithm.
- *Summary Building* - this activity performs a sentence ordering to generate the final summary containing both textual and multimedia content.

## 4 Preliminary Experimental Results

Summarization technologies are usually evaluated by measuring the agreement of their extracted summaries to human-generated summaries (*ground truths*). Many evaluation metrics have been proposed to measure the performance of a summarization approach and, according to the most common vision, these metrics are classified into three main categories: *recall-based*, *sentence-rank-based* and *content-based*.

In [6], we presented an evaluation of our textual summarization technique based on ground truth summaries produced by human experts, using recall-based metrics as the summary needs to contain the most important concepts coming from the considered documents. In particular, in order to compare the performances of our system with respect to the other ones, we employed the automatic summarization evaluation package ROUGE[7] using the data set of documents related to the *DUC* workshops.

**Table 2** Quality of the DUC 2007 summaries

| Rank | Summarizer ID | Grammaticality | Non Redundancy | Referential Clarity | Focus | Structure and Coherence | Average | R OUGE-2 Position | R OUGE-SU4 Position |
|------|---------------|----------------|----------------|---------------------|-------|-------------------------|---------|-------------------|---------------------|
| 2 | iWIN | 5 | 5 | 5 | 4 | 4 | 4,7 | 16 | 15 |
| 1 | 5 | 4 | 5 | 5 | 4 | 4 | 4,4 | 24 | 26 |
| 3 | 12 | 3 | 3 | 4 | 5 | 4 | 3,8 | 26 | 25 |
| ... | ... | ... | ... | ... | ... | ... | ... | ... | ... |
| 33 | 19 | 1 | 4 | 1 | 1 | 1 | 1,6 | 27 | 24 |

---

[7]*Recall-Oriented Understudy for Gisting Evaluation*, http://haydn.isi.edu/ROUGE/.

We observed that even if our systems has not been designed for a *topic-based* approach as in DUC 2007 requirements, the obtained performance is comparable with the best summarizers.

Here, we want to evaluate the *quality* of produced multimedia summaries. To this aim, we asked 50 students to read the summaries generated by our summarizer about 20 different topics and to answer several apposite questions provided by NIST for DUC 2007. Such evaluations are compared with other (only textual) summarizers officially provided by NIST. The obtained results are listed into Table 2.[8] The results show the very good quality of our summaries.

# 5 Conclusions

In this paper, we described a novel approach for generating multimedia summaries from web documents based on semantic extraction and description of documents. The semantics of a document is captured and modeled by a set of triples, i.e. a subject, verb, object for textual information and by a set of keywords for multimedia one.

We then proposed a novel methodology based on cluster analysis of triples, thus obtaining a summary as the sequence of sentences that are associated to the most representative clusters' triples. Multimedia data are then associated to each cluster using a term frequency measure.

Based on this approach, we implemented that provides all the functionalities of a multimedia summarization tool. We tested quality of our summaries using some well-known data sets.

Future work will be devoted to improve the current research into main directions: (i) extend the proposed methodology to the query-based approach; (ii) comparing ore system with other multimedia summarizers.

**Acknowledgments** This research has been partially supported by the FLAG- SHIP InterOmics project (PB.P05, funded and supported by the Italian MIUR and CNR organizations).

# References

1. Adzic, V., Kalva, H., Furht, B.: A survey of multimedia content adaptation for mobile devices. Multimedia Tools Appl. **51**(1), 379–396 (2011)
2. Amato, F., Chianese, A., Moscato, V., Picariello, A., Sperli, G.: Snops: a smart environment for cultural heritage applications. In: Proceedings of the Twelfth International Workshop on Web Information and Data Management, pp. 49–56. ACM (2012)
3. Amato, F., Mazzeo, A., Moscato, V., Picariello, A.: Exploiting cloud technologies and context information for recommending touristic paths. In: Intelligent Distributed Computing VII, pp. 281–287. Springer (2014)

---

[8]For brevity, only the first three positions and the last one are shown.

4. Ankerst, M., Breunig, M.M., Kriegel, H.P., Sander, J.: Optics: ordering points to identify the clustering structure. In: Proceedings of the 1999 ACM SIGMOD International Conference on Management of Data, pp. 49–60. SIGMOD '99, ACM (1999)
5. Blair-goldensohn, S., Neylon, T., Hannan, K., Reis, G.A., Mcdonald, R., Reynar, J.: Building a sentiment summarizer for local service reviews. In. In NLP in the Information Explosion Era (2008)
6. d'Acierno, A., Moscato, V., Persia, F., Picariello, A., Penta, A.: iwin: a summarizer system based on a semantic analysis of web documents. In: Proceedings of the 2012 IEEE Sixth International Conference on Semantic Computing, pp. 162–169. ICSC '12, IEEE Computer Society (2012)
7. Ding, D., Metze, F., Rawat, S., Schulam, P.F., Burger, S.: Generating natural language summaries for multimedia. In: Proceedings of the Seventh International Natural Language Generation Conference, pp. 128–130. Association for Computational Linguistics (2012)
8. Ding, D., Metze, F., Rawat, S., Schulam, P.F., Burger, S., Younessian, E., Bao, L., Christel, M.G., Hauptmann, A.: Beyond audio and video retrieval: towards multimedia summarization. In: Proceedings of the 2nd ACM International Conference on Multimedia Retrieval, p. 2. ACM (2012)
9. Ferreira, R., de Souza Cabral, L., Freitas, F., Lins, R.D., de Franca Silva, G., Simske, S.J., Favaro, L.: A multi-document summarization system based on statistics and linguistic treatment. Expert Syst. Appl. **41**(13), 5780–5787 (2014)
10. Gillick, D., Favre, B.: A scalable global model for summarization. In: Proceedings of the Workshop on Integer Linear Programming for Natural Language Processing, pp. 10–18. ILP '09, Association for Computational Linguistics (2009)
11. Gupta, V., Lehal, G.S.: A survey of text summarization extractive techniques. J. Emerg. Technol. Web Intell. **2**(3), 258–268 (2010)
12. Hahn, U., Mani, I.: The challenges of automatic summarization. Computer **33**(11), 29–36 (2000)
13. Hennig, L.: Topic-based multi-document summarization with probabilistic latent semantic analysis. In: Proceedings of the International Conference RANLP- 2009, pp. 144–149. Association for Computational Linguistics, Borovets, Bulgaria (2009)
14. Luhn, H.P.: The automatic creation of literature abstracts. IBM J. Res. Dev. **2**(2), 159–165 (1958)
15. Mandreoli, F., Martoglia, R.: Knowledge-based sense disambiguation (almost) for all structures. Inf. Syst. **36**(2), 406–430 (2011)
16. McDonald, R.: A study of global inference algorithms in multi-document summarization. In: Proceedings of the 29th European conference on IR research, pp. 557–564. ECIR'07, Springer-Verlag (2007)
17. Mendoza, M., Bonilla, S., Noguera, C., Cobos, C., Lean, E.: Extractive single-document summarization based on genetic operators and guided local search. Ex-pert Syst. Appl. **41**(9), 4158–4169 (2014)
18. Money, A.G., Agius, H.: Video summarisation: a conceptual framework and survey of the state of the art. J. Vis. Commun. Image Represent. **19**(2), 121–143 (2008)
19. Moscato, V., Picariello, A., Persia, F., Penta, A.: A system for automatic image categorization. In: Proceedings of the 3rd IEEE International Conference on Semantic Computing (ICSC 2009), 14–16 September 2009, Berkeley, CA, USA, pp. 624–629 (2009), doi:10.1109/ICSC. 2009.25
20. Nenkova, A., McKeown, K.: A survey of text summarization techniques. In: Mining Text Data, pp. 43–76. Springer, US (2012)
21. Rusu, D., Fortuna, B., Grobelnik, M., Mladenic, D.: Semantic graphs derived from triplets with application in document summarization. Informatica (Slovenia) **33**(3), 357–362 (2009)

22. Takamura, H., Okumura, M.: Text summarization model based on maximum coverage problem and its variant. In: Proceedings of the 12th Conference of the European Chapter of the AC, pp. 781–789 (2009)
23. Wang, D., Li, T.: Weighted consensus multi-document summarization. Inf. Process. Manage. **48**(3), 513–523 (2012)
24. Wu, Z., Palmer, M.: Verb semantics and lexical selection. In: 32nd Annual Meeting of the Association for Computational Linguistics, pp. 133–138 (1994)

# c-Space: A Mobile Framework for the Visualization of Spatial-Temporal 3D Models

Bruno Simões, Matteo Marangon and Raffaele De Amicis

**Abstract** Three-dimensional data acquisition systems are becoming progressively more affordable, especially those that rely on photographic cameras and motion sensing input devices such as the Microsoft Kinect and PlayStation Eye. At the same time, mobile devices capable of rendering complex 3D graphics and with always-on broadband connectivity are becoming increasingly wide-spread. As a result, there is a potential opportunity for the development of novel mobile-oriented streaming mechanisms that can support the visualisation of large and rapid changing spatial-temporal datasets. In this work, we introduce a framework for the visualisation of spatial-temporal point cloud models on mobile devices. One advantage of our framework is that it eliminates the need to pre-downloading spatial-temporal models prior to their visualisation, hence avoiding the need of large storage requirements. Additionally, it streams spatial-temporal geometry using progressive levels of detail that are optimised to the mobile's rendering capabilities and network bandwidth. The representation of different levels of detail is not bounded to a particular geometry reduction algorithm and the streaming process is totally transparent to the user, who perceives remote 3D models as local ones. To evaluate the relevance and impact of this study, a use case is also presented.

## 1 Introduction

Today we are in the midst of another major technological shift: mobile broadband networks are becoming available at flexible prices [1] and mobile devices are set to overtake PCs as the device of choice to access the Internet. This is bringing a new perspective on how we can access and distribute information and is paving the way to new types of interaction. The way information is created is also changing, in this case, driven in part by high-resolution 3D data acquisition technologies, such as those that rely on electronic cameras and motion sensing input devices such as

B. Simões (✉) · M. Marangon · R. De Amicis
Fondazione Graphitech, Trento, Italy
e-mail: bruno.simoes@graphitech.it

© Springer International Publishing Switzerland 2015

E. Damiani et al. (eds.), *Intelligent Interactive Multimedia Systems and Services*,
Smart Innovation, Systems and Technologies 40,
DOI 10.1007/978-3-319-19830-9_4

the Microsoft Kinect and PlayStation Eye, that are becoming progressively more affordable. As a result, this new way of creating and accessing information is revolutionising the design of applications for spatial-temporal data visualisation [2], disaster management [3], urban planning [4], and navigation of large architectural and outdoor spaces [5], which has now the means required to overcome the lack of ubiquity of existent Desktop solutions.

Many of these applications aim at visualising very large datasets. In many scenarios, it is preferable to visualise subsets of data that is stored remotely due to computational and memory requirements. Similarly, the development of new streaming mechanisms plays a very important role in the visualisation of large spatial-temporal datasets on mobile devices. On the one hand, large datasets do not have to be completely downloaded prior to their visualisation. This comes handy when models are too big to fit the device's memory or too large to be stored locally. On the other hand, having a cloud-like environment is useful for scenarios where data has a dynamic nature. Not only this approach keeps data integrity, but it also ensures that clients always receive the most recent data available – an important requirement given the nature of our 4D reconstruction pipeline where the quality of large scenes and events is improved across time with the upload of new video streams.

This paper describes a framework for the visualisation of spatial-temporal point cloud models (that are automatically reconstructed from mobile video streams) on mobile devices. The advantage of our framework is that it eliminates the need to pre-downloading static spatial-temporal models prior to their visualisation. Additionally, it streams spatial-temporal geometry using progressive levels of detail that are optimised to the mobile's rendering capabilities and network bandwidth. The representation of different levels of detail is not bounded to a particular geometry reduction algorithm.

The rest of this work is structured as follows. The next section surveys relevant works on streaming of 3D point clouds, both to analyse if similar studies have been done, and to provide the framework from which to evaluate the relevance and impact of our study. The third section of our study introduces the methodology and implementation details. The fourth presents benchmarks based on a test case and explores the meaning of this study. Section fifth wraps up possible extensibility.

## 2 Related Work

This section surveys recent works on streaming techniques that aim at visualising large datasets. In the past years, the use of streaming techniques to effectively display large datasets (generated by high-resolution 3D data acquisition technologies and stored at remote repositories) was proposed. However, the challenge of visualising large amounts of data on mobile devices, which have limited network bandwidth, memory and computing resources, made practical implementations almost infeasible.

Kammerl proposed a real time compression technique for streaming points as a way to overcome the issue of visualising large point cloud models. Meng et. al. proposed the streaming of a view-dependent level of detail [6]. The geometry inside the field of view is visualised with higher resolution. Another solution, proposed by Rusinkiewicz and Levoy described Streaming QSplat [7], a view-dependent progressive transmission technique based on a pre-processed bounding sphere hierarchy representation of the dataset. Similarly, Rodriguez et al. introduced a progressive transmission technique for mobile devices [8]. Many authors used octree-based point cloud coding. For example, Botsch et al. used octrees to encode sets of points with less than 5 bits per point [9]. Compression techniques such as spanning trees [10] and predictive techniques [11] were also used to organise a not structured point cloud and to perform the downscaling of 3D models [12–14]. Briceno et al. suggested a technique to reorganise dynamic 3D objects into 2D images [15]. The approach can obtain high compression rates even when used with standard techniques for video coding, but the reverse mapping, from 3D to 2D space, is computationally too expensive for real-time applications. Another attempt to promote the idea of streaming geometry is proposed in [16]. However, his work defines only the communication protocol and not the technique used to optimise the data to be transferred.

Finally, another relevant work is the one of Willow Garage [17], which laid the foundations for the Point Cloud Library, an open source project that includes functions and algorithms to manage point cloud data. This library includes the Point Cloud Streaming to Mobile Devices with Real-time Visualization Module which was used in our solution. Yet, not of the above techniques were designed to stream animated point clouds.

## 3 Streaming Framework

In this section, we describe a streaming framework for the visualisation of large spatial-temporal 3D models on mobile devices. The framework is built on top of three processes that are transparent to the users: a process to generate levels-of-detail (LODs) of a 4D model, optimised for each mobile device; a transmission process that uses target-aware optimised compressed forms to avoid memory or storage limitations; and process that capitalises on a data structure to store and exploit the LODs transmitted. Most of our work concentrates on a solution for the last two processes, which we refer to as a device-aware and progressive level of detail geometry streaming.

In the description of our framework, we distinguish between Application Layer (AL), Service Layer (SL), and Data Access Layer (DAL). The Application Layer is responsible for presenting the data to the user. The Service Layer is responsible for processing, transforming, and encoding the application data. The Data Access Layer encapsulates all the methods needed to access and manage the data.

## 3.1 Service Layer

The encoding process takes into account the limited and heterogeneous capabilities of the mobile devices, and finds the balance between the performance and the amount of geometry that they can display. To efficiently stream large datasets, the service layer has to evaluate, in real time, the client Field of View (FOV) and the mobile average frame rate (FPS), see Fig. 1.

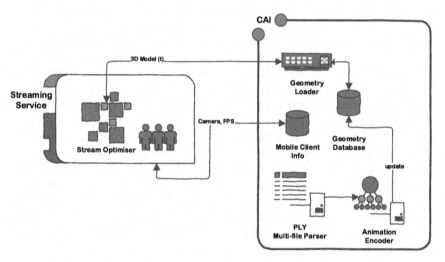

**Fig. 1** Server architecture

The FOV is used to sort the list of points that has to be sent, hence it does not need to be very accurate. The network bandwidth and the mobile average frame rate are used to encode the level of detail of the 3D model that has to be streamed. The sequence of request and responses is the following. The mobile client requests a specific model. The service assigns a univocal identifier to the client and creates its personal user structure. The structure stores the information about the model(s) chosen by the user, the last known number of frames per second, the last known camera view, as well as other variables that trace the user's activity. The Service Layer retrieves the model from the Data Layer and initiates the encoding process.

The encoding process is responsible for downscaling spatio-temporal 3D models and for defining all necessary streaming operations. A downscaled version of the 3D model is used during the first interactions to benchmark the network and the device capability. The downscaling factor is initially set to 512, which means that only 1/512 of the points of the original model are transferred. During the benchmarks, we use a uniform sample of the model defined by the indexes that are divisible by the downscaling factor.

At each iteration the server updates the downscaling factor to a smaller value (if possible) and then streams the new set of points. Since downscaled models do not share common points, the same point is never sent twice. The downscaling factor negotiation process is described in Fig. 2. The variable $n$ is the minimum frame rate considered acceptable.

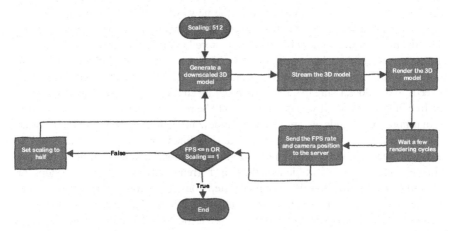

**Fig. 2** Benchmark workflow

The streaming of the animated model initiates once a stable downscaling factor is computed. The streaming of the models is encoded using three different blocks of information. The first block contains the list of points that are missing. The remaining two blocks contain a list of bits that are used to identify points that changed or that can be purged from the device's memory. We did not use a common data structure to create, read, update and delete (CRUD) geometric primitives because it would increase the size of the transfer up by 70 %. Additionally, our structure defines the transfer size to be directly proportional to the number of elements to be transferred, and zero when two or more consecutive frames are equal. Also, it does not require mobile devices to waste computational cycles managing the data, which instead are necessary to visualise large datasets.

## 3.2 Data Access Layer

The objective of this layer is to break down data as much as possible, while maintaining all its original properties, and then to provide access to it. That is, the objective of this layer is to parse the geometry files in one thread while calculating the differences between different frames in a second thread. A few optimisations were implemented to speed up the operation. When possible the parser takes each triplet of short type coordinates (x,y,z) and unsigned short type colours (r,g,b) and merges them into 64bit unsigned integers. Hence, point comparisons can be computed faster.

There are four distinct results when comparing two points. (1) They have the same coordinates and the same colour. In this case, we mark the last point as "not relevant". (2) We consider the last point as "relevant" if they have the same coordinates but not the same colour. We save the index of the first point to the output file, together with a flag ("c" as changed) and the value of the new colour. (3) If the point from the previous frame does not exist in the current frame, then it has to be removed. In this case, the index of the point to be removed is stored in the list of "free indexes buffer" and is then available to be re-used by new points without the need of allocating new memory. (4) If a point in the current frame is not present in the previous frame then it needs to be added to the list. First, the program checks if an index is available from the free indexes buffer. If yes, the first one will be removed from the list and assigned to that point. If no indexes are available, a new index is generated. A flag "a", the index of the point, its coordinates and colour (if defined) are saved to the intermediate storage format. If there are free indexes left after comparing the two frames, then we add their index to the output file with the flag "r" (removed). We repeat these steps for the remaining frames. Another necessary step consists in computing a comparison between the last and the first frame to complete the loop. This method allows us to optimise the index mapping, reducing as more as possible the index fragmentation.

The intermediate storage format is then loaded by the Service Layer into an octree that is used to generate models at a given resolution.

### 3.3 Application Layer

Our mobile application (Application Layer) integrates the following libraries: KiwiViewer Android application from VES, the VTK OpenGL ES Rendering Toolkit [18], and the Point Cloud Library. At the moment, the application runs only on Android devices. The Application Layer has two processes that run in parallel. The first process is responsible for establishing a connection with the server and for handling the stream. The second process is responsible for the visualisation, as well as for handling the user interaction with the 3D model. The information about field of view is stored in a shared memory with the first process and sent to the server when necessary. The information about the field of view is not a necessary requirement, but it is useful to optimise the stream. In fact, the information about the field of view is sent only after considerable changes.

The sequence of request and responses is the following. First the mobile application sends the information about the field of view to the server, as well as the id of the desired model. The information is then stored at the service level, as described in Sect. 3.1. After the initial negotiation, the service layer initiates the benchmark operation. The difference between a normal rendering operation and a benchmark operation is that during the benchmark the service layer has to wait a few a CPU cycles before sending additional data to the client, so it does not affect the frame rate.

The mobile application uses two different structures to store 3D points. The first one serves as a temporary buffer to store a dynamically allocated frame buffer. The second structure is used directly by the renderer thread for rendering purposes. The temporary structure can be manipulated in two different ways, depending on whether we receive the entire model or not. In the first case the old indexes are purged from the main structure and replaced with the new ones. If instead the streamed geometry belongs to a specific time frame, then we simply update the list of points. New points are stored in an empty location when no free indexes are available. All structural updates are performed in $O(1)$ because the local indexes match the ones being manipulated in the Service Layer. Even the indexes of the removed points have a correspondence in the final buffer (zero index), which allows the maintenance of a good performance without wasting memory or the necessity to rebuild the whole structure.

## 4 Evaluation

In the previous sections, we described how the streaming framework works. In this section, we present a use case to evaluate the relevance and impact of this study.

During the benchmarks tests, the streaming server ran on an Intel Core i7-2600 K @3.70 GHz CPU, 16 GB of RAM, OCZ Vertex 3 SSD. The mobile devices tested were the Nexus 7 v.2013, the Nexus 4, and the Galaxy S3, which were connected to the server by a Netgear DGN2200. We tested the mobile clients using a Wi-Fi connection at 300 Mbps and a LAN connection at 100 Mbps. Different network bandwidths were simulated by limiting the Wi-Fi network traffic from 300 to 145 Mbps and then to 54 Mbps.

To test the proposed framework, we created temporal animations for a few well-known datasets. One of the models used was the "Happy Buddha", created by the Stanford University Computer Graphics Laboratory [19]. This model includes a geometric representation of 543,652 points. The size of the uncompressed animated version of the model is around 547 MB. The size of the compressed version is 134 MB. Another model used for testing purposes is the uncompressed 175 MB "Armadillo", from the same repository, which contains about 172974 points. The size of the compressed version is 41.7 MB. We also tested models like the "Bunny" composed of 35947 points and the "Tea Pot" composed of 6468 points.

### 4.1 Measure of Quality

We devised an empirical measure to assess the overall quality of visualisation. The quality factor is calculated based on two key aspects of the overall experience: rendering quality and network utilisation. The rendering quality of a 3D model is defined as the ratio between the size of the entire 3D model and the maximum size that can be visualised interactively on the mobile device. At any instant of time, the rendering quality is given by $f(k) = n/(2^k)$, where $n$ is the number of point of a model and $k$ is the lowest value that provides at least a frame rate $t$. The parameter $t$ is a user defined value (Table 1).

**Table 1** Downscaling factors and streaming time costs

| Downscaling factor | Number of points | Time cost (in seconds) | |
| --- | --- | --- | --- |
| | | First iteration | Next iterations |
| 512 | 9012 | 4.21 | 0.44 |
| 256 | 18054 | 4.09 | 0.29 |
| 128 | 36108 | 4.21 | 0.29 |
| 64 | 72185 | 4.27 | 0.40 |
| 32 | 144400 | 4.38 | 0.66 |
| 16 | 288799 | 4.52 | 0.99 |
| 8 | 577627 | 4.95 | 1.66 |
| 4 | 1155253 | 5.97 | 3.15 |
| 2 | 2310506 | 9.37 | 6.24 |
| 1 | 4621042 | 15.68 | 12.08 |

The quality factor decreases in mobile devices with low computing capabilities because the frame rate depends also on time required to update the geometry and to maintain an optimised frame rate. Additionally, the quality factor decreases when the bandwidth is not enough to maintain an interactive visualisation.

**Table 2** Streaming benchmark on Nexus 7

| Model | Points | Data frames transferred | Average number of operations executed |
| --- | --- | --- | --- |
| Tea pot | 6468 | 30 | 1617 |
| Bunny | 35947 | 30 | 8986 |
| Armadillo | 172974 | 30 | 43243 |
| Happy Buddha | 543652 | 30 | 67956 |

Table 2 shows the number of points that were streamed, the number of data frames transferred, and the average number of points modified per frame. The data transmitted represents only a small fraction of the total size of the model in part due to the lower capabilities of mobile devices tested, emphasising once more the relevance of this framework to the mobile landscape.

Figure 3 depicts the average frame-rate for each model in Table 2. For testing purposed, we defined the minimum number of frames per second (in the negotiation phase) to be equal to 1. The objective was to keep the level of detail constant so we

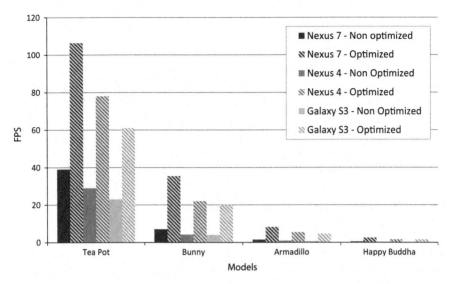

**Fig. 3** Performance comparison of optimised and non optimised models

could evaluate the impact of our streaming operations on different mobile devices. Our framework is proven to be more efficient from a memory and rendering perspective because in both tests the number of points streamed was the same.

The impact of the network bandwidth was also studied. Figure 4 shows how bandwidth can limit the number of points that can be transmitted per second. The limitation is quite visible for downscales smaller than 128.

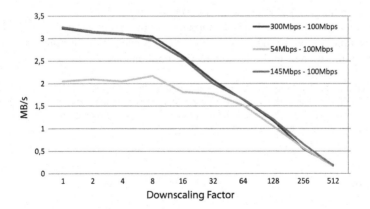

**Fig. 4** Happy Buddha - Nexus 7 2013. Transfer Rate (MB/s) - Scaling Factor

Lastly, Fig. 5 depict how the network speed and downscale factor affect the rendering quality.

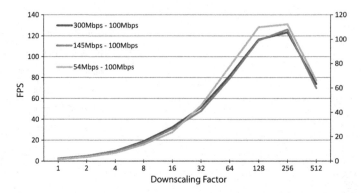

**Fig. 5**  Happy Buddha - Nexus 7 2013. FPS - Scaling Factor

## 5 Conclusion

This paper describes a framework for the visualisation of spatial-temporal point cloud models on mobile devices, which is built on top of three processes: a process to generate multiple levels of detail (LODs) of a 4D model, optimised for each mobile device connected; a transmission process that uses target-aware optimised compressed streams to avoid memory or storage limitations; and process that capitalises on a data structure to store and exploit the LODs transmitted.

The advantage of our framework is that it adapts to the mobile client's rendering capabilities, network bandwidth and user motion by automatically readjusting the level of detail of the 3D model, so the user experience while navigating the virtual environment is as fluid and jitter free. The results of our first benchmarks were promising both from a memory and rendering perspective.

Future improvements include the implementation of a downscaling factor based on the complexity of the 3D features, instead of a percentage of points to be sent, and the optimisation of multiple streams (in simultaneous) to the same device.

**Acknowledgments**  This research has been supported by the European Commission (EC) under the project c-Space. The authors are solely responsible for the content of the paper. It does not represent the opinion of the European Community. The European Community is not responsible for any use that might be made of information contained herein.

## References

1. Ericsson, A. Ericsson mobility report. http://www.ericsson.com/res/docs/2013/ericsson-mobility-report-june-2013.pdf, tech. rep., Ericsson AB, Tech. Rep., 2013
2. Nüchter, A., Lingemann, K.: Robotic 3d scan repository. http://kos.informatik.uni-osnabrueck.de/3Dscans/

3. Simões, B., Amicis, R.D.: User-friendly interfaces for web gis. In: The 2011 International Conference on Internet Computing, pp. 176–182. CSREA Press (July 2011)
4. De Amicis, R., Conti, G., Simões, B., Lattuca, R., Tosi, N., Piffer, S., Pellitteri, G.: Geo-visual analytics for urban design in the context of future internet. Int. J. Interact. Des. Manuf. (IJIDeM) 3(2), 55–63 (2009)
5. Beck, M.: Real-time visualization of big 3d city models. International Archives of the Photogrammetry Sensing and Spatial Information Sciences 34, (2003)
6. Meng, F., Zha, H.: Streaming transmission of point-sampled geometry based on view-dependent level-of-detail. In: Proceedings Fourth International Conference on 3-D Digital Imaging and Modeling (3DIM 2003), pp. 466–473. IEEE (2003)
7. Rusinkiewicz, S., Levoy, M.: Streaming qsplat: a viewer for networked visualization of large, dense models. In: Proceedings of the 2001 Symposium on Interactive 3D Graphics, ser. I3D '01, pp. 63–68. New York, NY, USA, ACM (2001). http://doi.acm.org/10.1145/364338. 364350
8. Rodriguez, M.B., Gobbetti, E., Marton, F., Pintus, R., Pintore, G., Tinti, A.: Interactive exploration of gigantic point clouds on mobile devices. In: VAST, pp. 57–64 (2012)
9. Botsch, M., Wiratanaya, A., Kobbelt, L.: Efficient high quality rendering of point sampled geometry. In: Proceedings of the 13th Eurographics Workshop on Rendering, ser. EGRW '02, pp. 53–64. Aire-la-Ville, Switzerland, Switzerland: Eurographics Association (2002). http:// dl.acm.org/citation.cfm?id=581896.581904
10. Merry, B., Marais, P., Gain, J.: Compression of dense and regular point clouds. In: Computer Graphics Forum, vol. 25, no. 4, pp. 709–716. Wiley Online Library (2006)
11. Gumhold, S., Kami, Z., Isenburg, M., Seidel, H.-P.:Predictive point-cloud compression. In: ACM SIGGRAPH 2005 Sketches, p. 137. ACM (2005)
12. Schnabel, R., Klein, R.: Octree-based point-cloud compression. In: SPBG, pp. 111–120 (2006)
13. Huang, Y., Peng, J., Kuo, C.-C.J., Gopi, M.: Octree-based progressive geometry coding of point clouds. In: Proceedings of the 3rd Eurographics/IEEE VGTC Conference on Point-Based Graphics, pp. 103–110. Eurographics Association (2006)
14. Pauly, M., Gross, M., Kobbelt, L.P.: Efficient simplification of point-sampled surfaces. In: Proceedings of the Conference on Visualization'02, pp. 163–170. IEEE Computer Society (2002)
15. Briceño, H.M., Sander, P.V., McMillan, L., Gortler, S., Hoppe, H.: Geometry videos: a new representation for 3d animations. In: Proceedings of the 2003 ACM SIGGRAPH/Eurographics Symposium on Computer Animation, pp. 136–146. Eurographics Association (2003)
16. Jovanova, B., Preda, M., Preteux, F.: Mpeg-4 part 25: a generic model for 3d graphics compression. In: 3DTV Conference: The True Vision-Capture, Transmission and Display of 3D Video, pp. 101–104. IEEE (2008)
17. Rusu, R.B., Cousins, S.: 3d is here: point cloud library (pcl). In: 2011 IEEE International Conference on Robotics and Automation (ICRA), pp. 1–4. IEEE (2011)
18. VTK, Ves, a vtk opengl es rendering toolkit. http://www.vtk.org/Wiki/VES
19. Stanford University: The stanford 3d scanning repository. http://www-graphics.stanford.edu/ data/3Dscanrep/

# The Method to Verify Facial Shape Model

Seonwoon Kim and Seokhoon Kang

**Abstract** This paper proposes a method to verify the facial shape model. The method can be used to improve or use of the shape model. We verify the shape model by analyzing the texture and shapes of landmarks that constitute the outline of the shape. It generates the facial face image of the face that is tracked for the invariant verification method to poses of the face. Next, it reduces the effect of environment by configuring a patch surrounding each landmark. The texture of the skin area to these patches is analyzed by the gray-value variance. Then, the result of analysis is corrected by the relationship between the landmarks of the shape. As a result, we identified the fitting result as true or false to the Multi-PIE database and obtained the accuracy of 83.32 %.

**Keywords** Gray-value variance · Shape model · Human-computer interaction

## 1 Introduction

The expectation for the Human Computer Interaction areas such as face recognition or interface has been increased for a long time. Accordingly, the study for the face tracking based on shape model, the base of the technique, has become more important because the fitting technique that is based on the shape model is a technique that can analyze faces in various conditions. The main study is alignment of shape model and the feature detection about landmark. We have determined that it is necessary to be verified before the shape model is utilized. For this, we propose

S. Kim
Mobile Solution Division, TecAce Solutions Inc, Incheon, Korea
e-mail: swkim@tecace.com

S. Kang (✉)
Department of Embedded System Engineering, Incheon National University, Incheon, Korea
e-mail: hana@inu.ac.kr

© Springer International Publishing Switzerland 2015
E. Damiani et al. (eds.), *Intelligent Interactive Multimedia Systems and Services*,
Smart Innovation, Systems and Technologies 40,
DOI 10.1007/978-3-319-19830-9_5

the method to determine whether a fitting of landmark is appropriate through the separate verification procedure from the tracking method.

This method applies the procedure to the contour landmarks that constitute a face shape. The outline of a face is a clear distinction element common to all faces. Since it is the side that contacts with the background, it is relatively easy to navigate and could be accurate. Because the shape model goes through the process of alignment with the relationship of all landmarks, by the verification of landmarks that constitute the outline, the success or failure of the fitting of shape model can be determined. We use a gray-value variance for the verification of landmarks constituting outlines. This method is appropriate to analyze the texture of skin.

In this paper, the proposed method will be described in detail. First, in Sect. 2, the technique of shape fitting and the similar studies to the present study are described. In Sect. 3, we explain the techniques we used in detail. In Sect. 4, the performance of our method is verified, and the conclusion is finally discussed in Sect. 5.

## 2 Relative Work

### 2.1 Face Shape Model and Fitting Method

Shape Model is a very useful method for tracking a deformable shape for an object. This technique has been firstly started from the Active Shape Model (ASM) [1]. In this study, the fitted shape is generated by

$$\mathbf{x} = \bar{\mathbf{x}} + \mathbf{P}_s \mathbf{b}_s, \tag{1}$$

where $\mathbf{x}$ be generated shape, $\bar{\mathbf{x}}$ is mean shape, $\mathbf{P}$ is a first eigenvector of deviation from mean shape and $\mathbf{b}_s$ is a vector of weights. Theses parameters are trained from training set. After movement vectors of each landmark using a method of detecting feature, the varying is calculated by

$$d\mathbf{b} \approx \mathbf{P}_s^T d\mathbf{x}, \tag{2}$$

where $d\mathbf{b}$ is shape parameter adjustments, $d\mathbf{x}$ is obtained movement vector. Then a parameter such as scale, translation rotation is calculated. This procedure is iterated until the parameters are converged. Here, calculating the parameter by approximate calculation can be seen. The basic shape model uses a method that applies each weight to variation modes of trained shape. This is a method that stably maintains the shape by maintain the shape and at the same time has a limitation of variation by the form of untrained shape. Further, since it searches the weights as the method that is approximating as much as possible, it is difficult to expect a perfect match.

Based on this, the Active Appearance Model (AAM) that even considers the appearance has been proposed [2]. The model of AAM uses the method of combining the shape model in ASM and the gray model. The gray model is defined as

$$\mathbf{g} = \bar{\mathbf{g}} + \mathbf{P}_g \mathbf{b}_g, \tag{3}$$

where $\mathbf{P}_g$ is set of orthogonal modes of variation, $\mathbf{b}_g$ is set of gray-level parameters. Therefore, the vector that combines shape model parameter $\mathbf{b}_s$ and gray-level parameter $\mathbf{b}_g$ is defined as a vector $\mathbf{b}$. Then a model is obtained using a principal component analysis. Therefore, vector $\mathbf{b}$ is defined as

$$\mathbf{b} = \mathbf{Qc}, \tag{4}$$

is obtained from principal component analysis. $\mathbf{Q}$ is eigenvector, and $\mathbf{c}$ is vector of appearance parameter that controls the shape and gray-levels. In the iterative process, this vector to be updated is calculated as

$$\delta \mathbf{c} = \mathbf{A}\delta \mathbf{g}, \tag{5}$$

where $\delta\mathbf{g}$ is evaluated error vector with normalized gray-levels, $\mathbf{A}$ is a relationship that is obtained by difference between the vector of gray-value in image and the gray-level values model displacements $\delta\mathbf{c}$ using multivariate linear regression. In this process $\mathbf{c}$ is repeatedly updated, the repeated is finished when the error vector $\delta\mathbf{g}$ is converged. It is repeated until it is converged adjusting the gray-level appearance and searches parameters that are closely converged.

And constrained local model (CLM) uses a strategy that constructs response maps with patches [3]. The response maps are generated for each patch. It uses the Nelder-Meade simplex algorithm that induces that parameter of shape model by maximizing the sum of the responses of each landmark. This method also finds an approximate result.

Now these techniques are being developed through the improved researches. In addition to the typical method, various shape model techniques are also proposed. The reviewed methods can not generate all of the shape that is used training. Also, fitting procedure uses approximate solution to linear or non-linear. Therefore, the verification of the landmarks can assist in improving performance and utilization.

## 2.2 Most Related Approaches

In [4], pose-invariant face recognition has been proposed. Here, the boundary of the face is extracted using seam carving and it goes through the process of updating the shape generated by the view-based active appearance model. The seam is a patch of pixels connected with minimum energy cost. The energy function values a pixel by measuring its contrast with its neighbor pixels. In this study, a path is defined as an optimal curve and the boundary of the shape is updated. This method compensates the part that the shape model cannot be transformed by using a separate technique.

# 3 Verifying Shape Model

The appearance to the around landmark for the shape model is verified. We check on the smooth texture of the inner side of the face and the edge of border that connects the landmarks.

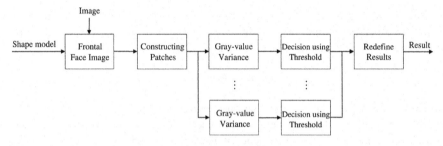

**Fig. 1** Flow of Verifying the facial shape model

As shown in Fig. 1, mean shape, tracking image, and points of shape are used, which are the trained information of techniques being tracked. First, using the triangulation method, tracking image is projected on the mean shape. As a result of this, the image similar to the frontal face can be obtained. In this image, the patches are constructed that contain the pixels around each points of the mean shape. In these patches, the gray-value variance is calculated and the texture of the smooth skin is identified. Furthermore, the result is adjusted by the relationship of landmarks which constitute the face.

## 3.1 Frontal Face with Patches

In this section, it goes through the process of getting the right information to verify the fitting in each landmark. In the pose invariant face model, as 2d image is tracked by the information of the 3d rotation. Also, the image will include various factors such as light direction, mustache, and so on. We use the frontal face image and patches in order to reduce the influence of the information. The whole process uses the transformed image to gray-levels. In order to compare under the same conditions for images applied by pose or scale, it uses the image that is projected in the same shape. We used the mean shape used in the tracking method. Also, it designates a rectangular area around the points as the patch. Reviewing a portion can reduce the influence on the gradient by light, or the images in which a part is covered.

## 3.2 Verifying Texture Using Gray-Value Variance for Patches

The patch that is constituted by the landmarks in the center of the boundary of a face is configured with skin and background. Here, we use the mask used in the projection of the frontal face in the previous triangulation method. As a result, we will have only the patch of the skin area. We use the method that can distinguish between the skin texture.

The gray-value variance means the distribution to the gray-value. Therefore, we determines by a smooth texture calculated from the gray-value variance. Basically, the calculation of variance to the gray-values set is calculated by

$$\sigma = E\left(\mathbf{g}^2\right) - E^2(\mathbf{g}),\qquad(6)$$

where $E$ is function that calculates the expected value, and $\mathbf{g}$ is the listed set of gray-value in the patch in 1-dimension. As noted above, the background area should be excluded through the mask $\mathbf{m}$. Therefore, the expected value is defined where the mask value $m_i$ is 0 as

$$EM(\mathbf{g}, \mathbf{m}) = \sum_{i=0}^{N} \mathbf{g_i m_i},\qquad(7)$$

where $m_i$ the set of values which are transformed the mask $\mathbf{m}$ of 2-dimension to 1-dimension that has the value of 0 and 1. The gray-value variance that excludes the area of mask $m_i$ is 0 using Eq. (8) is calculated by

$$\sigma' = EM\left(\mathbf{g}^2, \mathbf{m}\right) - EM^2(\mathbf{g}, \mathbf{m}).\qquad(8)$$

We decide a threshold for this value and determine the landmark with the gray-value variance higher than the threshold value is not correctly aligned.

## 3.3 Considering Relation Between Landmarks

In the previous process, we calculate the gray-value variance except the mask and checked for skin texture. This method cannot acquire a perfect result like other landmark feature detection methods because those still have the problem getting the wrong gray-value variance by the environment like light or beard. We correct it by the relationship between the landmarks of the face model.

Since the outline of the shape model is curve, only one result among the results of neighbor cannot be different. On the other hands, there are several points that out of the boundary with low value of gray-value variance. This phenomenon occurs when a landmark is placed on a simple background. To solve this case, we calculate the gray-value variance of the patches which includes the mask. We create another frontal image for this. Then, we configures the patches as same as the previous step and gets the gray-value variance that does not exclude the mask by the difference between the previous Eq. (6) and the Eq. (8) as

$$d = \sigma - \sigma'. \tag{9}$$

If $d > 0$, it is because that it includes the background. In the case of the complex background, since the gray-value variance is measured as high in the previous step, it is assumed as already being classified. On the other hand, if $d \leq 0$, it is defined that it is placed on the inside of the skin or the solid background. Based on this, the weakness of the method that distinguishes using the calculation of the gray-value variance is made up.

## 4   Experiment

The performance of the method we propose is evaluated on the CMU Multi-PIE database [5]. This database has various poses of human faces from different positions of cameras. Also, it varies the position of the sources in the 20 direction in the same photo. It consists of the faces of 346 people like that. We use 75,360 photos included in the set of the three directions of the camera position. Ground truth landmarks construct the shape model with 68 points.

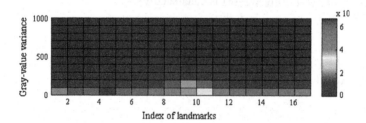

**Fig. 2** Gray-variance of the patch around the ground truth landmark in the Multi-PIE

We verify the validity of this method using the Ground truth landmarks. The result of measured gray-value variance is shown in Fig. 2.

As shown in the picture, the gray-value variance of 87.52 % of all the landmarks has a value of less than 200. Through this result, the method we propose showed its possibility. In addition, the result that considers relationship between landmarks is shown. As a result, the gray-value variance of 93.15 % of the landmarks has a value less than 200.

Here we use the fitting method [7] in order to measure the threshold of the appropriate gray-value variance. If the distance error between ground truth landmark and landmarks fitted shape is lower than 3 pixel, we decided the fitting result is success. Table 1 is the result of measuring the receiver operating characteristic only using the gray-value variance as threshold. As seen in the table, when the threshold of gray-value variance is 228, the accuracy is measured highest as 83.32 % (Fig. 3).

**Table 1** Results of verifying outer landmarks using g according to the threshold in the Multi-PIE

| Threshold | Total | Sensitivity | Specificity | PPV[a] | NPV[b] | ACC[c] |
|---|---|---|---|---|---|---|
| 100 | 1,277,091 | 51.90 % | 80.53 % | 87.46 % | 39.02 % | 59.82 % |
| 200 | 1,277,091 | 81.68 % | 75.48 % | 89.71 % | 61.16 % | 79.96 % |
| **228** | **1,277,091** | **87.54 %** | **71.84 %** | **89.07 %** | **69.09 %** | **83.32 %** |
| 300 | 1,277,091 | 89.39 % | 66.63 % | 87.52 % | 70.79 % | 83.17 % |
| 400 | 1,277,091 | 93.33 % | 23.81 % | 76.21 % | 57.70 % | 74.10 % |

[a] Precision or positive predictive value
[b] Negative predictive value
[c] Accuracy

For the experiment in the sequence of video, by applying fitting method [7] in the FGNet talking face database, our method is checked on. The FGNet talking face database consists of face images with various facial expressions and natural movement and the ground truth landmarks [6]. Similarly, the fitting method [7] is

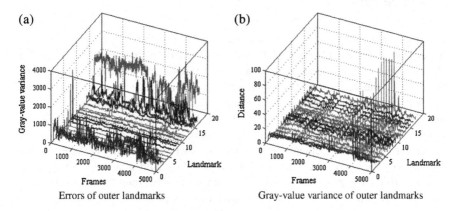

(a)  Errors of outer landmarks

(b)  Gray-value variance of outer landmarks

**Fig. 3** Each landmark error and gray-value variance in the FGNet database

applied in this database. As a result, the figure shows the error between the landmarks consist of face boundary of fitting method and the landmarks consist of face
boundary of database, and gray-value variances that corresponding each landmark.
Most of the landmarks are tracked with the error of around 5 pixel. Accordingly, the
gray-value variance is also stably measured. However, the 0th and 17th landmark
can be seen that their value of the gray-value variance is unstable by the hair. This
result can be compensated by applying the dealt method in Sect. 3.3.

**Fig. 4** Examples of the result of verifying landmarks in the complicated background. The red
point represents the failure of fitting and the green point represents the success (Color figure
online)

Finally, the Fig. 4. shows the face shape model fitting in the complex background
and its consequence landmark verifying results. The fitting method in a complex
background generates more cases that stay in the background due to the features of
the background which could be easily misunderstood. In tracking with simple
background, the technique of verifying landmarks considering relationship with
landmarks more necessary. Generally, the function of verifying landmarks shows the
better result in the previous measurement results of the database. If it is determined
that the landmark is misaligned with our method, it is shown as a red dot.

## 5 Conclusion

In this paper, we proposed the method that verifies the fitting result of landmarks
that configure the face boundary of the face shape model. The tracking based on the
shape models used the shape fitting method and alignment method. In addition to
this process, the verification process is subjected to a further process. We analyze
the texture of the skin by gray-value variance, and studied the process that considers
the relationship between the landmarks of the shape. The gray-value variance is a
method that measures the softness of the texture and has a characteristic that is
proper to recognize the human skin. In addition, since the boundary of the facial has
the shape of the curve, it can be used complementing with gray-value variance.

We verified the proposed method by measuring the ground truth to the face
image of the Multi-PIE and that the gray-value variance has a possibility. In
addition, we checked out the performance by applying the fitting method about
whether the fitting result is appropriate or not. The fitting method was applied to

these pictures and the suitability of performance is confirmed. In addition, in a series of images of the FGNet talking face database, it showed the features of the results for each landmark and the relationship between the error and gray-value variance. Finally, by showing the verification result in the actual complex image, the result in the actual use was presented as a figure.

The proposed method is useful to determine whether the shape model is proper to use in the systems using the facial shape model. In addition, it checks the landmarks of the outside face through our method, and could be used to search for a better point.

# References

1. Cootes, G.T.F., Taylor, C.J., Cooper, D.H., Graham, J.: Active shape models-their training and application. Comput. Vis. Image Underst. **61**(1), 38–59 (1995)
2. Timothy, F.C., Gareth J.E., Christopher J.T.: Active appearance models. IEEE Trans. Pattern Anal. Mach. Intell. **23**(6) (2001)
3. Cristinacce, D., Cootes, T.F.: Feature detection and tracking with constrained local models. Proc. British Mach. Vision Conf. **3**, 929–938 (2006)
4. Asthana, A., et al.: Fully automatic pose-invariant face recognition via 3d pose normalization. In: IEEE International Conference on, Computer Vision (ICCV) (2011)
5. Gross, R., et al.: Multi-PIE. In: 8th IEEE International Conference on Automatic Face and Gesture Recognition, 2008. FG'08. IEEE (2008)
6. Cootes, T.F., et al.: Groupwise construction of appearance models using piece-wise affine deformations. BMVC, vol. 5 (2005)
7. Jason M.S., Simon L., Jeffrey, F.C.: Deformable model fitting by regularized landmark mean-shift. Int. J. Comput. Vision, pp. 200–215 (2011)

# Effective Visualization of a Big Data Banking Application

**Luigi Coppolino, Salvatore D'Antonio, Luigi Romano, Ferdinando Campanile and Alexandre Valle de Carvalho**

**Abstract** Data analysis and monitoring is currently carried out within enterprises using Business Intelligence tools that are subject to major limitations (as outlined in the state of the art analysis that we perform). Effective visualization support is a very much needed feature in Big Data applications. In this paper we examine the visualisation requirements of a real world banking application, and identify generic visualisation tasks that are essential for doing effective analysis of a complex process that produces amazingly large amounts of data. The requirements for the visualization support that we propose are modelled using an application wireframe that acts a story-board. The effectiveness of the visualization facilities that we propose is demonstrated through their application to the Big Data banking use-case.

**Keywords** Visualization · Big data · Cloud computing · Banking fraud · Business intelligence · Direct debit

L. Coppolino (✉) · S. D'Antonio · L. Romano
Department of Engineering, University of Naples "Parthenope", Naples, Italy
e-mail: luigi.coppolino@uniparthenope.it

S. D'Antonio
e-mail: salvatore.dantonio@uniparthenope.it

L. Romano
e-mail: luigi.romano@uniparthenope.it

F. Campanile
Sync Lab S.r.l, Naples, Italy
e-mail: f.campanile@synclab.it

A.V. de Carvalho
FEUP/INESC Porto, Porto, Portugal
e-mail: alexandre.carvalho@inescporto.pt

© Springer International Publishing Switzerland 2015                    57
E. Damiani et al. (eds.), *Intelligent Interactive Multimedia Systems and Services*,
Smart Innovation, Systems and Technologies 40,
DOI 10.1007/978-3-319-19830-9_6

# 1   Introduction

Data analysis and monitoring is currently carried out within enterprises using Business Intelligence (BI) tools that are subject to major limitations. Big Data applications pose challenges that cannot be coped with via existing BI technologies. In this paper we examine the visualisation requirements of a real world banking application. The application deals with the electronic alignment of Direct Debit (DD) transactions, which is a key business function of modern banking systems. In a DD financial transaction, one party withdraws funds from another party's bank account (both parties can either be a company or a person). The party receiving the funds is called "the payee", the one being charged is called "the payer". To set up the process, the payer must advise his/her bank that he/she authorizes the payee to directly draw the funds from a specified bank account belonging to him/her (for this reason, DD is also called pre-authorized debit (PAD) or pre-authorized payment (PAP)). DDs are available in a number of countries, where they are made under each country's specific rules, and are usually restricted to domestic transactions. In Italy, DDs must comply to the rules set out by SEPA DD [12] and CBI-STD-001 [13]. DDs are massively used for recurring payments, and in particular utility and credit card bills. A major risk factor lies in the fact that the authorisation that is given - i.e. the circumstances in which the funds can be drawn - are a matter of agreement between the payee and the payer, of which the bankers are not concerned. The difficulty of enforcing controls is exhacerbated in particular by the fact that, since payment amounts vary from one payment to another, the payer typically grants authorizations that are based on a "worst case" estimation of the potential transaction volumes. Even in countries where a number of controls are enforced on the authorization set up process, the problem of direct debit fraud is extensive. A recent research by Liverpool Victoria Insurance [14] reveals that over 97,000 Britons have fallen victim to criminals setting up fraudulent direct debits from their accounts, and that this trend is increasingly sharply. The pre-condition for a DD fraud is – almost invariably – an identity theft. As such, it is typically detected, eventually. Thus, once the fraudster has successfully set up a DD authorization from the victim's account, he/she would use the newly created fraudulent authorization for irregular payments. An additional problem is also represented by cancelled and/or obsolete DDs being wrongfully or maliciously revived or re-implemented. In both the aforementioned misuse cases, the fraud (or the system failure, respectively) are typically characterized by an anomalous increase in the frequence/amount of transactions. With current computing technology (typically, BI), although no specific figures are - to the best of our knowledge - available, it appears that a substantial number of people lose considerable amounts of money annually because of fraudulent and revived/re-implemented DD authorizations, since frauds/failures go unnoticed for a relatively long time. In order to timely spot frauds, computing and visualization facilities are needed, that allow fraud analysts to extract anomalous transactions from the Big Data flows in real-time [6], and inspect them in detail.

Based on a requirement analysis of the case study application, via a generalization process, we identify generic visualisation tasks that are essential for doing effective analysis of a complex process that produces amazingly large amounts of data [3]. The requirements for the visualization support that we propose are modelled using an application wireframe, that acts a story-board. The effectiveness of the visualization facilities that we propose is demonstrated through their application to the Big Data banking use-case.

The rest of the paper is organized as follows. Section 2 provides a brief yet right to the point treatment of the state of the art of currently available BI technologies. Section 3 describes the banking application that is used as a reference for extracting the requirements of the visualization support that we propose in Sect. 4. Finally, Sect. 5 concludes the paper with lessons learned and final remarks.

## 2 Business Intelligence: State of the Art and Limitations

Business Intelligence (BI) platforms aim at supporting data scientists to do data retrieval, processing and analysis on computing platforms [9]. In this section we provide a brief yet right to the point state of the art analysis of current BI offerings, and we pin point their main limitations, with respect to the new challenges posed by Big Data applications [1]. BI tools provide a front-end user-interface that allows users to write structured queries, view query results, generate charts and construct reports. Traditionally BI [2] has focussed on generating descriptive statistics of data that has been pre-processed by an ETL tool (Extract, Transform and Load) and stored in a data mart. The schemas for the data mart are pre-defined based on business goals resulting in fast query responses but with limited flexibility. Increasing data volumes and the need for more flexible querying led to the development of Massively Parallel Processing (MPP) [4] database architectures and distributed storage and processing frameworks such as Hadoop [5]. Many BI platforms are providing API's to make use of these technologies.

The rise of the Big Data phenomenon is disrupting the BI market with a number of emerging vendors providing "Advanced Analytics" platforms. Gartner describes Advanced Analytics as, "the analysis of all kinds of data using sophisticated quantitative methods (for example, statistics, descriptive and predictive data mining, simulation and optimization) to produce insights that traditional approaches to business intelligence — such as query and reporting — are unlikely to discover" (source – Gartner [10]). New trends are starting to emerge within the BI market [11]. These include:

- **Agile self-service BI**: Many advanced analytics platforms include visual query builders that support drag and drop query workflow building and execution. This codeless querying will become more widespread.
- **Improved Visualization**: Extended charting libraries with improved composition capabilities are expected.

- **Broader device support**: Trends around increasing self-service BI and Visual improvements open up possibilities for touchscreen and gestural query composition.
- **Everything in the Cloud**: Fully browser-based GUI's offer improvements for limiting data transfer, collaboration and device support.

In our state of the art analysis we select a number of BI tools and projects that provide analytics functionality. We focus on Open Source tools, as these may be of particular interest to readers engaged in research projects that aim at leveraging existing technologies by adding additional functionality.

**RapidMiner** (https://rapidminer.com/) is a software platform that provides an integrated environment for machine learning, data mining, text mining, predictive analytics and business analytics. RapidMiner uses a client/server model with the server offered as Software as a Service or on cloud infrastructures. Functionality can be extended with additional plugins and the Rapid Miner Extensions marketplace provides a platform for developers to create data analysis algorithms and publish them to the community. RapidMiner is distributed under the AGPL open source license. RapidMiner is very popular and has a wider user group across domains. The workflow element of the software is part of RapidMiner Studio, a desktop application developed in Java and data must be downloaded to the client before workflows can be built.

**KNIME** (https://www.knime.org/) is an open source data analytics, reporting, and integration platform. KNIME allows users to visually create data flows (or pipelines), selectively execute some or all analysis steps, and later inspect the results, models, and interactive views. KNIME is written in Java and based on Eclipse and makes use of its extension mechanism to add plugins providing additional functionality. As of version 2.1, KNIME is released under GPLv3. As with RapidMiner, KNIME uses a client/server model with the expectation that data will be loaded onto the client before workflows are constructed.

**BIRT** (http://eclipse.org/birt/) is an open source technology platform used to create data visualizations and reports that can be embedded into rich client and web applications. BIRT has two main components: a visual report designer within the Eclipse IDE for creating BIRT Reports, and a runtime component for generating reports that can be deployed to any Java environment. The BIRT project also includes a charting engine that is both fully integrated into the report designer and can be used standalone to integrate charts into an application. BIRT is released under an Eclipse Public License.

**Chorus** (http://alpinenow.com/) is a collaborative platform for data science. It exists as both a commercial offering by Alpine Data Labs and as an OpenSource project name OpenChorus. The commercial version of the platform includes a Workflow tool that allows you to design an analytic process visually. However, this functionality is not available in the open-source version. Unlike other platforms, Chorus was developed as a browser based tool and has a JavaScript front-end with a ruby on rails backend. This opens up potential for collaboration and improved performance with big data, as it supports moving the modelling to the data rather than forcing data transfer.

**Jaspersoft** (https://www.jaspersoft.com) provides a suite of tools that bring benefits both to business and to IT audiences, enabling self-service BI for organizations of all sizes. The Jaspersoft BI Suite provides full-featured reporting, dashboards, and analytics that can be embedded in a web-based application. The platform is available on several editions: Community, Express, AWS, Professional and Enterprise. The Community Edition is the only one that is open-source, but it is stripped of the major features relevant for our purposes, such as dashboards, interactive visualizations, embedded data visualizations and reports inside web applications and storing interactive views.

The **Pentaho** (http://www.pentaho.com/) solution aims on leveraging innovation, participation and cooperation. This solution has a solid Enterprise Edition. However, the community edition focuses on enabling a jump-start of the development, in order to innovate and experiment, until a solution is found, and then upgrade to the Enterprise Edition for production environments. The open-source version lacks some important features for our purposes, such as aggregation designer and schema workbench.

In conclusion, all of the existing solutions examined have major limitations, with respect to some important requirements of applications that handle large amounts of data in real-time (which is typical of Big Data applications). Based on this analysis, in the next section we present the design of a visualization platform that provides effective visualization support to Big Data applications.


# 3 The Banking Application

The Single Euro Payments Area (or "SEPA" for short) is where more than 500 million citizens, over 20 million businesses and European public authorities can make and receive payments in euro under the same basic conditions, rights and obligations, regardless of their location.

The overall gains expected from SEPA for all stakeholders has been evaluated at €21.9 billion per year by PWC in 2014 confirming a Cap Gemini study of 2008 evaluating these benefits at €123 billion cumulated over 6 years.

SEPA adopts the ISO 20022 standard, a multi-part International Standard prepared by ISO Technical Committee TC68 Financial Services. It describes a common platform for the development of messages, using:

- A modelling methodology to capture in a syntax-independent way financial business areas, business transactions, and associated message flows;
- A central dictionary of business items used in financial communications;
- A set of XML and ASN.1 design rules to convert the message models into XML or ASN.1 schemas, whenever the use of the ISO 20022 XML or ASN.1-based syntax is preferred.

In Italy, by the 1st February 2014 domestic credit transfers, banking (RID) and postal direct debits will have to be replaced by the corresponding SEPA instruments.

SEPA migration requirements are set by European and national regulations which have to be fully respected without exception. In Italy the SEPA standard adopted is slightly different from the canonical one (ISO 20022 standard). In particular, the standard adopted for the SDD request is: "CBIBdySDDReq.00.01.00".

The recent adoption of the Single Euro Payments Area (SEPA), which follows the European Union (EU) payments integration initiative (http://ec.europa.eu/finance/payments/sepa), has moved more attention to the mechanisms to avoid frauds in banking/financial transactions.

As far as an SDD (SEPA Direct Debit) transaction is concerned, the SEPA standard has simplified a lot the payment process, while moving the consequences of a fraud from the user to the bank. In particular in an SDD transaction one person/company withdraws funds from another person's bank account. Formally, the person who directly draws the funds ("the payee or creditor") instructs his or her bank to collect (i.e., debit) an amount directly from another's ("the payer or debtor") bank account designated by the payer and pay those funds into a bank account designated by the payee.

Typically examples of SDD transactions are services that require recurrent payments, such as pay per view TV, energy distribution, credit card etc.

To set up the process, the creditor has to acquire an SDD mandate from the debtor and advise his/her bank about that mandate.

Each time it will be needed, the Creditor sends a direct debit request (with amount specification) to his/her bank that will start the process to request the specified amount on the Debtor's bank account.

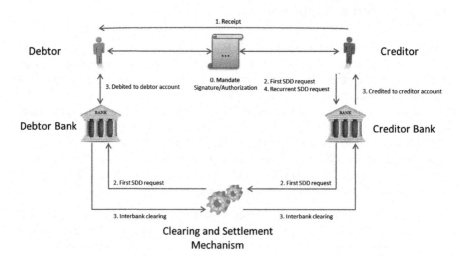

**Fig. 1** SEPA Direct Debit process

The debtor only has to provide the signature of the mandate and the debtor could not receive communications about the SDD request. The debtor has no prior acknowledgement about the direct debit being charged to his bank account. Typically, the creditor sends a receipt to the debtor using a best effort service, so no guarantee is provided about delivery time and delivery itself. The debtor will only have access to the direct debit after the transaction has been completed. He/she can identify an unauthorized SDD amount only when it receives its bank statement (Fig. 1).

Of course this exposes the debtor to a number of possible frauds. For this reason, with SEPA, in case of error/fraud with an SDD, a debtor can request for a refund until: 8 weeks after the SDD deadline or 13 months for unauthorized SDD (no or revoked mandate). The SDD process is shown in.

The Financial Banking application monitors the debit request to try to recognize unauthorized debit, thus providing an anti-fraud system helping the financial institution to reduce costs due to frauds. Unauthorized debits are typically due to an Identity Theft that occurs at the beginning of the process, when the SDD mandate is being created.

The SDD mandate is authorized not by the debtor, but by someone impersonating the debtor. For instance, this identify thief will benefit from the product/service being provided and charged by the creditor but will not pay for the service.

The debtor will be charged for a product service that did not acquire and, will detect the fraud after the direct debit is performed.

To allow the detection of possible unauthorized SDD, the Financial Banking application will correlate different information coming from different sources (social, questionnaires, personal info etc.) [8] to create a debtor profile that summarizes the habits and interests of the debtor. The debtor profile will be a list of preferences/attributes with the relative weight and also personal information such as address, social account etc.

## 4 Visualization Support for the Banking Application

Visualization supports two main activities comprised within the banking application and described in the following sections: Sects. 4.1 and 4.2 address visualization support for online detection and validation of possible fraud matches. This helps fraud analysts to be informed on possible fraud matches and helps to perform a root cause analysis of possible matches in order to validate/refute them. Section 4.3 addresses visualization support for both online and offline analytical processing of data, where fraud analysts are required to gain a more insight of data for various reasons such as improve the anti-fraud system with more accurate rules.

### 4.1 Online Visualization of Possible Frauds Matches

The system uses an online data stream of SDD transactions which are tested against the set of anti-fraud rules created by the fraud analyist. When a match occurs it is displayed on a dashboard from which the user can choose among several display views. Other linear-structure type visualizations can be used and are to be further researched. For instance, space-filling techniques and linear/hexagonal binning can help a user to understand the amount of possible fraud matches as the system computes the bin size according to the number of items at a particular interval of time. Visual variables such as color, texture, pattern and size can perform an important role in the presentation of possible fraud items in order to visually accumulate other information such as severeness level, type of fraud, status (if it is being managed by a fraud analyst), etc. Aditionally, sorting of these items according to a user-defined criteria is also usefull. As the number of possible fraud matches may get too high, undesired visual clutter may occur in the visual representation of the items. Visual clutter reduction techniques [7] regard appearance (opacity, sampling and clustering), spatial distortion (topological distortion, space-filling) and time can be used to overcome this issue (animation). Furthermore, interaction techniques such as zoom + panning can also be used.

## 4.2 Validation of a Possible Fraud

After detection and presentation, confirmation is required. The goal is to validate if the possible match constitute a fraud or, instead, is a false positive. To do so, the fraud analyst selects a possible match and inspects the SDD transaction data, the anti-fraud rule matched, the degree of severeness the anti-fraud system placed on this possible fraud. This inspection provides the idea of why it can be a possible fraud, but the validation process requires additional data regarding the profile of the debtor (as a last resort, the validation processs may require to contact the debtor). The profile is composed of information such as address and the set of interests, such as sports, culture, football, etc. for each interest associated to a particular debtor, a plausibility factor is calculated.

The inspection process is inefficient if the analyst is required to read this amount of data for each possible fraud match. Furthermore, this task should be performed very promptly and efficiently. A visual representation of this heterogeneous and multi-source data is required, which can be effectively perceived by the analyst. To do so, the visualization system provides a force-directed graph-based visualization of the dataset. This approach is currently adopted by other entities within the same domain [16, 17]. In this type of visualization (example in Fig. 2), nodes constitute entities or pieces of information and the edges denote relations between the nodes they connect. The analyst can further inspect details by selecting nodes/edges. Alternatively hierarchical graph visualizations are used in order to allow the selection of a node and the presentation of a new level of data.

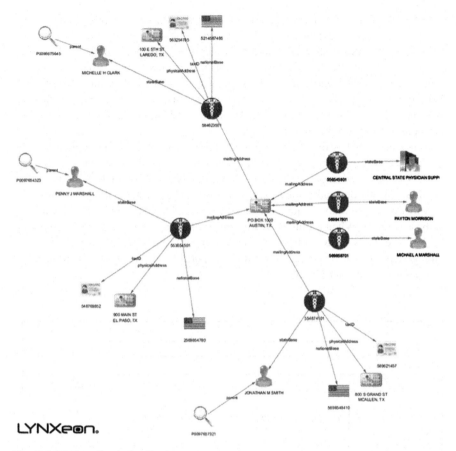

**Fig. 2** LYNXeon fraud visualization system

## 4.3 Exploratory Data Analisys

The last two sections address visualization support for online data, regarding inspection of possible fraud matches. However fraud analysts are also required to do analytical processing of data for several reasons, being the most important, to better understand data and behaviours, to look for new patterns of frauds and to improve anti-fraud rules. To do so, a framework is provided to allow users to inspect datasets, perform queries using an easy, drag and drop and visual process workflow, inspect results, associate charts to results, create and manage dashboards and scrutinize data by means of charts.

Client components of this system are the front-end user interfaces that a user interacts with, delivered as HTML, JavaScript and CSS to browsers on desktop and mobile devices following responsive design principles. The visualization framework is composed of the following main components: (a) The Workflow Builder

allows users to do visual composition of queries, execute data processing workflows Fig. 3) and preview results It includes a library of operators containing
common operators for selection filtering, cluster, join of datasets, binarization, map,
etc.; (b) The Chart Builder allow users to associate results coming from a workflow
execution to several chart types. Charts are selected from a predefined library and
their configuration relies on the chart type, the results metadata, more specifically,
the data types of primary variables; (c) The dashboard builder allow users configure
dashboards based on the existing charts.

Hence the user has the ability to build queries that, once executed, their results
can be visually inspected and periodically updated when assigned to a dashboard
view. The provided support, helps the fraud analyst to formulate hypothesis on
potential correlations and behaviors, to express the hypothesys with the help of one
or more queries, to map the queries against the available chart library for visual data
analysis (Fig. 4). When required, charts can be associated to dashboards. With this
support, the analyst can call a particular dashboard and visually inspect the containing set of visual representations which is periodically updated, according to a
user-defined time span per chart.

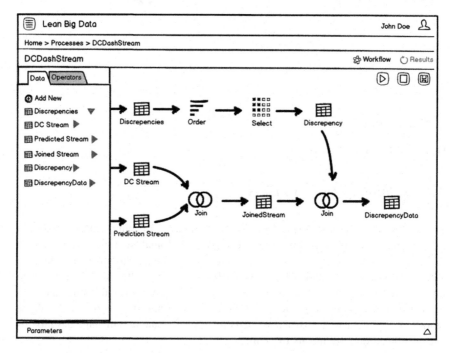

**Fig. 3** Example of process workflow GUI to generate Bank Branch Frauds data table

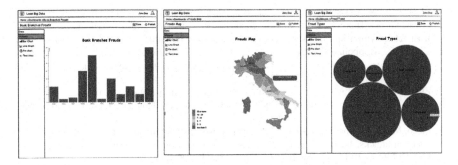

**Fig. 4** Visualization support regarding queries from analytical processing

## 5 Conclusions and Future Work

In this work we have examined the visualisation requirements of a real world banking application, and identified generic visualisation tasks that are essential for doing effective analysis of a complex process that produces amazingly large amounts of data (which is typically the case for Big Data applications). The requirements for the visualization support that we have proposed were modelled using an application wireframe that acts a story-board. The effectiveness of the visualization facilities that we have proposed have been demonstrated through their application to a Big Data banking use-case from the real world. The main avenues for future research will be: (i) the development of additional functionalities in the visualization framework, and (ii) the execution of a thorough experimental campaign, to validate the proposed solution under realistic operation conditions.

**Acknowledgments** The research leading to these results has received funding from the European Union's Seventh Framework Programme for research, technological development and demonstration under grant agreement no 619606 (LeanBigData Project).

It has been also supported by the Italian Ministry for Education, University, and Research (MIUR) within the framework of the Project of National Research Interest (PRIN) "TENACE", and by the Regione Campania within the framework of the project "Embedded Systems in critical domains".

## References

1. Big Data Application Architecture Q&A: A problem - solution approach (expert's voice in big data) paperback – by Nitin Sawant 17 Dec (2013)
2. Watson, H.J., Wixom, B.H.: The current state of business intelligence, Computer **40**(9)
3. Campanile, F., Cilardo, A., Coppolino, L., Romano, L.: Adaptable parsing of real-time data streams. In: 15th EUROMICRO International Conference on, Parallel, Distributed and Network-Based Processing, 2007. PDP '07. pp. 412–418, 7–9 Feb 2007. doi:10.1109/PDP. 2007.16

4. Metropolis, N.: Massively parallel processing. J. Sci. Comput. **1**(2), 115–116 (1986)
5. Shvachko, K., Hairong, K., Radia, S., Chansler, R.: The hadoop distributed file system. In: IEEE 26th Symposium on, Mass Storage Systems and Technologies (MSST) (2010)
6. Cicotti, G., Coppolino, L., Cristaldi, R., D'Antonio, S., Romano, L.: QoS monitoring in a cloud services environment: the SRT-15 approach. Lect. Notes Comput. Sci. **7155**, 15–24 (2012). doi:10.1007/978-3-642-29737-3_3
7. Ellis, G., Dix, A.: A taxonomy of clutter reduction for information visualisation. In: IEEE Trans. Visual. Comput. Graphics, vol. 13(6), pp. 1216–1223, Nov–Dec (2007). doi:10.1109/TVCG.2007.70535
8. Ficco, M., Coppolino, L., Romano, L.: A weight-based symptom correlation approach to SQL injection attacks. In: Fourth Latin-American Symposium on, Dependable Computing, 2009. LADC '09. pp. 9–16, 1–4 Sept 2009. doi:10.1109/LADC.2009.14
9. Coppolino, L.D'., Antonio, S., Garofalo, A., Romano, L.: Applying data mining techniques to intrusion detection in wireless sensor networks. In: Eighth International Conference on, P2P, Parallel, Grid, Cloud and Internet Computing (3PGCIC). Compiegne 28–30 Oct 2013
10. Gartner Report: Magic quadrant for advanced analytics platforms
11. http://www.slideshare.net/johblom/the-new-normal-in-business-intelligence
12. SEPA Direct Debit Core Rulebook: Version 6.1, European payments council (EPC), Nov 2011
13. Criteri e regole generali - CBI - Standard tecnici: http://www.querciacb.info/1399.pdf
14. Sharp increase in direct debit fraud: Gill Montia, 19 Nov 2010
15. A taxonomy of clutter reduction for information visualisation. Trans. Visual. Comput. Graphics, (6), Nov–Dec 2007. IEEE (2007)
16. http://neo4j.com/
17. http://www.21ct.com/products/lynxeon/
18. http://keylines.com

# A Study About the Comprehensibility of Pictograms for Order Picking Processes with Disabled People and People with Altered Performance

Andreas Baechler, Liane Baechler, Peter Kurtz, Georg Kruell, Thomas Heidenreich and Thomas Hoerz

**Abstract** This study evaluates, the effectiveness of pictogram meanings for order picking processes with disabled people and people with altered performance. Pictograms were created for guidance and feedback of process steps in order picking. The comprehensibility and colors of the pictograms were evaluated through a survey with 45 normal performance people and 26 disabled people. The interviewees responded to a standardized questionnaire with at least three different types of pictograms for each process step.

**Keywords** Human-computer-interaction · Pictogram · Order picking · Adaptive assistance systems disabled people · People with altered performance · Manual processes · Demographic change

A. Baechler (✉) · G. Kruell · T. Hoerz
Faculty of Mechanical Engineering, University of Applied Sciences Esslingen,
Esslingen, Germany
e-mail: andreas.baechler@hs-esslingen.de

G. Kruell
e-mail: georg.kruell@hs-esslingen.de

T. Hoerz
e-mail: thomas.hoerz@hs-esslingen.de

L. Baechler · T. Heidenreich
Faculty of Social Work, Health Care and Nursing Sciences,
University of Applied Sciences Esslingen, Esslingen, Germany
e-mail: liane.baechler@hs-esslingen.de

T. Heidenreich
e-mail: thomas.heidenreich@hs-esslingen.de

P. Kurtz
University of Technology Ilmenau, Ilmenau, Germany
e-mail: peter.kurtz@tu-ilmenau.de

© Springer International Publishing Switzerland 2015
E. Damiani et al. (eds.), *Intelligent Interactive Multimedia Systems and Services*,
Smart Innovation, Systems and Technologies 40,
DOI 10.1007/978-3-319-19830-9_7

# 1  Introduction and Related Work

## 1.1  Pictograms

"A pictogram is a stylized figurative drawing that is used to convey information of an analogical or figurative nature directly to indicate an object or to express an idea" [1]. Pictograms are part of our daily lives and they can fulfill a number of functions. They are used in transportation, medication, the textile industry, computers and many others to represent directions, written instructions, actions and objects. There are different advantages of the usage of pictograms, for example the information is processed quickly (e.g. traffic signs), the comprehension of information is independent from language (e.g. non-natives) and it can be understood with limited linguistic ability and visual problems (e.g. people with little education, mental disability and older people) [1]. But the main objective of pictograms is to get the user's attention and to convey a certain information as fast as possible. Multiple studies display that information given by images are interpreted quicker and more accurately than information by words [2]. Other studies show that pictograms can be a helpful instrument of conveying information to people with limited verbal skills or to people with different native languages [3].

Based on these results, pictograms, which are specifically designed for the needs of people with disabilities, can be represented as a supportive and appropriate means of inclusion of these people in industrial processes.

## 1.2  Theoretical Backgrounds

Because of globalization and the ongoing demographic change there are significant changes in Germany's industrial production today. As contemporary production trends move more and more away from the mass production with a long production life and ever more towards the production of products with a high variety and shorter product lives, the proportion of small series with small numbers greatly increases [4]. Due to these developments, automated processes are no longer viable in the order-picking and the assembly area. Therefore manual order- picking becomes increasingly more important. Manual processes require, however, due to high product intervals and -variety a highly qualified staff.

In addition to that, current figures show that the average age of employed persons working in Germany, and thus the proportion of workers who are subject to a performance change, increases. The Federal Statistical Office of Germany predicts a nine percent decline of the population level in Germany from 2000 to 2050, the most relevant group in production—the 20 to 60 year olds – is predicted to decrease from 45.5 to 35.4 million. The proportion of 60-year-old people, for example, is predicted to increase from 19.4 to 27.5 million people [5].

Furthermore, results of a market study with 130 industrial enterprises showed that the proportion of employees with altered performance[1] in German companies occupy a substantial height of up to 20 % of the total workforce.

Due to the lack of young and skilled workers, companies are not able to find enough suitable candidates in each field of activity [8]. Therefore, it is required - not only because of social, but also because of economic aspects - to employ staff of advanced age (50 years and older) and staff with disabilities in the industrial sector [4].

The increasing demands on sheltered workshops cause, especially for disabled people[2] and people with altered performance, new ways of employment. Therefore, new ways of support must be found [10]. In addition the effort of the industry for an inclusive labor market for these people increases [8].

## 1.3 Order Picking

Manual order picking[3] is one of the most important manual activities in the industry. This area represents the core of today's warehouse logistics, in particular of the intra-logistics. The client orders are put together under the factors of quality and time, causing a successful customer loyalty and competitiveness of companies [12].

As a component of the intra-logistics, order picking is one of the most complex and labour-intensive sectors in the supply chain [13]. The importance of order picking is also reflected in its high proportion of logistic costs. Despite numerous developments in recent years of purely manual systems to fully automated solutions, manual designs remain the most widespread in practice, sort of an order picking system [14]. Based on the low level of investment, the cognitive skills and the flexible tactile- and gripping ability, humans are not replaceable by machines in order picking processes in most cases [15].

Manual order picking is usually carried out according to the principle "picker to part". In this principle the parts are static, usually provided in racks and the picker moves. The required information for the picker is usually provided by a paper list (pick-by-paper), an illuminated container indicator (pick-by-light), using an audio wizard (pick-by-voice) or using a mobile terminal with display (pick-by-display).

---

[1]Employees with altered performance are people that are unable to exercise their original work activities with the appropriate requirements and burdens for a period of time or permanently. Here mainly employees with acquired disabilities are concerned as a result of illness, accident or symptoms of old age. However, at an adapted workplace, these employees can realize their full potential [6, 7].

[2]To counteract stigmatization and inappropriate connotation, the people with impairment of functional health are referred as "disabled people" in this document. This term refers to a limitation of the performance, thus it is meant the maximum power level of a person regarding a task, or action under test, standard or hypothetical conditions [9].

[3]Order picking is the process of collecting items from stock and transporting them to a specific location [11].

## 1.4 Objectives of an Assistance System in Order Picking

In the future changes and innovations are needed in the area of guidance and control for the employees to perform reliably in order picking. In this case, research is being conducted, especially in the field of technical support by assistance systems for disabled people and people with altered performance, but also for people without restrictions. The main objective of the assistance system is to compensate the functional impairments of the employees, so that they are balanced on the one hand and on the other hand existing skills are promoted. Furthermore the system should also reduce the effort and complexity for the training of employees with different levels of performance and technical background. In addition, the flexibility for the use of employees with different performance levels or different lot size is supposed to be increased [16].

Similarly the work ability and motivation of older and disabled people should be improved or maintained and the number of picking errors should be reduced [17]. Active participation of motivating elements and ergonomic aspects shall support healthy work behaviour and prevent musculoskeletal disorders.

Another aspiration of such a solution is to support and facilitate the inclusion of disabled people. The work assistance in the area of order picking reflects an important aid for disabled and performance altered people, which is in the extended sense regulated by law in §33 para. 8 No. 3 and § 102 para. 4 of the German Social Code IX (2004).

Despite manual activities in order picking the competitiveness of German companies and sheltered workshop should be further improved compared to the global market [8]. These goals can only be achieved if the people get instructions during the teaching and performing of work processes. This information must be communicated quickly and in a way, which is easily accessible to understand and overcome educational, language and cultural barriers.

## 2  Experimental Design

### 2.1  Layout Order Picking System

An assistance system for order picking is developed and evaluated. The current status of the project is structured as follows (Fig. 1). In a flow rack different items in containers are provided. Across to the rack a height adjustable picking cart is coupled via linear sliding rails with a mobile assistance unit including two projectors, two infrared cameras, a height-adjustable scale and a touchscreen monitor. On the touchscreen monitor a picking order can be selected and the compilation of the order starts. The picker can take either individual items or entire containers. The removal of the items or containers will be guided through light signals, emitted from a projector. An infrared camera and a scale make sure the parts are removed correctly and in the right amount. The removal position and required amount are mapped on projected symbols. The subsequent deposition position in a container on the scale or on the picking cart is also shown by projected pictograms. After proper removal and storage of a component the picking cart is moved to the next removal position. This position is illustrated by icons, which are displayed on the touch-screen monitor.

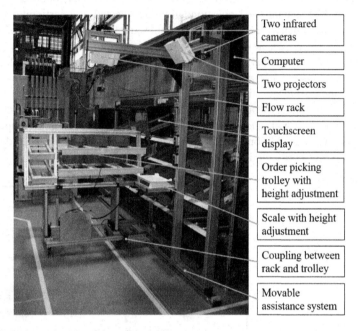

**Fig. 1** Experimental setup of the order picking system

The pictograms are shown as light signals, which represent and instruct the affected objects (hand, container, picking cart) and the performed procedures (removal, storage, moving). At the same time the pictograms are also used for the confirmation of correctly and incorrectly performed actions. Thereby pictograms shall indicate the type of error as precise as possible, to perform a fast and reliable troubleshooting. The pictograms are derived from a flow chart of the picking process. They are designed for the following process steps:

– Guidance of removal operations
– Guidance of storage operations
– Guidance of moving the picking cart to the next removal position
– Feedback about correct or incorrect performed operations.

## 3  Methodological Approach

At least three pictograms in iconic, symbolic and hybrid form were designed and evaluated for six different action cases. The design of the pictograms is mainly influenced by a simple representation, the use of striking and closed objects and by a uniform design.

To find the most suitable variety of pictograms, a study was conducted in form of a survey. The standardized questionnaire was selected as questioning method after performing a pairwise comparison and subsequent cost-benefit analysis. This type of survey recorded, next to an e-mail questionnaire, a free interview and an open online survey, especially in the criteria of the response rate, the intelligibility and representativeness as the most suitable variant.

In the standardized questionnaire associations with colors and pictograms with 45 normal performance people were polled (Fig. 2). In order to obtain representative results people with and without prior technical knowledge in the population were interviewed. In addition, the questionnaire has been carried out to 26 disabled people to ensure representativeness.

**Fig. 2** Abstract of the questionnaire

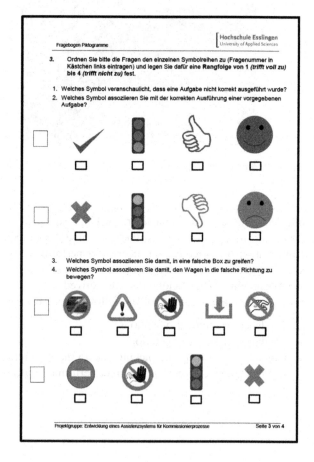

In a first step, as a control function for conscientious filling of the questionnaires, it is needed to select the right row (from two possibilities) before answering the questions. This control element (inverse design) was used to avoid random ticking of the responses.

## 4 Experimental Results

### 4.1 Survey with Normal Performance People

A first survey took place in May 2014 and covers a sample of 45 participants, 24 male and 21 female, aged 17–60 years (average = 33.4 years, standard deviation = 12.4). A randomly selected sample of normal performance people with technical knowledge (N = 11, 24.4 %) and non-technical (N = 34, 75.6 %) was adopted to conduct this survey. The results of the survey through the questionnaire clearly show that as a color guide for removal operations at the rack the color green appears most suitable (N = 26, 57.8 %). As feedback for correct removing operations the color green has, according to participant's assessment, also proved to be most intelligible (N = 42, 93.3 %). Most of the participants associate the color red with the feedback of an incorrect removal (N = 41, 91.1 %). As a color to guide the back setting of an incorrectly removed part, the color yellow appears as the most suitable (N = 32, 71.1 %). For the feedback that the picking cart is moving into the wrong direction, according to the participants estimate, the color red makes the most sense (N = 37, 82.2 %).

**Fig. 3** Symbols for a correct execution of a given task

In the second part of the questionnaire different pictograms were compared. The results of the normal performance people are indicated in the following figures by a blue frame, the results of the disabled people are marked with a green frame. Thereby a "confirmation hook" presented as the most appropriate symbol (N = 27, 60 %) for the feedback of the correct execution of a given task (Fig. 3).

As the most convincing symbol for feedback of an incorrectly performed task, a red "X" was most commonly selected (N = 24, 53.3 %), see Fig. 4.

**Fig. 4** Symbols for an incorrect execution of a given task

The results of the pictograms for the feedback of grasping into the wrong container were ambiguous. Symbol 1 with a "photographed hand" (N = 20, 44.4 %) and symbol 5 with a "caricatured hand" (N = 19, 42.2 %) were the most meaningful pictograms (Fig. 5).

**Fig. 5** Grasp in the wrong box

When comparing the pictograms for the feedback of pushing the picking cart into the wrong direction, the "one-way street" pictogram (N = 33, 73.3 %) turned out to be the most suitable variant (Fig. 6).

**Fig. 6** Cart is moved into the wrong direction

For instruction how far the picking cart has to be moved until it is at the removal position, the illustration of a simplified cart and a finishing flag is the most understandable solution (N = 25, 55.5 %), see Fig. 7.

**Fig. 7** Moving cart

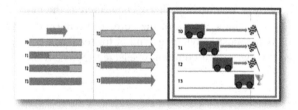

The instructions for the number of parts, which have to be taken is best illustrated on the third pictogram with a green background "European Uppercase number" (N = 24, 53.3 %), see Fig. 8.

**Fig. 8** Number of parts

As a symbol for the removal of an object from a container symbol 3 (N = 23, 51.1 %) is the best suitable variant (Fig. 9).

**Fig. 9** Removal parts from a container

## 4.2 Survey with Disabled People

A second survey using the same questionnaire, but with a target group of disabled people also took place in May 2014 and covered a sample of 26 participants, 10 male and 16 female, aged 16–54 years (average = 34.1 years, standard deviation = 10.4). For the survey disabled people of a sheltered workshop, which could potentially work on an assistance system, were arbitrarily selected. The survey was conducted by the pedagogical specialists of the sheltered workshop in interview form.

The results clearly show that as a color guide for removal operations at the rack the color green is the most appropriate (N = 14, 53.8 %). Also, as a feedback for the correct removal, the color green (N = 17, 65.4 %) and as a false feedback removal, the color red (N = 18, 69.2 %) is most suitable. As a color to guide the back setting of an incorrectly removed part the disabled people associate, however equally red and yellow as appropriate colors (each N = 11, 42.3 %). According to the participants assessment, the feedback that the picking cart is moved into the wrong direction, shows the color red (N = 13, 50 %).

The results of the second part show the symbol "thumbs up" (N = 10, 38.5 %) as well as the symbol "Smiley" (N = 9, 34.6 %) as appropriate pictograms for the feedback of the correct execution of a given task (Fig. 3). Similarly, these picto-grams are also best suited in the negated form for feedback of an incorrectly performed task (N = 10, 38.5 % and N = 8, 30.8 %) as shown in Fig. 4. The comparison of the pictograms for the feedback of grasping into the wrong container led to an inconclusive result. Both, symbol 1 "photographed hand" (N = 10, 38.5 %) and symbol 3 "halt stop" (N = 9, 34.6 %) were rated equally high (Fig. 5). For the understanding of the feedback of pushing the picking cart into the wrong direction symbol 2 "halt stop" (N = 11, 42,3 %) and symbol 1 "one-way street"(N = 7, 26.9 %) turned out to be the most suitable (Fig. 6). An instruction of how far the picking cart has to be moved until it is at the correct removal position is given according to the assessment of the disabled people by the pictogram of a simplified cart and a finishing flag (N = 14, 53.8 %). An instruction for the number of parts, which have to be taken are best illustrated on pictogram 1 (N = 12, 46 %) and pictogram 3 (N = 10, 38.5 %) with the "European Uppercase number" (Fig. 8). To guide the removal of an object from a container, the photographic image with the additional arrow symbol is the most suitable pictogram (N = 20, 76.9 %) see Fig. 9.

## 5   Discussion

The results of the surveys display clearly which colors are best suited for which activities. The only distinction in the color association between the disabled and normal performance people is the color for the back setting of an incorrectly removed part. Hereby the survey of the disabled additionally results with the color red. It can be assumed that for this target group, red is considered as a signal color, especially for deviant activities and that due to the limited education an uncon-ventional color understanding is present among disabled people and people with altered performance. Furthermore, differences occur at the feedback of a correctly and incorrectly performed task for the evaluation of these pictograms. The normal performance people associate these two activities mostly with a "confirmation hook" or a red "X", whereas the disabled people emphasize pictograms with a human face, gestures and photographic images, such as "Smileys", "Thumbs up or down" and photos. The same phenomenon is also found in the results for the understanding of grasping into the wrong container.

The pictographic symbols for the feedback of pushing the picking cart into the wrong direction indicate significant different results, between the two user groups. The favorite of the normal performance user group is the road sign "one-way street", whereas for the disabled user group the sign "halt stop" proves to be easier to understand. This difference may result from the fact that the normal performance people possess a driving license, and therefore have a significantly different understanding of symbols.

It is observed that the survey of the disabled people (in contrast to normal performance people) often results in inconclusive results. This wide variation in results may be explained with the fact that this group has different performance levels or degrees of disability.

# 6  Future Work

The discussed results are characterized mainly by distinctive differences between the two user groups. It is therefore necessary to ensure that two separate operating systems are developed and that these systems are evaluated within the appropriate user groups. This aspect must also be taken into account with the further development of adaptive systems.

Furthermore, in a subsequent evaluation with other pictograms, special attention should be given to the design with human faces, gestures and photographic images to be able to respond adequately to the target group of disabled people. Additionally, the use of avatars could be investigated.

In such an evaluation the ambiguous results of this questionnaire could be re-examined, by taking the factors described into account. Previous feedback and suggestions from users such as the mapping of numbers with the corresponding number of articles to be picked or images of cube numbers could also be investigated.

In addition, the practical understanding must be proofed in another study with reference to an experimental setup. So in a case of actual use it can be ensured, that the understanding by the pictograms is given. It is even more important to carry out this practical study with the target group of disabled people, to obtain representative results for the assistance system in order picking.

In a final study the complete assistance system (including the selected pictograms) is evaluated in relation to the benefits (time and error rate) and stress with the current standard versions of picking systems (paper pick list, Cart-Mounted Display, pick-by-light).

# References

1. Tijus, C., Meunier, J.-G., Cambon de Lavalette, B., Barcenilla, J.: Chapter 2: the design, understanding and usage of pictograms. In: Terrier, P., Alamargot, D., Cellier, J.M. (eds.) Written Documents in the Workplace, pp. 17–31. Brill, Netherlands (2007)
2. Department of Trade and Industry (ed.): The role of pictograms in the conveying of consumer safety information. Government consumer safety research (2000)
3. Dowse, R., Ehlers, M.: Pictograms for conveying medicine instructions: comprehension in various South African language groups. In: South African Journal of Science **100**(11 +12):687–693 (2004)

4. Lotter, B.: Überlegungen zum Montagestandort Deutschland. In: Lotter, B., Wiendahl, H.-P. (eds.) Montage in der industriellen Produktion, 2nd edn. pp. 389–396. Springer, Heidelberg (2012)

5. Statistisches Bundesamt: Bevölkerung Deutschlands bis 2060 - Begleitheft zur Pressekonferenz am 18. Nov 2009

6. Adenauer, S.: Die (Re-) Integration leistungsgewandelter Mitarbeiter in den Arbeitsprozess. Das Projekt FILM bei FORD Köln. Angewandte Arbeitswissenschaft, **181**:1–18 (2004)

7. Jahn, H.-P.: Datenerfassung und -verarbeitung bei der ergonomischen Gestaltung von Arbeitsplätzen -mehrere Jahre nach Abschluss eines HdA-Projektes für Leistungsgewandelte: Herbstkonferenz GfA, 12 Oct 2001

8. Stratmann, A.: Aktuell. Die Firmenzeitschrift der GWW (127). http://www.gww-netz.de/tl_files/gww/Aktuell_127_2014_08.pdf. Checked on 9 Nov 2014

9. Schuntermann, M.F.: Einführung in die ICF. Grundkurs, Übungen, offene Fragen. 2., überarb. Aufl. Landsberg/Lech: ecomed Medizin (2007)

10. Töpffe, M., Maas, O.: In einer inklusiver werdenden Gesellschaft darf das Berufsleben für behinderte Menschen nicht exklusiv gestaltet sein. Pressemeldung. Diakonie Rheinland-Westfalen-Lippe (2014)

11. Baumann, H.: Order picking supported by mobile computing (2013). http://elib.suub.uni-bremen.de/edocs/00102979-1.pdf. Checked on 9 Oct 2014

12. Pulverich, M., Schietinger, J. (eds.): Handbuch Kommissionierung. Effizient picken und packen, 1st edn. Vogel, München (2009)

13. Hompel, M., Sadowsky, V., Beck, M.: Kommissionierung. Materialflusssysteme 2 – Planung und Berechnung der Kommissionierung in der Logistik. Springer, Berlin (2011)

14. Straube, F.: Trends und Strategien in der Logistik. Ein Blick auf die Agenda des Logistik-Managements 2010. Hamburg: Dt. Verkehrs-Verl (2005)

15. Arnold, D., Furmans, K.: Materialfluss in Logistiksystemen. Springer, Berlin (2009)

16. Bächler, L., Bächler, A., Kölz, M., Hörz, T., Heidenreich, T. (awaiting publication): Über die Entwicklung eines prozedural-interaktiven Assistenzsystems für leistungsgeminderte und -gewandelte Mitarbeiter in der manuellen Montage. In: Soeffke Uni Magdeburg (Hg.): Tagung Mensch Maschine. Magdeburg (2014)

17. Rammelmeier, T., Galka, S., Günthner, W.A.: Fehlervermeidung in der Kommissionierung. In: Proceeding (2012). doi:10.2195/lj_Proc_rammelmeier_de_201210_01

# Mining Popular Travel Routes from Social Network Geo-Tagged Data

Carmela Comito, Deborah Falcone and Domenico Talia

**Abstract** On line social networks (e.g., Facebook, Twitter) allow users to tag their posts with geographical coordinates collected through the GPS interface of smart phones. The time- and geo-coordinates associated with a sequence of tweets manifest the spatial-temporal movements of people in real life. This paper aims to analyze such movements to discover people and community behavior. To this end, we defined and implemented a novel methodology to mine popular travel routes from geo-tagged posts. Our approach infers interesting locations and frequent travel sequences among these locations in a given geo-spatial region, as shown from the detailed analysis of the collected geo-tagged data.

**Keywords** Social networks · Trajectory pattern mining · Semantic location detection

## 1 Introduction

Social media become very popular in recent years and is receiving an always increasing attention from the research community as through the user-generated data it embeds precious information concerning human dynamics and behaviors within urban context. The ability to associate spatial context to social posts is a popular feature of the most used on line social networks. Facebook and Twitter exploit the GPS readings of users phones to tag user posts, photos and videos with geographical coordinates.

C. Comito (✉)
ICAR-CNR, Calabria, Italy
e-mail: ccomito@dimes.unical.it

D. Falcone · D. Talia
DIMES-University of Calabria, Calabria, Italy
e-mail: dfalcone@dimes.unical.it

D. Talia
e-mail: talia@dimes.unical.it

© Springer International Publishing Switzerland 2015
E. Damiani et al. (eds.), *Intelligent Interactive Multimedia Systems and Services*,
Smart Innovation, Systems and Technologies 40,
DOI 10.1007/978-3-319-19830-9_8

According to this view, people travelling and visiting a sequence of locations generate many trajectories in the form of geo-tagged tweets. The time- and geo-references associated with a sequence of tweets manifest the spatial-temporal movements of Twitter users. This paper aims to analyze those trajectories to discover people and community behavior, i.e. patterns, rules and regularities in moving trajectories. The basic assumption is that people often tend to follow common routes: e.g., they go to work every day traveling the same roads. Thus, if we have enough data to model typical behaviors, such knowledge can be used to predict and manage future movements of people.

In particular, the objective of this study is to provide the top interesting locations and frequent travel sequences among these locations, in a given geo-spatial region. For interesting locations we mean the culturally important places, such as National Gallery or Buckingham Palace in London (i.e. popular tourist destinations), and commonly frequented public areas, such as shopping malls/streets, restaurants, cinemas and bars.

To this aim we defined and implemented a methodology to mine popular travel routes from geo-tagged posts of an urban area such as those collected from Twitter. The proposed methodology consists of various phases allowing to collect tweets, detecting locations from tweets, extracting travel routes through those locations and, then, mine *frequent travel routes* using sequential pattern mining. We, then, derive spatial-temporal information for each frequent travel route to capture the factors that may drive users' movements. In particular, for all the frequent patterns, we compute a set of daily snapshots concerning the visited places, the movements among them, and the duration of the visits at each location.

The rest of the paper is organized as follows. Section 2 overviews related works. Section 3 presents the methodology. Section 4 describes the statistical analysis of frequent travel routes. Finally, Sect. 5 concludes the paper.

## 2 Related Work

The rapid rise of social network has motivated a lot of studies on human mobility and its implications in the social relationships and location-based services [2, 3, 10]. Although, to the best of our knowledge, few previous studies have been carried out on the extraction of trajectories by exploiting only the geo- references of tagged social posts. Our work shares the same aim as [11] of identify the most frequent travel routes and the top interesting locations in a given geo-spatial region. However, the authors in [11] mine trajectory patterns from photos on Flickr. They associate semantics to the locations on the basis of associated tags that are contributed by users for each photo. In contrast, the geo-tagged tweets we analyzed are lacking of such information. Moreover, we deal with a huge quantity of data compared to them: their most frequent trajectory is traveled by just 21 users, whereas in our case the same trajectory is traveled by 176 users.

Other approaches studied mobility patterns in a city during exceptional events by observing micro-blog posts. As an example [5, 8, 9] used Twitter as a sensor to detect natural phenomena. However, compared to our proposal they perform a more coarse grain analysis of typical trajectories by considering a fixed set of areas in the city.

A lot of research work has been done in trajectory pattern mining [1, 6]. Also with respect to this branch of works, our approach presents elements of novelty that could be considered as an improvement in the field. First, we deal with the more unpredictable and irregular data of the Twitter social network, whereas they use GPS traces of mobile devices or synthetic datasets. In addition, they lack some semantics about the type of place in the travel sequences, which would allow a better understanding of users' patterns.

## 3 Methodology

Figure 1 outlines the main steps of the methodology we used to find people trajectories. The first step is the *collection of geo-tagged data*; in our case we used as data source Twitter. The second step is *semantic location detection*. Locations are detected through dense clustering that allows to cluster GPS coordinates into specific places and associate them to Foursquare categories when available. After locations detection we *generate trajectories or travel routes*. We can then *mine frequent travel routes* using sequential pattern mining. If a trajectory pattern repeats often, we consider it as a *frequent travel route*. Finally, we extract *spatial-temporal features* so as to capture the factors that may drive users' movements. In the following of the section we will describe in detail each of the steps of the proposed methodology.

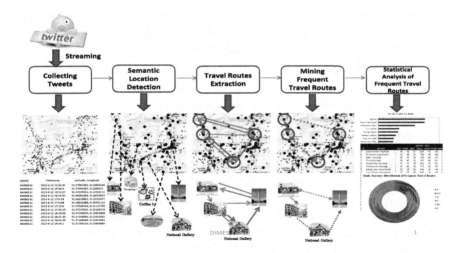

**Fig. 1** Methodology

## 3.1 Collecting Tweets

We built a multi-threaded crawler to access the Twitter Streaming API. The crawler collects the geo-tagged tweets filtered by location and processes the results to obtain a dataset representing a sequence of daily snapshots.

The geo-located data mined in this work is a Twitter dataset of 7,424,112 tweets issued by 292,195 mobile users in 6,098,148 distinct locations of the city of London from June 2013 to November 2013, with an average number of tweets per day greater than 40,000. We define a geo-tagged tweet as follows.

**Definition 1** *A geo-tagged tweet is defined as a triple* $tw = (u, l, t)$, *where* $u$ *is the user id,* $l$ *is the location expressed with latitude and longitude,* $l = (x, y)$, *and* $t$ *is the timestamp of the tweet.*

## 3.2 Semantic Location Detection

The *semantic location detection* step consists of two main phases: (i) location detection and (ii) semantic category association to the detected locations so as to obtain *semantic locations*.

**Location detection** Due to imprecision of GPS, a specific place in the city might be represented by slightly different GPS coordinates. Thus, it is necessary to *cluster the locations* of the geo-tagged tweets so that each place is identified by a single pair of geographic coordinates. To this aim, we used a dense clustering algorithm. The result of the clustering is a model composed of a set of clusters; each of such cluster corresponds to a geographic region that actually is a *dense region* visited by many users. A dense region can be defined as follows:

**Definition 2** *A dense region stands for a geographic area of points that are densely visited by users, meaning that many tweets have been posted from that geographic area. It is, thus, characterized by a group of nearby GPS points and is represented by the GPS coordinated of its centroid point.*

Using a density-based clustering algorithm, we hierarchically cluster this geo-tagged data into some geo-spatial regions (set of clusters) in a divisive manner. The close geo-locations in the tweets from various users would be assigned to the same clusters on different levels. Given this hierarchical structure, a node on a level of the tree (that is a cluster of locations) represents a larger region that can be used to represent its descendant nodes obtaining this way a lower granularity in the location identification process. In general, these regions covered by clusters on different level of the hierarchy might stand for various semantic meanings, such as a venue, a suburb and a district. To detect significant locations in a city, the algorithm parameters are set such as to aggregate the geo-tagged locations at the extend of venues in the city. To this aim we used a clustering neighborhood radius of about 100 m as it is reasonable to take this value for the spatial extension of typical locations in city environments. This setting of the cluster radius parameter enable us to find out some significant places such as museums, restaurants, shopping malls, etc., while ignoring geo-regions without semantic meaning, like the places where people wait for traffic lights or meet congestion. The evolution of the clustering algorithm is shown in Fig. 2.

**(a)**                         **(b)**                         **(c)**

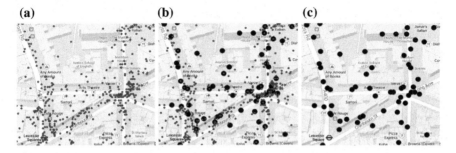

**Fig. 2** Evolution of Clustering Algorithm: the tweets geo-locations (**a**) are grouped by the algorithm so that each cluster respects spatial distance constraints (**b**). Each cluster of geo-locations is represented by its centroid (**c**)

**Category association** The semantic category of a place is a significant information that can be added to the traditional latitude-longitude coordinates used by online social networks to represent a location. Once the clusters are detected we identify place semantics. We formulate the problem as a supervised learning task where the input vectors are designed according to: (i) a set of spatial-temporal features such as stay duration, time of day, place popularity, extracted from irregular social posts; and (ii) a Foursquare dataset retrieved by the Foursquare API. Specifically, we use the Foursquare categories as the ground truth class attribute of the category classifier. As compared to a raw GPS point, each location detected with our methodology carries a particular semantic meaning, such as the shopping malls we accessed or the restaurants we visited, etc. Moreover, we made a more fine grain classification, retrieving by the Foursquare API, the precise name of places, e.g. London Eye, Buckingham Palace, National Gallery and so on. Details about place semantic category association can be found in [4].

*Example 1* Figure 3 shows an example of semantic location detection obtained through the proposed hierarchical dense clustering approach.

Analysing the Twitter dataset of London, with the semantic location identification phase we automatically labeled 70177 locations.

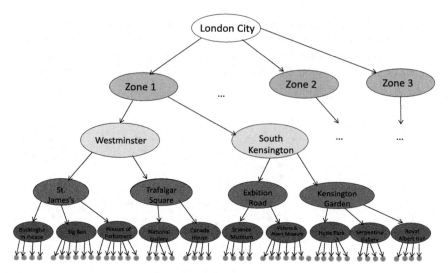

**Fig. 3** Example of hierarchical clustering

## 3.3 Travel Routes Extraction

A travel route is a temporally ordered sequence of places visited by users. In a day d a user u might visits one or more locations in a city. For each user we extract his daily travel routes. A daily travel route is defined as follows:

**Definition 3** *A travel route is a sequence of locations visited by users during the same day according to temporal order. It is represented as a sequence of n pairs in the form $(l_i, t_i)$ where $l_i = (x_i, y_i)$ and $t_i$ is the timestamp. We can, thus, define a travel route as a spatial-temporal sequence of visited place:*

$$TR_{u,d} = v_{l_0,t_0} \longrightarrow v_{l_1,t_1} \longrightarrow \cdots \longrightarrow v_{l_n,t_n}$$

According to the tree model defined for the location identification process, we also obtain a hierarchy of travel routes. In particular, we connect the clusters of the same level according to the extracted trajectories: if consecutive places on one route are individually contained in two clusters, a link would be generated between the two clusters in a chronological direction according to the time serial of the two places.

*Example 2* Referring to the example shown in Fig. 3 we can derive the following travel routes with different granularity detail:

$TR_1$ : *BigBen → NationalGallery → ScienceMuseum → HydePark*

$TR_2$ : *St.James's → TrafalgarSquare → ExbitionRoad → KensingtonGarden*

$TR_3$ : *Westminster → SouthKensington*

Considering the finer granularity detail we derive the travel route $TR_1$ among venues; the travel route $TR_2$ expresses the same travel path of $TR_1$ but here the geo-coordinates are aggregated at the extend of suburbs; in the travel route $TR_3$ the aggregation is at the extend of districts in London.

To the aim of this paper we extract travel routes at the finest granularity detail, so travel routes among venues in the city. We extracted 455422 travel routes with an average cardinality of 3.31.

## 3.4 Mining Frequent Travel Routes

The extraction of spatial-temporal patterns from travel routes is a key point of our methodology. Thus, the problem of trajectory pattern mining can be modeled as an extension of traditional sequential pattern mining and association rule mining. The idea is to determine sequences of places, which occur together frequently in the data, and with similar transition times. The sequential pattern mining paradigm can be extended to this case by incorporating temporal constraints into successive elements of the sequence. Given the travel routes generated at the previous step, the sequential pattern mining algorithm will find all the sequential patterns whose frequencies are no smaller than the minimum support. We refer to the mined frequent patterns as *trajectory patterns* or *frequent travel routes*.

We adopt a two-phase approach for mining popular travel routes: (i) the first phase consists of applying sequential pattern mining on the location sequences; (ii) the second one consists of extracting the most frequent maximal sub-sequences from all the frequent sequences obtained with the sequential pattern mining algorithm. This second step is necessary in order to ensure that trajectories with large segments in common are not reported simultaneously. To this aim we modify the well-known PrefixSpan [7] algorithm to obtain only maximal frequent patterns.

In the following is described a toy example of frequent travel routes mining.

*Example 3* Let's suppose to extract from the tweets of London the following travel routes:

$T1$ : $BigBen \rightarrow TheLondonEye \rightarrow TowerBridge \rightarrow TateModern$

$T2$ : $BuckinghamPalace \rightarrow MadameTussauds \rightarrow BigBen \rightarrow TheLondonEye \rightarrow TowerBridge \rightarrow WestminsterBridge \rightarrow ParliamentSquare$

$T3$ : $BigBen \rightarrow TheLondonEye \rightarrow TrafalgarSquare \rightarrow TempleofMithras$

$T4$ : $BigBen \rightarrow TheLondonEye \rightarrow TowerBridge \rightarrow WestminsterBridge \rightarrow ParliamentSquare$

$T5$ : $BuckinghamPalace \rightarrow MadameTussauds \rightarrow TowerBridge \rightarrow TateModern$

Given the above 5 travel routes the maximal frequent pattern mining algorithm determines the following frequent travel routes with a minimum support fixed at 2:

$BigBen \rightarrow TheLondonEye$ (frequency 4)

$TheLondonEye \rightarrow TowerBridge$ (frequency 3)

$TowerBridge \rightarrow TateModern$ (frequency 2)

$WestminsterBridge \rightarrow ParliamentSquare$ (frequency 2)

$BuckinghamPalace \rightarrow MadameTussauds$ (frequency 2)

Concerning our real evaluation of the London tweets, we mined 11959 patterns and then we extract only the maximal frequent patterns with a minimum support of 25, obtaining this way just 923 maximal frequent patterns to which we refer to as frequent travel routes. Of the 923 mined frequent travel routes, 451 concern routes traveled by multiple users whereas 461 are routes traveled by just one user.

## 3.5 Spatial-Temporal Features of Frequent Travel Routes

This section describes how to obtain daily snapshots and statistics on the mined travel routes.

We define a set of features that exploit different information dimensions about users' movement such as historical visits or social ties. The features answer to several questions allowing to characterize the dynamics of human behavior and of their activity together with the featuring of cities and urban areas. Among such questions the following ones also included: (1) how popular is a travel route? (2) is a travel route frequent? (3) how many popular locations are included in the route? (4) how many people visit the location? (5) how long/ (6) when/ (7) how is the place visited?

We denote the set of frequent travel routes in the Twitter dataset as $\mathcal{FTR}$, the set of category $C$ and the set of users $\mathcal{U}$.

**Number of journeys**. The number of journeys along a travel route is the overall number of times that the route is traveled; it corresponds to the support of the frequent travel route $ftr$ and can be defined as follows:

$$\text{nJourney}(ftr) = |\{ftr_i \in \mathcal{FTR} : ftr_i = ftr\}| \tag{1}$$

**Number of distinct travelers**. The number of people who travel a route is indicative of its *popularity*. The number of travelers of a frequent travel route $ftr$ can be expressed as follows:

$$\text{nTraveler}(ftr) = |\{u \in \mathcal{U} : \mathcal{J}_{ftr,u} \neq \emptyset\}| \tag{2}$$

where $\mathcal{J}_{ftr,u}$ is the set of journeys in the travel route $ftr$ by user u. According to that we define a *popular travel route* as follows:

**Definition 4** *A frequent travel route is popular when the number of distinct users traveling it is higher than 20 % of the total number of users.*

**Number of journeys per traveler**. This feature describes the periodicity of users behavior along the trajectory. The feature is formalized as follows:

$$\text{JourneyPerUser}(ftr) = \frac{\text{nJourney}(ftr)}{\text{nTraveler}(ftr)} \tag{3}$$

**Travel route category**. This feature characterizes the travel route in terms of the category of the locations crossed in the route. Formally can be expressed as follows:

$$C(\text{ftr}) = \{c : \forall l \in \mathcal{L}_{\text{ftr}}, \text{Category}(l) = c\} \qquad (4)$$

where $\mathcal{L}_{ftr}$ is the set of locations in the travel route $\text{ftr}$ and *Category(l)* represents a category of the location $l$.

**Entropy of a travel route**. This feature tells us how a travel route $\text{ftr}$ is visited, whether users tend to travel regularly the route at usual times or they transit in it without any regularity. For this purpose we use the Shannon Entropy:

$$H(X_{ftr}) = - \sum_{u=1}^{n} p(x_u) log p(x_u), \quad where \ p(x_u) = f(ftr, u) \qquad (5)$$

$f(\text{ftr}, u)$ describes the distribution of the movements across the users, in other words it is the user's proportion of journeys at the route $\text{ftr}$ and is defined as:

$$f(\text{ftr}, u) = \frac{|\mathcal{J}_{\text{ftr},u}|}{|\mathcal{J}_{\text{ftr}}|} \qquad (6)$$

where $\mathcal{J}_{ftr}$ is the whole set of journeys in the travel route.

We expect a small entropy for routes involving Professional or College & University places in which people tend to have more stable and periodic behavior. In contrast, a higher entropy value implies that many users do journeys along the route $\text{ftr}$, but they have very few journeys. This user behavior is typical along routes involving Arts & Entertainment or Nightlife & Spot places.

## 4 Statistical Analysis of Frequent Travel Routes

In this section we present a statistical study on the twitter dataset described in Sect. 3.1. In particular, we focus on the 451 routes traveled by multiple users obtained after the mining phase, as detailed in Sect. 3.4. Specifically, we describe the daily snapshots and statistics obtained on the mined travel routes by exploiting the spatial-temporal features introduced in Sect. 3.5. The snapshots concern routes dynamics such as the traveled paths, the visited places, the movements among them, the duration of the visits of each location.

The overall analysis of traffic flows across time and space reveal that the majority of the frequent travel routes are tourist movement patterns, as the visited locations are London's most important historical and cultural sites. Moreover, the spatial-temporal information featuring the travel routes, like the number of users traveling the route (popularity), the number of journeys (frequency) along the route, and the regularity of users behavior (entropy), outline the profile of touristic trajectories.

The average length of the analyzed frequent travel routes is 2.4. This number is also confirmed in many studies in literature as reported in the study [11] where the average length of touristic travel routes is 2, and in the study [12] where the authors explicitly say that it is not necessary to find out trajectories with long length as people would not visit many places in a trip and for this reason in that paper they work with 2-length trajectories. The number of users traveling a specific route is indicative of its *popularity*. Figure 4a shows the probability distribution function (PDF) of the number of distinct users per frequent travel route, highlighting that about 86 % of travel routes are traveled by a number of distinct users greater than 10, with a peak when the number of user is 23. Among them, about 10 % of trajectories is visited by more than 60 users.

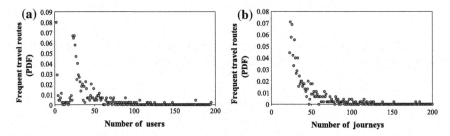

**Fig. 4** Probability Distribution Function of number of distinct users (**a**) and number of journeys (**b**) per frequent travel route

Figure 4b shows the PDF of the number of journeys per travel route. This value gives information about the *frequency* of travel routes. The graph highlights a trend similar to the previous distribution. In fact, more than 80 % of the travel routes are traveled a number of times ranging from 10 to 60, while there is a few number of travel routes (about 16 %) with a very large number of journeys; these last trajectories cross the most popular tourist attractions.

Even if the volume of journeys and users per travel route is quite high, we observe that the number of journeys per user is 1 or 2 for the 85 % of travel routes. This depends on the fact that those travel routes are touristic and, thus, users occasionally travel across them, as shown for some trajectories in Table 1.

In Table 1 we list the 20 top frequent travel routes. We can notice that most of the frequent routes concerns visits to world famous historic sights, including Houses of Parliament, Big Ben, Buckingham Palace, and other London attractions like the London Eye and Harrods. Table 1 contains 16 travel routes, referred to as *touristic travel routes*, characterized by a high number of distinct users traversing them (popular routes) and a high number of journeys along them. Important information is given by the high entropy values, indicating that the behavior of visitors is not regular, in fact they travel the routes on average only once, as confirmed by values in column *Journeys/Users*.

We observe that the district of Westminster presents the highest tourist density with travel routes covered by more than 170 users.

A different profile is presented by 4 trajectories in Table 1: T10, T12, T13 e T14, referred to as *residential travel routes*. These routes are traveled by 2-3 users who

**Table 1** The TOP 20 frequent travel routes

| ID | Frequent travel route | Tweets | Users | Journeys | Journeys/user | Entropy | Time (min) | Space (km) |
|----|----------------------|--------|-------|----------|---------------|---------|-----------|-----------|
| T0 | Houses of Parliament → The London Eye | 581 | 235 | 244 | 1 | 7.85 | 122.34 | 0.43 |
| T1 | The London Eye → Houses of Parliament | 540 | 217 | 229 | 1 | 7.72 | 94.06 | 0.43 |
| T2 | Big Ben → The London Eye | 490 | 195 | 205 | 1 | 7.58 | 97.96 | 0.46 |
| T3 | The London Eye → Big Ben | 434 | 176 | 184 | 1 | 7.43 | 99.37 | 0.46 |
| T4 | Buckingham Palace → The London Eye | 430 | 176 | 179 | 1 | 7.45 | 195.04 | 1.57 |
| T5 | The Tower of London → Tower Bridge | 377 | 147 | 152 | 1 | 7.18 | 87.79 | 0.25 |
| T6 | Buckingham Palace → Houses of Parliament | 318 | 137 | 138 | 1 | 7.09 | 187.28 | 1.26 |
| T7 | Houses of Parliament → Buckingham Palace | 302 | 125 | 127 | 1 | 6.96 | 134.06 | 1.26 |
| T8 | Harrods → Hyde Park | 299 | 120 | 126 | 1 | 6.88 | 211.24 | 1.09 |
| T9 | Tower Bridge → Tower of London | 334 | 117 | 123 | 1 | 6.84 | 77.12 | 0.25 |
| T10 | Clapham Junction → ASDA Clapham Junction | 533 | 2 | 119 | 60 | 0.07 | 114.56 | 0.24 |

(continued)

**Table 1** (continued)

| ID | Frequent travel route | Tweets | Users | Journeys | Journeys/ user | Entropy | Time (min) | Space (km) |
|---|---|---|---|---|---|---|---|---|
| T11 | Trafalgar Square → National Gallery | 276 | 116 | 118 | 1 | 6.85 | 60.04 | 0.08 |
| T12 | Red Lion → The Courtyard Theatre | 374 | 2 | 115 | 58 | 0.07 | 173.33 | 0.22 |
| T13 | ASDA Clapham Junction → Clapham Junction | 507 | 2 | 115 | 58 | 0.07 | 90.60 | 0.24 |
| T14 | 06 St. Chad's Place → Burger King | 483 | 3 | 114 | 38 | 0.14 | 93.84 | 0.29 |
| T15 | Hyde Park → Harrods | 257 | 96 | 112 | 1 | 6.51 | 200.37 | 1.09 |
| T16 | Buckingham Palace → Trafalgar Square | 274 | 107 | 111 | 1 | 6.72 | 142.34 | 1.22 |
| T17 | The London Eye→ Buckingham Palace | 280 | 107 | 110 | 1 | 6.73 | 181.46 | 1.57 |
| T18 | The London Eye → Tower Bridge | 254 | 103 | 105 | 1 | 6.68 | 195.84 | 3.08 |
| T19 | Tower Bridge → The London Eye | 247 | 101 | 102 | 1 | 6.65 | 182.68 | 3.08 |

repeat often the paths, in fact the number of journeys is high, consequently they are frequent but not popular. In this case, the entropy is low because visitors run through these routes in a periodic way, returning on different days. For instance, T10 and T13 represent the movement from a railway station to a grocery store and vice versa. The users could be two employers of the store that arrive to work and come back to home by train.

Table 2 shows the 20 popular locations. As we expected, it contains the London's most famous tourist attractions. These locations are characterized by a high number of distinct visitors and by a high number of visits. The *FrequencyFTR* column

**Table 2** The TOP 20 popular locations

| Popular location | Users | Visits | VisitPerDay | Frequency FTR | Reachability | Incoming |
|---|---|---|---|---|---|---|
| The London Eye | 1594 | 3323 | 21.58 | 52 | 24 | 25 |
| Buckingham Palace | 1284 | 2784 | 18.20 | 52 | 29 | 22 |
| Harrods | 996 | 1898 | 12.82 | 45 | 23 | 22 |
| Houses of Parliament | 922 | 1728 | 13.09 | 27 | 14 | 12 |
| Trafalgar Square | 890 | 1879 | 12.28 | 39 | 20 | 19 |
| Tower Bridge | 765 | 1419 | 9.72 | 27 | 12 | 15 |
| Big Ben | 737 | 1431 | 15.39 | 25 | 11 | 13 |
| Hyde Park | 650 | 1089 | 7.83 | 25 | 13 | 12 |
| The Tower of London | 539 | 988 | 7.06 | 22 | 12 | 10 |
| Piccadilly Circus | 501 | 894 | 11.92 | 21 | 9 | 12 |
| Oxford Street | 458 | 892 | 9.91 | 22 | 12 | 10 |
| Westminster Abbey | 438 | 815 | 5.82 | 16 | 9 | 7 |
| St Paul's Cathedral | 374 | 655 | 4.85 | 17 | 10 | 7 |
| Selfridges & Co | 355 | 607 | 4.71 | 13 | 9 | 4 |
| M & M's World | 337 | 493 | 3.79 | 15 | 5 | 10 |
| Leicester Square | 332 | 513 | 3.83 | 15 | 6 | 9 |
| Frankie & Benny's | 330 | 605 | 6.24 | 16 | 10 | 6 |
| Tate Modern | 302 | 433 | 3.52 | 11 | 6 | 5 |
| London Pavilion | 300 | 503 | 6.53 | 16 | 7 | 9 |
| Westminster Bridge | 278 | 434 | 3.29 | 9 | 5 | 4 |

highlights that the locations can belong to travel sequences of different frequent travel routes (ftr). For instance, The London Eye and Buckingham Palace are in 52 distinct travel sequences. For each location, the table shows the number of other locations that can be reached from the one (*Reachability*), and the number of places from where the specific location can be reached (*Incoming*). In agreement with *FrequencyFTR*, The London Eye and Buckingham Palace are the places that allow to reach and are accessed by the highest number of places. We determine place semantics according to Foursquare classification. Figures 5a, b show the categories distribution to which the frequent travel routes and the visited locations belong, respectively. Both the distributions highlight that the prominent activities of users are spending leisure time in the outdoor (e.g., park, square, mountain and river) or visiting landmarks (e.g. historic buildings, monuments, castles and gardens). In fact 42 % of routes has

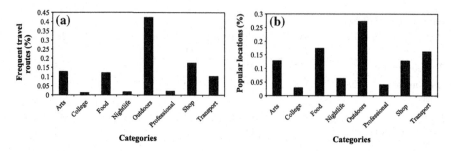

**Fig. 5** Categories distribution of the frequent travel routes (**a**) and of visited locations (**b**)

travel sequences among *Outdoor & Recreation* locations (Fig. 5a). Those locations represent 27 % of the overall visited places (Fig. 5b). Over half of the travel routes (53 %) visits places belonging to the categories *Food, Shop & Service, Arts & Entertainment*, and *Travel & Transport*. Only 5 % of routes visits work, educational and night-life locations. A similar pattern arises when considering the categories distribution of visited locations in Fig. 5b. This evidence reinforces our hypothesis that most of the travel routes are touristic movements.

## 5 Conclusion

In this paper we presented a novel methodology to extract and analyze the time- and geo-references associated with social data so as to mine information about human dynamics and behaviors within urban context. We performed a fine grain analysis of unpredictable and irregular information coming from geo-tagged tweets. In particular, we extracted a set of daily trajectories and we used a sequential pattern mining algorithm to discover frequent travel routes. We then, defined a set of spatial-temporal features over such routes and, accordingly, performed a statistical characterization of patterns, rules and regularities in moving trajectories. Future work include the integration of such methodology in a recommender system for trip planning, personalized navigation services, and location-based services useful for urban planning and management.

## References

1. Cesario, E., Comito, C., Talia, D.: Towards a cloud-based framework for urban computing, the trajectory analysis case. In: CGC, pp. 16–23 (2013)
2. Crandall, D.J., Backstrom, L., Huttenlocher, D., Kleinberg, J.: Mapping the world's photos. In: WWW, pp. 761–770 (2009)
3. Cranshaw, J., Toch, E., Hong, J., Kittur, A., Sadeh, N.: Bridging the gap between physical location and online social networks. In: Ubiquitous, Computing, pp. 119–128 (2010)

4. Falcone, D., Mascolo, C., Comito, C., Talia, D., Crowcroft, J.: What is this place? inferring place categories through user patterns identification in geo-tagged tweets. IEEE Comput. Soc. MobiCASE (2014)
5. Gabrielli, L., Rinzivillo, S., Ronzano, F., Villatoro, D.: From tweets to semantic trajectories: mining anomalous urban mobility patterns. In: Citizen in Sensor, Networks, pp. 26–35 (2014)
6. Giannotti, F., Nanni, M., Pinelli, F., Pedreschi, D.: Trajectory pattern mining. In: KDD, pp. 330–339 (2007)
7. Pei, J., Han, J., Mortazavi-Asl, B., Pinto, H., Chen, Q., Dayal, U., Hsu, M.-C.: Prefixspan: mining sequential patterns efficiently by prefix-projected pattern growth. In: ICDE, p. 215 (2001)
8. Sakaki, T., Okazaki, M., Matsuo, Y.: Earthquake shakes twitter users: real-time event detection by social sensors. In: WWW, pp. 851–860 (2010)
9. Takahashi, T., Abe, S., Igata, N.: Can twitter be an alternative of real-world sensors? In: Human-Computer Interaction. Towards Mobile and Intelligent Interaction, Environments, pp. 240–249 (2011)
10. Wakamiya, S., Lee, R., Sumiya, K.: Urban area characterization based on semantics of crowd activities in twitter. In: GeoSpatial Semantics, pp. 108–123 (2011)
11. Yin, Z., Cao, L., Han, J., Luo, J., Huang, T. S.: Diversified trajectory pattern ranking in geo-tagged social media. In: SDM, pp. 980–991 (2011)
12. Zheng, Y.-T., Zha, Z.-J., Chua, T.-S.: Mining travel patterns from geotagged photos. ACM Trans. Intell. Syst. Technol. 3(3), 56:1–56:18 (2012)

# Enhancing Workspace Composition by Exploiting Linked Open Data as a Polymorphic Data Source

Giuseppe Desolda

**Abstract** In the last decade, the World Wide Web has been evolving as a data infrastructure, where a wide variety of resources is increasingly being made available as Web services. This trend is pushing the researchers to investigate approaches like composition platforms, aimed at empowering end users to access, compose and use these services. Despite the wide availability of data sources, due to the specific and diverse end users' information needs often no data source can satisfy these needs. This limits the adoption of composition platforms in real contexts and everyday use. In order to overcome this limitation, this paper presents a polymorphic data source that exploits the wide availability of information structured in the Linked Open Data cloud. To build this data source, a semi-automatic annotation algorithm is presented that creates semantic annotations for services available in a composition platform. An implementation of this approach in a mashup platform is described.

**Keywords** Mashup · Linked open data · Semantic web

## 1 Introduction and Motivations

Over the past years, we have been facing a growing amount of heterogeneous data sources available on the Web. When writing this paper, the site programmableweb.com lists more than 12000 API to retrieve data or exploit functionalities. This paper is about data retrieval APIs that can be classified into cross-domain (Wikipedia, YouTube, Google, etc.) or domain specific APIs (Government, Life Science, Music, etc.). This huge amount of information available on the Web and the opportunities offered by Web 2.0 are pushing researchers to investigate new

G. Desolda (✉)
Dipartimento di Informatica, Università degli Studi di Bari Aldo Moro,
Via Orabona, 4, 70125 Bari, Italy
e-mail: giuseppe.desolda@uniba.it

© Springer International Publishing Switzerland 2015
E. Damiani et al. (eds.), *Intelligent Interactive Multimedia Systems and Services*,
Smart Innovation, Systems and Technologies 40,
DOI 10.1007/978-3-319-19830-9_9

methodologies, technologies and mechanisms to allow laypeople, i.e., end users without expertise in programming, to access and manipulate data sources by exploiting visual mechanisms. In the last 10 years, different platforms with various composition paradigms have been proposed [1, 2]. Typically, they implement visual mechanisms to access, create, compose, modify and use data sources usually available through the APIs. They are often known as mashup platforms.

According to [3], different features affect the mashup quality, for example, the *data quality* dimension, characterized by *accuracy, timeliness, completeness* and *availability*. Regarding this dimension and in particular the completeness, the data sources available nowadays describe a portion of a domain and often do not include many details. It is sometimes possible to overcome this limitation by composing different data sources, but in some cases, when the end users' information need is more specific, no data source could provide the useful information. This is a vast limitation in exploiting mashup platforms in real contexts. In fact, although the current platforms allow laypeople to easily use and compose data sources, often they cannot benefit from a composition platform due to a lack of data sources if used in real contexts.

To overcome this lack of information and better satisfy the end users' information needs, this paper presents a new *polymorphic* data source built upon the Linked Open Data cloud. It is called polymorphic because it provides mutable information with respect to the data sources of which it is composed.

The remainder of this paper is structured as follows. Section 2 describes the polymorphic data source, the use of Linked Open Data to build this data source and the integration of the polymorphic data source in a mashup platform. Section 3 describes the annotation algorithm and its performance evaluation. Section 4 reports related works. Section 5 concludes the paper and also outlines future work.

## 2  Polymorphic Data Source: a Source for Many Purposes

To explain the idea of a polymorphic data source, let us consider the following scenario that refers to a typical situation when laypeople want to exploit a platform to mashup services according to their needs, but at a certain point they leave it because they do not find useful data sources available through the platform. "John is using a mashup platform. He adds the SongKick service to his workspace to find upcoming musical events in his city. He also needs to retrieve, for each event artist, a list of related videos. For this purpose, John composes SongKick artist attribute with YouTube. Now, John has two widgets in his workspace: SongKick and YouTube; the first allows him to search upcoming musical events and the second automatically performs a search (with the artists' name) each time John clicks on a specific musical artist in the list of upcoming events. Afterwards, he wants to know, for each artist, details such as genre, starting year of activity and artist photo. *Searching for useful services on the composition platform, John does not find any service that satisfies his needs. Thus, John is not supported anymore by the platform and has to go to the Web for a usual (manual) search for the specific information".

Let us now look at a scenario that is the same as the previous one until the asterisk, but it goes on with the following. "To retrieve the desired information, John decides to expand the SongKick artist attribute with the polymorphic data source. When he chooses the polymorphic data source, the platform shows a list of new properties related to the concept of musical artist. Thus, John decides to create the new data source with the genre, the starting year of activity and the artist photo properties. Henceforward, John can find a list of upcoming events on SongKick and can visualize the additional artist's information on the polymorphic data source by clicking on a specific artist on SongKick (Fig. 1)".

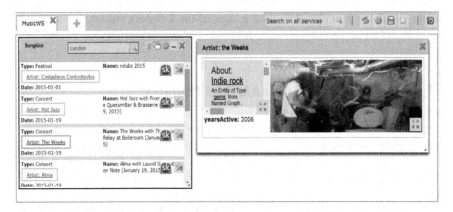

**Fig. 1** Composition of DBpedia polymorphic data source with SongKick artist attribute

In the previous scenario, John could continue to compose SongKick with the same polymorphic data source starting from other SongKick attributes. For each attribute that John decides to expand, the polymorphic data source provides different properties related to the semantics of the starting attribute (for example, for the SongKick place attribute, properties like borough, census, year and demographics should be shown). Thus, this type of data source is considered polymorphic because it can provide different information (properties) according to the data source attribute that is selected. On the contrary, the classic data sources (YouTube, Wikipedia, etc.) provide the same properties independently of the selected attribute.

## 2.1 Linked Open Data as a Basis for the Polymorphic Data Source

The polymorphic data source is built by exploiting the huge amount of information available in the Linked Open Data (LOD) cloud. In 2009, Tim Berners-Lee defined Linked Data as "a set of best practices for publishing and connecting structured data on the Web" [4]. The goal of the Linked Data project is to publish data so that they are readable by a human and an automatic agent. The LOD are Linked Data distributed under an open license that allows its reuse for free. At the time of this paper, there are more than 1000 KB datasets published in the LOD cloud.

Nowadays, one of the biggest KBs in the LOD cloud is DBpedia (the structured version of Wikipedia). The DBpedia English version describes 4.58 million things, out of which 4.22 million are classified in its ontology. In the DBpedia ontology, there are 685 classes. Thanks to the availability of this huge amount of information and its semantics structured in the DBpedia ontology, DBpedia has been chosen as the starting point to create the polymorphic data source.

An annotation algorithm has been developed to annotate each attribute of each service (available in a platform) with a class of the DBpedia ontology. Each class has to be semantically similar to the attribute.

## 2.2 An Implementation in a Mashup Platform

The algorithm and the polymorphic data source have been implemented in the mashup platform described in [5]. This platform provides a composition paradigm elicited during users' studies [6] and evaluated during field studies in cultural heritage and technology enhanced learning domains [7]. The platform is implemented by using Primefaces, an open source User Interface (UI) component library for JavaServer Faces (JSF) based applications. It is deployed on a remote Apache Web server and a Mysql database is used for the user account management. It provides different mashup mechanisms to access, compose and use services. In particular, the end user can add services to his own workspaces and can compose services by using two mashup mechanisms called *join* and *union* [6].

In this paper only the join composition is considered. As reported in the scenarios, with the join function the user composes a service A with a service B, in order to expand the results of service A with details provided by service B. This composition is assisted by a wizard procedure that the user activates by clicking on a gearwheel button in the upper right corner of the service widget. By clicking on this icon, the user (1) selects the attribute that he wants to expand, (2) selects the data source from which to gather information (DBpedia in our case), (3) chooses a visual template to visualize the new results and finally (4) uses drag&drop to map a subset of service attributes into the visual template. The fourth step is the most interesting for the aim of this paper. In fact, while composing a service with another 'traditional' service the list of attributes in step 4 is always the same, by choosing the polymorphic data source the list of attributes is different in relation to the semantics of the selected attribute during step 1.

In the current implementation, the service that provides details is shown as a window only when the user clicks on a specific item (e.g., click on "the Weeks" artist in Fig. 1). This visualization emerged as requirement during the users' study described in [6].

# 3 An Algorithm for Data Source Annotation

This section describes a semi-automatic algorithm that creates semantic annotations for services available in a composition platform. Its performance evaluation is also reported.

## 3.1 Generation of a Set of Candidate Classes to Annotate Attributes

In order to create this polymorphic behaviour, a mapping step is required between all data source attributes registered in the platform and the DBpedia ontology classes. In general, this problem falls in the ontology matching area. In order to start from consolidated approaches, the methodologies surveyed in [8, 9] were investigated with the aim of creating an ad hoc solution based on the literature, without the pretension of building a new ontology matching methodology. Furthermore, a great deal of specific literature in the semantic web area has already been produced for the problems of semantic annotation of a service [10]. These approaches have been taken into account to design the proposed algorithm.

The proposed solution can be classified as a semi-automatic and instance-based annotation algorithm. As is described in the following, it is defined as semi-automatic because a user has to provide a set of example queries (about 10) when a data source is registered in the platform. Furthermore, it is defined as instance-based because it infers a set of candidate classes to annotate the attributes starting from the results (instances) of the queries. The main goal of the algorithm is to annotate each attribute of each service with a DBpedia class that is semantically similar to the attribute. The algorithm is reported in Table 1. The criterion for choosing the most important class, as specified in line 12 of Table 1, is illustrated in Sect. 3.2.

In order to understand the algorithm, an example of its execution on a real data source is here reported. Let us consider SongKick data source $s$ and a set $Q$ of queries on it (e.g., London, Liverpool, Rome). This set Q is manually provided only once at the time of service registration in the platform. Each instance of the SongKick results is characterized by the set of attributes $A = \{artist, place, event\_type, event\_name, date\}$. The algorithm starts by executing all the queries in Q on SongKick and by collecting all the results in set $I$. For each attribute $a_i$ in A, the algorithm considers all the instances selecting only the $a_i$ attribute values. For example, after having queried SongKick with queries in Q, for the artist attribute the algorithm creates set $I_{artist} = \{Ligabue, U2, One Direction, Taylor Swift,...\}$, that is a set of musical artists who will perform at places stored in Q. The algorithm uses each instance of $I_{artist}$ to query DBpedia. The aim is to find the same instance of each element in $I_{artist}$ as a DBpedia thing and add its classes in set $L\_temp$. Obviously, although the instances of each attribute have the same meaning (for example, all artist instances are singers), not all the retrieved DBpedia things are instances of the same class. Thus, at the end of the

**Table 1**  The instance-based semi-automatic annotation algorithm

|   | |
|---|---|
| | **Input:** Set $T$ of Triples $t=(s,A,Q)$, $s$ is a data source, $A$ is a set of attributes $a$ for s, and $Q$ is a set of queries on $s$ |
| | **Output:** Set $R$ of results $r=(s, L)$, $s$ is a data source, $L$ is a set of $<a_i,l_i>$ where $a$ is an attribute of $s$ and $l$ is its label |
| 1: | **for** each $t \in T$ **do** |
| 2: | **create** $I$ as empty set of instance results for queries and an empty set M |
| 3: | **for** each $q \in Q$ **do** |
| 4: | **query** $s$ by using $q$ and collect instance results into $I$ |
| 5: | **end for** |
| 6: | **for** each attribute $a$ of $s$ **do** |
| 7: | **create** $L\_temp$ as empty set of labels for $a$ |
| 8: | **for** each $i \in I_a$ **do** |
| 9: | **query** DBpedia by using value of $i$ and obtain a set $C$ of classes |
| 10: | **put** values of $C$ into $L\_temp$ |
| 11: | **end for** |
| 12: | **calculate** most important class $l$ in $L\_temp$ and put $<a,l>$ into L |
| 13: | **end for** |
| 14: | **end for** |

execution on all the attributes, the results appear as shown in Table 2, where the *Class* column indicates the DBpedia classes inferred for each attribute and the % column indicates the frequency of the classes at the end of DBpedia queries. Furthermore, when the algorithm queries DBpedia, sometimes the retrieved things are wrongly classified. For example, when DBpedia is queried with the 'Ligabue' string, three thing instances of different classes are retrieved: one instance of Artist (Luciano Ligabue, singer), one instance of Agent (Antonio Ligabue, painter) and one instance of Italian_opera_singer (Ilva Ligabue, opera singer). Obviously, the second and third things are false positives that create noise in the set L_temp. However, it is empirically observed that this noise represents only a tiny percentage (typically less than 4 %). For this reason, no classes less than 4 % are considered in the rest of the algorithm.

Until now, the algorithm has collected a set of promising (candidate) classes to annotate attribute data sources. The easiest annotation solution could be to select the most frequent class (In Table 2 *Agent* for Artist, *Event* for EventType and *Place* for Location), but the performance of this solution is improved by the second step of the proposed algorithm that takes into account both the class frequency and the ontology tree structure.

## 3.2  Choosing the Best Class from the Set of Candidates

The starting point for choosing the best class for each data source attribute is the set of promising classes that the algorithm has built in the previous step. The goal of the next step is to assign to each class in Table 2 a rating that takes into account both the class frequency and the ontology tree structure. In the end, the class with the highest score is used to annotate the service attribute. To explain why the tree structure is important, consider the DBpedia sub-tree in Fig. 2. In that figure the

**Table 2** Candidate classes for annotable attributes of the SongKick data source; the *Class* column indicates the DBpedia class associates; the % coloumn indicates the frequency of each class

| Artist | | EventType | | Location | |
|---|---|---|---|---|---|
| *Class* | *%* | *Class* | *%* | *Class* | *%* |
| Agent | 14 | Event | 25 | Place | 17 |
| Work | 13 | FilmFestival | 25 | Settlement | 11 |
| MusicaWork | 12 | Organization | 13 | PopulatedPlace | 11 |
| Organization | 7 | Television | 13 | Work | 6 |
| Album | 7 | Agent | 12 | Album | 6 |
| Person | 7 | TelevisionShow | 12 | MusicalWork | 6 |
| Band | 6 | | | Agent | 6 |
| Artist | 6 | | | Organization | 6 |
| Single | 5 | | | Building | 6 |
| MusicalArtist | 5 | | | Architectural | 6 |
| ... | ... | | | ... | ... |

sub-tree of the DBpedia ontology is depicted; it has been built by considering, for an easy explanation, the first ten candidate classes of the attribute Artist in Table 2. If the algorithm annotates the Artist attribute with the most frequent class, then the Agent class (14 %) will be chosen. However, by looking inside the semantics and the properties that characterize this class, it is evident that the Agent class is too general for the concept of the musical artist of the SongKick Artist attribute. We need to choose a more specific class. There are two aspects to consider, in order to annotate data sources with the best classes: the *class coverage* and the *number of properties*. The class coverage is the percentage of retrieved DBpedia instances covered by each class (node percentages in Fig. 2) and it is calculated as sum of class percentage with all its sub-class percentages. This value is higher in the ontology top level classes and vice versa, because each class also cover the sub-class instances. On the other hand, the more specific is the class, the more properties the user can choose when creating the polymorphic data source. For this reason, the proposed algorithm tries to find a trade-off between the class coverage and the number of properties.

**Fig. 2** Sub-tree of the DBpedia ontology built by using the SongKick artist attributes. For each node, the percentage indicates the class coverage

In order to take into account these aspects, the *x-value* is introduced. It is an index that quantifies the semantic similarity between a class and an attribute considering both the frequency and the ontology structure. In particular, to consider the ontology structure, the algorithm starts by generating all the combinations of the sub-trees built with the classes in the list of candidates. The length of these groups ranges from 2 up N (where N is the number of candidate classes of an attribute). For each class in these sub-trees, the algorithm calculates the x-value. At the end of the computation, each class has many x-values, but only the highest value of each class is considered for the final ranking. The generation of these groups is performed to explore all possible paths in the ontology tree, as also performed in [11]. In particular, when the algorithm generates all sub-trees, new classes could be added to the candidate list. For example, let us consider the Fig. 2 and suppose that the MusicalWork class is not generated in the candidate list of Table 2. During the generation of all sub-trees, when the Single and Album classes are used to create a sub-tree, also MusicalWork is considered because it is their common ancestor. Thus, MusicalWork is added in the candidate list and its x-value is calculated. Choosing the MusicalWork, or an ancestor in general, the coverage could be higher than its sub-class coverage, maintaining a good number of properties that describe the semantics of the considered data source attribute.

The formula of an x-value that estimates the power of a class into each sub-tree is:

$$x - value = (classAncestorRatio + parentPower + nodePower)\%class.$$

Let us analyse in detail each component. The first one is *classAncestorRatio*.

$$classAncestorRatio = \frac{\%class * nLevelScore}{\%commonFather * nLevelScore}$$

This value takes into account the coverage of the current class with respect to the coverage of the first common ancestor in the sub-tree. To penalize classes in the higher levels (such as the ancestors), the numerator and denominator are multiplied by the *nLevelScore*, a number that ranges from 1 to 100 and it is calculated as:

$$nLevelScore = (100/OntologyDepth) * classLevel$$

In this formula, the *OntologyDepth* indicates the maximum depth of the ontology (7 in DBpedia), while *classLevel* specifies the depth of the considered class. It is evident that nLevelScore is high in deeper levels and low in higher levels. In this way, in the classAncestorRatio, the classes at higher levels are penalized.

The second component in the x-value is the *parentPower*. It quantifies the impact of the common ancestor with respect to all the classes at the same level.

$$parentPower = \frac{\%TotRoot}{\%TotRootLevel} \qquad classPower = \frac{\%class}{\%allClasses}$$

In this component, $\%TotRootLevel$ is the sum of the coverage rate of all classes at the same level of the sub-tree root. The $\%TotRoot$ is the coverage rate of the sub-tree root. This component is introduced in the x-value to solve the problem of the class sparsity in a tree. Let us consider in the Fig. 2 the Work and Agent classes. When sub-trees with Work or Agent as ancestors are generated, this component has a high value in Agent sub-classes. In this way, the x-value rewards sub-classes in the Agent branch instead of those in the Work branch.

The third component is the *classPower*, which indicates the weight of the considered class in the sub-tree. In classPower, the $\%class$ is the percentage of the class considered in the sub-tree and $\%allClasses$ is the sum of the percentage of all classes in the considered tree.

At the end of the generation of all trees and the calculation of the x-values, the list of candidate classes is expanded with all the new ancestor classes of the generated sub-tree. Each class has several ratings, one for each group in which it appears. The class with the highest score is selected to annotate the service attributes.

### 3.3 Performance Evaluation of the Annotation Algorithm

To the best of our knowledge, no datasets with data sources attributes exist annotated with DBpedia class. Thus, to establish the performance of the algorithm, two experts created and manually annotated a set of 7 services (for a total of 18 annotable attributes) by using DBpedia classes. In fact, not all the services attributes can be annotated with a DBpedia class (i.e. URL attributes). Furthermore, due to the nature of the algorithm, the numerical attributes (ticket price, temperature, humidity, height, weight, etc.) cannot be annotated because it is impossible to infer the classes from numerical values. As described in the Future Work Section, this limit can be overcome by combining the proposed approach with natural language processing of the attribute name [12, 13].

To evaluate the performance of the automatic annotation algorithm, a new metric called *Accuracy* is introduced. First, a score has been associated with each attribute comparing its automatic annotation to the manual one (AAA stands for Attribute Automatically Annotated; AMA stands for Attribute Manually Annotated). In particular:

- 10 points if AAA = AMA;
- 8 points if AAA is at the same level of AMA (not the same, but very similar semantic);
- 7 points if AAA is 1 level up/down from AMA as a sub-class or a super-class;
- 5 points if AAA is 2 levels up/down from AMA as a sub-class or a super-class;
- 0 points in all other cases.

The *Accuracy* of the overall automatic annotation is calculated as:

$$Accuracy = \frac{\sum_i^N A_i}{\sum_i^N MAXscore}$$

In the Accuracy formula the numerator is the sum of the accuracy of all the attributes, instead the denominator is the sum of the MAXscore that is the maximum accuracy that an attribute can have (10 in our case). The final accuracy ranges from 0 to 100. The Accuracy is calculated both for the annotation performed by associating the most frequent classes in Table 2 (baseline) and for the annotation performed by associating the classes with the proposed algorithm.

| **Table 3** Accuracy comparison between the baseline and the algorithm | | **Accuracy** |
| --- | --- | --- |
| | **Baseline** | 56 % |
| | **Algorithm** | 91 % |

As shown in Table 3, it is evident how important it is to consider the ontology structure during the automatic annotation procedure. In fact, in the latter (91 %) the accuracy is clearly improved.

This metric is quite different with respect to the ones such as precision and recall used for the service semantic annotations [13]. In fact, the classic precision and recall consider true and false values, if the automatic annotation matches the manual one. However, in our case, we can also consider as good annotations classes like super-classes/subclasses, but penalizing them because they do not match exactly the manual annotation. The penalizing factor has been empirically established.

## 4  Related Work

The core of the entire paper is the new polymorphic data source, introduced in the composition platforms to overcome the problem of lack of data source. To build this type of data source, the services available in composition platforms need to be enriched with semantic annotations. To the best of our knowledge, no previous works have tried to solve this problem with similar data sources. However, much effort has been dedicated to enriching Web services with semantic annotations, as described in [10]. In general, the goal of semantic annotations is to improve the mashup platforms with mechanisms like service recommendation, to assist the users during composition [14] or service discovery [15]. This is a hard problem because the Web APIs lack explicit and sufficient semantic information. In fact, API providers usually offer details like input and output parameters in the form of unstructured text in their Web page. A system that tries to exploit this HTML information is SWEET that assists the user in manually annotating services with hRESTS and MicroWSMO formats with Web page information [16]. However, the weakness of this approach is the heterogeneity of the provider Web pages, the lack of information and the manual end user effort that limits the large-scale service annotation.

To overcome the limits of manual methods, different semi-automatic approaches have been proposed. For example, two different solutions reported in [12, 13] annotate services with DBpedia classes and their attributes with DBpedia properties based on syntactical matching and other natural language processing techniques. Although the fact that these approaches seem promising, they still require the intervention of an expert to set-up the system by adding low-level APIs details [12] and, mostly, the presence of WSDL service descriptors that are not available for all Web APIs and that still require a heavy manual effort.

In order to overcome the limitations identified in the literature, the annotation algorithm proposed in this paper has been designed to be: (1) usable by laypeople (no expertise is required to provide example queries), (2) fully automatic (except for the typing of a set of queries) and (3) not constrained by the presence of an HTML or a WSDL service description. It is also reusable, in order to annotate services with other ontology classes (e.g., Freebase or Yago).

## 5   Conclusions and Future Work

This paper describes a polymorphic data source, a solution that aims to address an important limitation that affects the use of a composition platform in real contexts. In order to build this new polymorphic data source, an annotation algorithm has been developed. The initial results reveal that the algorithm creates good annotations that reflect a good quality of the polymorphic data source.

Although the algorithm performance appears encouraging, the proposed algorithm does not aim to solve ontology matching problems. It is only based on the literature and is an ad hoc solution for the specific problem. Thus, one aspect that could be addressed in the future is the improvement of the algorithm by investigating techniques as, for example, NLP approaches [12, 13], to improve the annotation accuracy and to annotate non-annotable attributes (e.g., the numerical attributes). Finally, user studies are planned to evaluate the benefits of a polymorphic data source.

**Acknowledgments** This work is partially supported by the Italian Ministry of University and Research (MIUR) under grant PON 02_00563_3470993 "VINCENTE" and by the Italian Ministry of Economic Development (MISE) under grant PON Industria 2015 MI01_00294 "LOGIN". We also thank the student Vincenzo Lucente for contributing to system implementation.

# References

1. Cappiello, C., Matera, M., Picozzi, M., Sprega, G., Barbagallo, D., Francalanci, C.: DashMash: a mashup environment for end user development. In: Auer, S., Díaz, O., Papadopoulos G.A. (eds.) Web Engineering, vol. 6757, pp. 152–166. Springer, Berlin (2011)
2. Ardito, C., Costabile, M.F., Desolda, G., Lanzilotti, R., Matera, M., Piccinno, A., Picozzi, M.: User-driven visual composition of service-based interactive spaces. J. Visual Lang. Comput. 25(4), 278–296 (2014)
3. Yahoo! Inc. YahooPipes: Retrieved 22 Feb from http://pipes.yahoo.com/pipes/
4. Cappiello, C., Daniel, F., Matera, M.: A quality model for mashup components. In: Gaedke, M., Grossniklaus, M., Díaz, O. (eds.) Web Engineering, vol. 5648, pp. 236–250. Springer, Berlin (2009)
5. Bizer, C., Heath, T., Berners-Lee, T.: Linked data-the story so far. Int. J. Semant. Web Inf. Syst. 5(3), 1–22 (2009)
6. Ardito, C., Costabile, M. F., Desolda, G., Lanzilotti, R., Matera, M., Picozzi, M.: Visual composition of data sources by end-users. In: Proceedings of the International Working Conference on Advanced Visual Interfaces, pp. 257–260. Como, Italy (2014)
7. Ardito, C., Bottoni, P., Costabile, M.F., Desolda, G., Matera, M., Picozzi, M.: Creation and use of service-based distributed interactive workspaces. J. Visual Lang. Comput. 25(6), 717–726 (2014)
8. Hooi, Y., Hassan, M.F., Shariff, A.: A survey on ontology mapping techniques. In: Jeong, H.Y., Obaidat, M.S., Yen, N.Y., Park, J.J. (eds.) Advances in Computer Science and its Applications, vol. 279, pp. 829–836. Springer, Berlin (2014)
9. Shvaiko, P., Euzenat, J.: Ontology matching: state of the art and future challenges. IEEE Trans. Knowl. Data Eng. 25(1), 158–176 (2013)
10. Reeve, L., Han, H.: Survey of semantic annotation platforms. In: Proceedings of the 2005 ACM Symposium on Applied Computing, pp. 1634–1638. Santa Fe, New Mexico (2005)
11. Jain, P., Hitzler, P., Sheth, A.P., Verma, K., Yeh, P.Z.: Ontology alignment for linked open data. In: Proceedings of the 9th International Semantic Web Conference on the Semantic Web - Volume Part I, pp. 402-417. Shanghai, China (2010)
12. Saquicela, V., Vilches-Blázquez, L.M., Corcho, Ó.: Semantic Annotation of RESTful services using external resources. In: Daniel, F., Facca, F. (eds.) Current Trends in Web Engineering, vol. 6385, pp. 266–276. Springer, Berlin (2010)
13. Zhang, Z., Chen, S., Feng, Z.: Semantic annotation for web services based on DBpedia. In: Proceedings of the IEEE 7th International Symposium on Service Oriented System Engineering (SOSE), pp. 280–295 (2013)
14. Bianchini, D., De Antonellis, V., Melchiori, M.: A recommendation system for semantic mashup design. In Proceedings of the Workshop on Database and Expert Systems Applications (DEXA), pp. 159–163 (2010)
15. Talantikite, H.N., Aissani, D., Boudjlida, N.: Semantic annotations for web services discovery and composition. Comput. Stand. Interfaces 31(6), 1108–1117 (2009)
16. Maleshkova, M., Kopecký, J., Pedrinaci, C.: Adapting SAWSDL for semantic annotations of RESTful services. In: Meersman, R., Herrero, P., Dillon, T. (eds.) On the Move to Meaningful Internet Systems: OTM 2009 Workshops, vol. 5872, pp. 917–926. Springer, Berlin (2009)

# Benjamin Franklin's Decision Method is Acceptable and Helpful with a Conversational Agent

**Daniel Mäurer and Karsten Weihe**

**Abstract** In this paper, we show that rational decision-making methods such as Benjamin Franklin's can be successfully implemented as a text based natural language dialog system. More specifically, we developed a prototype, *vpino*, and conducted a user study. Vpino acts maieutically: the questions raised by vpino encourage the user to reflect about potential options and arguments and help her structure her thoughts. To maintain a real dialog, *vpino* unobtrusively attempts to keep control of the conversation at all times. Serious, motivated users evaluated acceptance and usefulness of *vpino* quite positively. Users that are more open to computer based decision support held better and more fruitful dialogs than those with a sceptical attitude. This quantitative result conforms well to our qualitative observation that *vpino* shows good human like behaviour whenever the user is serious and motivated. We also found that users with a more hypervigilant approach to decisions particularly benefit from *vpino*.

**Keywords** Decision support · Rational decision coaching · Conversational agent

## 1 Introduction

The goal of this work is to find out if a text based dialog system in natural language can effectively support humans in making decisions on a distinctly rational basis. We consider decisions of individuals only, not group decisions. Ideally, the user will forget she is working with a software system and believe she is chatting with a human. In addition to general effectiveness and user acceptance, we want to find out how the user's personality and her usual approach to decisions may affect success of the dialog.

D. Mäurer (✉) · K. Weihe
Department of Computer Science, TU Darmstadt, Hochschulstr. 10,
64289 Darmstadt, Germany
e-mail: maeurer@cs.tu-darmstadt.de

K. Weihe
e-mail: weihe@cs.tu-darmstadt.de

© Springer International Publishing Switzerland 2015
E. Damiani et al. (eds.), *Intelligent Interactive Multimedia Systems and Services*,
Smart Innovation, Systems and Technologies 40,
DOI 10.1007/978-3-319-19830-9_10

Of course, a successful conversation requires a reasonable level of motivation and cooperation from the user; users who do not take the system seriously or try to challenge it will not enjoy target-aimed dialogs. Therefore, we distinguished between motivated and cooperative users and other users in our evaluation.

In this work, we explore the potential of IT for simulated human coaching or consulting. More specifically, we developed *vpino*, a prototypical text based dialog system in German language that supports individuals in making a decision on a rational basis. The intended field of application for our system is decision problems from any domain, where a mathematical problem formulation is not accepted by the user or does not exist at all. In our user study, the participants could freely choose their decision problems. For example, many participants chose individual career decisions.

A human coach does not necessarily know anything about the domain and often has but a vague understanding of the decision to be made. However, such a coach may do a good job because she has a thorough understanding of structured decision-making. This observation gives a perspective for a tool such as vpino: A general-purpose tool cannot understand the problem in depth, but nevertheless, it can help the user to clarify the problem, determine obscurities, and guide the user to a solution.

Exactly that is the purpose of vpino. Our system is not an expert on specific problem domains. The intelligence/value of our dialog system is not in its ability in detailed understanding what the user says. Rather, it is an expert on structuring the user's implicit knowledge and merely assists the user like a human coach.

Therefore, the system does not make a decision, nor does it suggest a particular option. Instead, the dialog system guides and leads the user through a chat dialog while hiding the details of the underlying methods on decision-making. Behind the scenes, *vpino* systematically structures the user's decision problem in a rational way.

Various concepts for rational decision-making have been proposed throughout the last centuries. We base *vpino* on Benjamin Franklin's famous method, but extend his method to multiple options.

Technically, *vpino* is a mixed initiative dialog system, in which both the user and the system are free to send messages and take the initiative at all times. The key is that, nevertheless, *vpino* keeps control over the dialog in a way that will not be noticed by the user.

**Background and literature** Decision-making is commonly defined as a cognitive process with the goal to select a final choice from a set of options. This involves handling information, uncertainty and resources and is highly influenced by psychological factors. The range covered by *vpino* is very broad, from private decisions up to management decisions in companies. A large set of guidebook literature on decision-making has been published. Consulting a coach or counsellor is a more expensive alternative.

The subject of decision-making has been covered by many research disciplines like psychology [1, 6] and business economics [3, 16]. While the early software solutions were based on spreadsheets, recent systems usually come with web-based interfaces. Most of the decision-making software developed in the last decades are based on mathematical methods such as *multiple-criteria decision analysis* (*MCDA*).

In contrast to our work, the focus of MCDA lies on formal methods and mathematical models to calculate an optimal solution.

Our focus is completely different: rational decision-making in the spirit of Benjamin Franklin [4], often referred to as *pros and cons*, realised in a natural language dialog system (NLDS). Franklin's approach begins with framing the problem and building a list of pros and cons, which are weighted and rated regarding to risk, probability of occurrence and importance. In a reflection step, the ratings and weights are reviewed to form further conclusions about the possible options.

NLDS are now used in many domains. However, none of the current systems focuses on general decision support or coaching. The majority of NLDS are designed for rather specific topics in domains such as healthcare, education, gaming or sales and support [5, 7, 9, 11, 13, 15]. In [10], we conducted a user study using a dialog system as a true virtual coach for transferring newly acquired communication skills in everyday work. We found that using an NLDS can perform better than using an online diary, in which the users had to fill out a form every day, depending on the participants' personality trait "openness to new experience."

One of the first NLDS was ELIZA [17], a simulation of a Rogerian psychotherapist. ELIZA was implemented using simple pattern matching techniques. Responses are generated by substitution of key words, which were extracted from the user's utterances, into predefined phrases. Nevertheless, it was taken seriously by a substantial number of users. The most prominent examples for state of the art systems on natural language personal assistance, question answering or information retrieval are Apple's personal assistant Siri[1] and IBM's Watson [2]. In contrast to *vpino*, both are request-response systems with a focus on information retrieval, not holding a complete dialog.

## 2 Concept of Vpino

From the user's point of view, the system is presented as a web based html5 interface. It was designed to look familiar and similar to common messaging user interfaces as known from Facebook or other common chat programs. Our system does not reply to user input immediately. Instead, it waits for a short period of time before posting our response. This makes the interaction feel more natural and allows the user to add further text. On further input, the system's response is recalculated and finally presented to the user. In the following, we present our adaptation of Benjamin Franklin's decision analysis approach and the dialog plan from a non-technical perspective. All information gathered throughout the conversation is stored in a *structured knowledge memory* (*SKM*). Vpino's reactions follow a dialog plan, which is ad-hoc composed from pre-designed building blocks (subdialogs) during the dialog. Most subdialogs require that some specific pieces of information have already been gathered in the SKM. The next subdialog is chosen among all those for which this requirement is fulfilled.

---

[1]http://www.apple.com/ios/siri.

## 2.1 Structured Knowledge Memory (SKM)

Franklin's approach frames the decision problem into an either/or or yes/no or for/against decision. After that, the user basically sets up a list of pros and cons and rates each argument regarding the risk and chance. In contrast to Franklin, we allow more than two options (not more than five to keep the dialog focused). We keep an SKM, which is organised in a tree-like structure. The root represents the decision problem as a whole. The descendants of the root are the individual options. Each option has two types of descendants: the pro arguments and the con arguments. Throughout the dialog, the tree grows dynamically and the options are added according to *vpino*'s questions and the user's answers. Each of these nodes holds a set of attributes. The attributes are set step-by-step throughout the dialog. For each attribute, *vpino* asks a specific question requesting that particular piece of information. The user's answer is always stored in the SKM as plain text, mainly for reference resolution and for extracting specific information at later stages (e.g. to present a list of all options or a list of all pros to the user later on). For an attribute that is boolean or numerical or a reference to another attribute, this piece of information is extracted from the text and stored in addition for later access.

- *Attributes of the decision problem*: The SKM stores the list of options and the user's general description of the problem, the users's short-term and long-term goals, and whether a compromise is possible and desired. Furthermore, the SKM stores the user's criteria for a hypothetic/theoretical optimal solution, and whether or why these criteria are realistic goals.
- *Attributes of an option*: For each option, besides the list of pro and con arguments, the SKM stores the user's answers on the following aspects: a description of that option, a shortcut description in a few words and finally risk and chance of that option.
- *Attributes of an argument*: Arguments include the user's description of that argument and a boolean flag indicating whether or not this argument is rated as particularly important by the user.

Besides the information about the decision problem given by the user, the SKM also stores internal information about the dialog state, which is used for selection of the subdialogs during the conversation.

## 2.2 Dialog Plan

The resulting dialog plan consists of three stages, which the dialog system follows in a strict order. Each stage consists of a set of specific questions that the user needs to answer before she is silently guided to the next stage:

*1. Problem framing and goal setting* This is called problem framing in Franklin's approach. Beyond Franklin, we aim to manage the user's expectations and set goals for the conversation. More specifically, we clarify what the dialog will do and what it will not do, especially that *vpino* will not suggest any option. This stage covers all attributes for the *Decision Problem* in the SKM.

*2. Options and arguments collection* In the second stage, we let the user describe her options and the pros and cons for each option, step-by-step. This stage includes questions for all attributes of *Options* and *Arguments* in the SKM. Furthermore, the user may identify particularly important pros and cons. The approach suggested by Franklin rates each pro and con separately with regard to risks, their probability of occurrence and importance of that argument. In contrast, the user of *vpino* has to rate each option as a whole instead of each argument separately. We did this to keep the dialog shorter and avoid boring the user with repeated requests on every single argument. The definition of risk is left to the user. This is the right way for a system such as *vpino*, which only supports the user in reflecting on her thoughts (in contrast, a system that indeed computes and suggests a decision, must come to a consensus with the user on the exact meaning of 'risk'). Finally, the user is asked to identify a favourite option.

*3. Review and next steps* After collecting options and arguments and considering them separately, the goal of this stage is to elaborate a more differentiated view on the options by relating them to each other. As mentioned before, *vpino* does not suggest a solution at any point. Instead, it always tries to encourage the user to reflect about the options to come to a conclusion by herself. Vpino verbalises/paraphrases the options based on the available information in the SKM from a pre-formulated set of text snippets. In particular, based on a combination of the risk/chance ratio, the number of arguments and whether there were more important pros or cons, *vpino* creates a positive, negative or neutral paraphrasing of that option. The following is an actual example for the resulting text generated by *vpino* (translated from German):

> *In the beginning, you said you cannot point out a favourite option yet. Option 'moving to berlin' yields a high chance with a low risk. On the other hand, your cons seem slightly overweight the pros here. In contrast to that, your option 'hamburg' seems rather risky with a low possible gain. Furthermore, cons seem to outweigh the pros here. When comparing these two, what are the reasons to still consider option 'hamburg' as a solution?*

In most dialogs in our user study, the user named three or fewer options. After generating a paraphrase for each option, *vpino* selects a pair of options, preferably including the user's favourite and options with interesting risk/chance or pro/con ratio. An option is interesting if it is either biased, for example a strong overweight of pros or cons: or if it is ambivalent, for example, a very high risk combined with a very high chance, or many important pros and many important cons simultaneously. The user is asked to compare the selected pair of options. Thereby, we try to reveal problematic or obvious options the user has potentially not taken into account yet. Of course, pairwise comparison could be repeated for all other combinations of options as well. However, for our study, we limited the comparison to two options to avoid that the conversation becomes lengthy and tedious.

At the end of this stage, the user should be able to point out a favourite option. The user is asked to identify her favourite option and justify/reason why this option is favoured. Finally, the user is asked to sum up the results and make a plan for the next steps before *vpino* finishes the dialog politely.

## 2.3 Behind the Scenes

To refine our three-stage decision support dialog, we have developed a hierarchical dialog model, called the dialog plan. The highest level of the dialog plan consists of the three stages from Sect. 2.2. The stages are subdivided into subdialogs, where each subdialog represents the part of the dialog that is devoted to a specific detail in the SKM (a new option or argument or the value of an attribute). More precisely, each subdialog models all possible dialog turns needed to more or less successfully answer *vpino*'s question for a specific detail.

A subdialog is initiated by a question from *vpino*, which targets the information on that specific SKM detail. For our dialog system to hold a reasonable conversation, *vpino* has to understand at least what type of dialog act [14] the user performed. However, *vpino* does not understand the actual content of that message. A subdialog handler defines *vpino*'s reactions to all user dialog acts types for that subdialog. For example, *vpino* asks a question and the user answered with "I don't know". *Vpino* classifies this user response as a *passing* and selects its response from a set of valid pre-formulated text snippets. After a user's *passing*, linguistically valid response types would be *insist*, *restate question* or *accept*. Depending on the current state of the subdialog, *vpino* chooses one of these response types and writes a pre-defined response. After finishing a subdialog, the user messages are parsed once again for content and the relevant piece of information is stored in the SKM. Due to lack of space, we cannot go into further detail here.

## 3 User Study

We conducted a study to evaluate user acceptance and effectiveness of the *vpino* dialog system. Our study was conducted as a field experiment with 128 (38 female) participants, who voluntarily signed up. On average, participants were 25.14 years old (SD = 6.04). The study was conducted via a web-interface that allowed the participants to choose time and location freely. We offered a 10€ incentive for successful participation.

In a pre-test, we tested for the participant's approach to decisions, the Big Five personality traits and their attitude towards decision coaching dialog systems. The participant's decision type was tested with the Melbourne Decision Making Questionnaire [8] on a five-point Likert scale (0 = disagree strongly to 4 = agree strongly). The attitude (*atti*) towards computer based decision coaching was measured by a scale based on 5 items (Cronbach's $\alpha = 0.744$, *A chatbot can help me with solving my problems in general, A chatbot can help me with personal problems, A chatbot can help me with decision problems, I would discuss private topics with a chatbot, I would discuss career-related topics with a chatbot.*) on a five-point Likert scale (0 = disagree strongly to 4 = agree strongly). We tested the Big 5 personality traits openness, conscientiousness, extraversion, agreeableness and neuroticism with the Big 5 inventory by [12].

The dialog phase follows the pre-test immediately. In this phase, users talked to *vpino* on a rational decision problem of their own choice.

The dialog was immediately followed by a post-test to evaluate effectiveness of the decision problem, general user acceptance and self-perception of the participant's own work with the dialog system. The effectiveness of *vpino* on decision-making was measured by 7 items (Cronbach's $\alpha = 0.913$, *I have reached my goal, I have made a decision, I realised what next steps I have to take, I have have gained clarity on my situation, I feel emotional relieved, I feel satisfied, The dialog was motivating*) in the post-test on a five-point Likert scale (0 = disagree strongly to 4 = agree strongly). The user's self-perceived work and cooperation (*spwc*) was measured by 3 items ($\alpha = 0.694$, *I reacted the same as I would have with a human chat partner, I was respectful, I am satisfied with my own work during the dialog.*). Pearson correlations between *spwc*, *atti* and the evaluation results will be presented in Table 2. The general overall evaluation of user acceptance and usefulness (*eval*) were measured by 5 items (Cronbach's $\alpha = 0.923$, *Working with the decision helper was fun, The conversation was motivating, I was positively surprised by the dialog system, I would recommend the conversation with vpino to others.*) on a five-point Likert scale (0 = disagree strongly to 4 = agree strongly) and *How would you rate your experience with the dialog overall?* (4 = very good to 0 = bad). The items *decision* and *eval* are strongly intercorrelated ($\alpha = 0.936$), we also calculated a *combined* item including all evaluation questions from *decision* and *eval*.

## 3.1 Results

The average length of the chat dialog was about 23:15 min (SD = 695 sec) with an average amount of 55.65 (SD = 15) user messages sent. Overall, users evaluated our system quite positively. Table 1 shows the ratings for the most relevant evaluation questions.

51.2 % of the participants evaluated their experience (a) with *vpino* as *very good* or *good*, compared to 29.5 % *acceptable* and 19.4 % *not so good* or *bad*. The mean rating was 2.36 (SD = 1.01). Vpino also evaluated fairly positively on the other relevant items: (b) 39.5 % users agreed to have reached their goal. (c) 55 % realised what their next steps should be. (d) 48 % agreed to have gained clarity on their situation. (e) The evaluation on how motivating the system was is rather neutral. (f) 41.8 % of the participants would recommend *vpino* to others compared to 31.8 % that would rather not.

As expected, talking to *vpino* did not work equally well for all participants. With our study, we also want to find out for which users *vpino* worked particularly well. We investigated correlations for our measures from the pre-test and the post-test. The statistical significance is denoted by *p*. A *p*-value of less than .01 is commonly regarded as a strong level of significance. The most relevant results are shown in Table 2.

**Table 1** Results on evaluation of the dialog in percentage. (a) *The overall user experience* (0 = bad to 4 = very good). Questions on the decision process (0 = disagree strongly to 4 = agree strongly): (b) ... *I have reached my goal.*, (c) ... *I realised what next steps I have to take.*, (d) ... *I have have gained clarity on my situation.* Questions on overall evaluation: (e) *The conversation was motivating,* (f) *I would recommend the decision dialog to others*

|             |   | 0    | 1    | 2    | 3    | 4    | avg  |
|-------------|---|------|------|------|------|------|------|
|             | a | 5.4  | 14.0 | 29.5 | 41.9 | 9.3  | 2.36 |
|             | b | 8.5  | 20.2 | 31.8 | 24.8 | 14.7 | 2.17 |
| All         | c | 14.0 | 11.6 | 19.4 | 37.2 | 17.8 | 2.33 |
| N=128       | d | 10.1 | 19.4 | 21.7 | 35.7 | 13.2 | 2.22 |
|             | e | 11.6 | 24.8 | 18.6 | 35.7 | 9.3  | 2.06 |
|             | f | 13.2 | 18.6 | 26.4 | 27.1 | 14.7 | 2.12 |

**Table 2** Pearson correlations for attitude, hypervigilance, self-perceived work and cooperation with evaluated results on decision, overall rating an combined rating ($p < .001$ for all entries)

|          | att   | hyp-vig | spwc  |
|----------|-------|---------|-------|
| decision | 0.441 | 0.274   | 0.348 |
| eval     | 0.335 | 0.322   | 0.553 |
| combined | 0.419 | 0.319   | 0.480 |

In general, the user's Big Five personality traits did not seem to affect success of the conversation. In contrast to [10], we found no relation to openness as a general personality trait. However, we did find a relation to the user's attitude/openness on dialog systems in particular. More specifically, participants who believed that dialog systems can support them in making a decision were more likely to reach their goals ($0.441, p < 0.001$) and also evaluated the system more positively ($0.335, p < 0.001$). Besides that, our results also suggest that participants with a higher rating on their self-perceived work and cooperation with the system were more likely to give a higher rating for overall evaluation ($0.553, p < 0.001$) and decision efficiency ($0.348, p < 0.001$). Figure 1 displays the *combined* evaluation results in relation to self-perceived work (*spwc*) and the participant's attitude (*atti*). Participants that were more cooperative and have a more open attitude to dialog support systems evaluated significantly higher than participants that lack one of them or both.

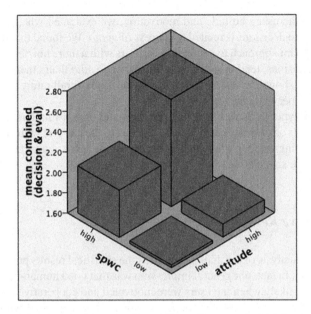

**Fig. 1** Mean *combined* evaluation ratings by median split groups for high/low attitude towards dialog systems and high/low self-perceived work and cooperation (spwc)

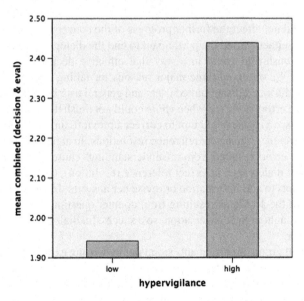

**Fig. 2** Mean for *combined* evaluation ratings by median split groups for high/low hypervigilance

Apart from the user's attitude and motivation, we evaluated whether the user's usual approach to decisions is related to success of *vpino*. We found that users with a more hypervigilant approach to decisions, i.e. users with a more hurried and anxious approach to decisions, tend to profit more from using *vpino* than others ($0.319$, $p < 0.001$). Figure 2 shows the difference in mean results on overall rating for users with high and low hypervigilance (split by median).

Apart from hypervigilance, both buck-passing and procrastination did not show significant effects on effectiveness. This may be due to the fact that with taking part in our study and using *vpino*, participants were yet past the point of avoidant behaviour, but this is a mere speculation.

## 3.2 Qualitative Results

Our qualitative analysis of the dialogs confirms the statistical results presented in the previous section. In fact, our general impression was that good human-like conversations were more likely when the users were motivated and cooperative and followed *vpino*'s lead. The best results were obtained when users responded with short but precisely formulated answers that meet the questions exactly.

For a qualitative evaluation, we define a successful conversation as a natural-looking dialog with human like behaviour at all times or a dialog with only minor problems that did not affect the further progress of the conversation. A dialog failed when a problem either caused the participant to end the dialog or somehow affected the user's relationship to *vpino* in a way that effective decision support was no longer possible. We observed three major reasons for failing dialogs: dialog reference errors, disillusioned/disappointed users and general user behaviour/motivation. Dialog reference errors occurred when *vpino* could not finish the dialog properly due to user initiatives, e.g. a user's attempt to correct a previous answer or jumping back to an earlier question, or incorrect reference resolutions. In about half of these cases, dialog reference errors resulted from misunderstandings caused by *vpino* (e.g. mis-classifications of dialog acts, incorrect reference resolution), the other half resulted from user attempts to add information or revise her answers. In contrast to reference failures, most of the problems resulting from counter questions could be smoothed out by *vpino* throughout the conversation, so a successful dialog was finally possible despite this error.

A rather small group of participants stood out by writing extremely detailed messages with three up to ten sentences per message. Of course, these users got disappointed when *vpino* did not respond to their specific problem. Some users accepted the limitations of the system and the conversation still led to positive results, while other users, although originally motivated, stopped their efforts on continuing the dialog in a serious and motivated way. They radically changed their answering behaviour to minimalist answers or even stopped the conversation.

A few failures resulted from challenging or uncooperative users. Due to our observation, for many of them, the motivation to use our system was curiosity and mischief rather than trying to solve a decision problem.

An interesting observation was the fact that users tend to avoid a clear no when disagreeing. While agreement is uttered straight forward and directly, most of the disagreements contained modifiers that weakened their disagreement. As a result, our dialog act classifier did not classify some of them correctly and *vpino* had to pump users (*"So does that mean you agree?"*).

## 4 Conclusion

We have built the natural language dialog system *vpino* to support humans in making rational decisions. A user study showed that the system was generally well accepted. Vpino was able to help many of them on either making a decision or clarifying their situation. The effectiveness of decision support was affected from the user's attitude towards computer based dialog systems: Users that are more open to using a coaching dialog system were more likely to hold an effective conversation than those who were more sceptical. As expected, a successful conversation requires a reasonable level of motivation and cooperation by the users. However, that is also true for working with a human coach or using any other kind of decision support. Besides, more hypervigilant participants seemed to profit in particular from using the system.

Through our user study, we have gained a lot of new insights on how users interact with the system. With the lessons learned, we plan to further improve our dialog plan and the system as a whole. We are also planning a couple of user studies, which address various target groups and focus on different aspects.

**Acknowledgments** Special thanks to Anna Bruns for her expert advice on evaluation of our study.

## References

1. Clemen, R. (ed.): Making Hard Decisions: An Introduction to Decision Analysis. Duxbury Press, Boston (1996)
2. Ferrucci, D., Brown, E., Chu-Carroll, J., Fan, J., Gondek, D., Kalyanpur, A.A., Lally, A., Murdock, J.W., Nyberg, E., Prager, J., et al.: Building watson: an overview of the deepqa project. AI Mag. **31**(3), 59–79 (2010)
3. Figueira, J., Greco, S., Ehrgott, M.: Multiple criteria decision analysis: state of the art surveys, vol. 78. Springer, Boston (2005)
4. Franklin, B.: Letter to joseph priestly. Fawcett, Reprinted in The Benjamin Franklin Sampler. New York (1956)
5. Graesser, A.C., Chipman, P., Haynes, B.C., Olney, A.: Autotutor: an intelligent tutoring system with mixed-initiative dialogue. IEEE Trans. Educ. **48**(4), 612–618 (2005)
6. Janis, I.L., Mann, L.: Decision Making: a Psychological Analysis of Conflict, Choice, and Commitment. Free Press, New York (1977)

7.  Latorre-Navarro, E.M., Harris, J.G.: A natural language conversational system for online academic advising. In: The Twenty-Seventh International Flairs Conference (2014)
8.  Mann, L., Burnett, P., Radford, M., Ford, S.: The melbourne decision making questionnaire: an instrument for measuring patterns for coping with decisional conflict. J. Behav. Decis. Making **10**(1), 1–19 (1997)
9.  Mateas, M., Stern, A.: Façade: an experiment in building a fully-realized interactive drama. In: Game Developers Conference. pp. 4–8 (2003)
10. Mäurer, D., Bruns, A., Weihe, K.: Be open to computer based coaching. In: Proceedings of SemDIAL (2013)
11. Morbini, F., DeVault, D., Sagae, K., Gerten, J., Nazarian, A., Traum, D.: Flores: a forward looking, reward seeking, dialogue manager. In: Natural Interaction with Robots, Knowbots and Smartphones, pp. 313–325. Springer (2014)
12. Rammstedt, B., John, O.P.: Measuring personality in one minute or less: a 10-item short version of the big five inventory in english and german. J. Res. Pers. **41**(1), 203–212 (2007)
13. Rizzo, A., Sagae, K., Forbell, E., Kim, J., Lange, B., Buckwalter, J., Williams, J., Parsons, T., Kenny, P., Traum, D., et al.: Simcoach: an intelligent virtual human system for providing healthcare information and support. In: The Interservice/Industry Training, Simulation and Education Conference (I/ITSEC), vol. 2011. NTSA (2011)
14. Searle, J.: Speech Acts. Cambridge University Press, Cambridge (1969)
15. Shawar, B.A., Atwell, E.: Chatbots: are they really useful? LDV Forum. **22**, 29–49 (2007)
16. Simon, H.A.: Rational decision making in business organizations. The American economic review pp. 493–513 (1979)
17. Weizenbaum, J.: Eliza - a computer program for the study of natural language communication between man and machine. Commun. ACM **9**(1), 36–45 (1966)

# Towards Model-Driven Assessment of Clinical Processes

Flora Amato, Giovanni Cozzolino, Alessandra D'Alessio, Stefano Marrone, Nicola Mazzocca, Gianluca Mele and Roberto Nardone

**Abstract** E-Health organisations have seen in these years a rapid growth in the complexity and criticality of the processes they manage. This paper defines an approach for modelling clinical workflows based on Model-Driven principles; in particular the modelling activity is supported by the Dynamic State Machine (DSTM) formalism, that is a well-formed graphical language able to represent state based systems. The main advantage of using such a language resides in obtaining formal models of clinical workflows (whose semantics is strong and precise), with an high level of usability. While the focus of the paper is clearly on modelling, the application of Model-Driven principles allows a tight integration between the control flow of the clinical processes and the information that can be extracted from informal documentation. The approach is shown by applying it to the case study of a real world treatment process of bipolar and mood disorders.

**Keywords** Dynamic state machine · Workflow management · Workflow modelling · Semantic processing · E-health organisations

F. Amato (✉) · G. Cozzolino · A. D'Alessio · N. Mazzocca · G. Mele · R. Nardone
Università di Napoli "Federico II", DIETI, (Italy)
e-mail: flora.amato@unina.it

G. Cozzolino
e-mail: giovanni.cozzolino@unina.it

A. D'Alessio
e-mail: alessandra.dalessio@unina.it

N. Mazzocca
e-mail: nicola.mazzocca@unina.it

G. Mele
e-mail: gianluca.mele@unina.it

R. Nardone
e-mail: roberto.nardone@unina.it

S. Marrone
Seconda Università di Napoli, Dip. di Matematica e Fisica, (Italy)
e-mail: stefano.marrone@unina2.it

© Springer International Publishing Switzerland 2015
E. Damiani et al. (eds.), *Intelligent Interactive Multimedia Systems and Services*,
Smart Innovation, Systems and Technologies 40,
DOI 10.1007/978-3-319-19830-9_11

121

# 1 Introduction and Related Work

The E-Health domain implies a proper massive document processing. In particular, knowledge management activities must be performed in reliable, effective and error-free way. Hence, E-Health organizations needs to be supported with approaches aimed at assessing clinical guidelines, and supporting their correct and efficient execution. For this reason, there is a need of combining both formal approaches with automatic elaboration of medical results: these goals can be achieved respectively by adopting Model-Driven principles and automatic on-line and/or off-line functionalities, aimed at analysing clinical documents.

In the last decades, many works have been proposed with the objective of improving the efficacy and efficiency of E-Health process [3, 20], some of which have been actually implemented in clinical application: these works are founded on the modelling of clinical processes through workflows, expressed in proper formalisms. These approaches introduce many advantages, including, among others, the coordination of multiple contributions, the reuse of assessed portions in new projects, the possibility of verifying the correctness by adopting Model-Based approaches, and so on. Unfortunately, many of these approaches do not scale well with the dimensions of the processes since there is a lack of high-level modelling primitives in these formalisms. To overcome the application limit of Model-Based approaches, Model-Driven methodologies have been proposed in other approaches [10–12, 15, 16] based on high-level models, formal representations automatically generated by means of transformation chains and verification of specific properties [13, 14].

From the other hand semantic analysis of documents has been also considered in many works during last years. In fact, as choice blocks in a workflow can be used to evaluate the content of the clinical records, it is possible to perform semantic processing procedures in the context of such evaluations by the adoption of semantic techniques. The objective is the analysis of texts and questionnaires. Specifically, most of the medical data is unstructured and can reside in multiple different places as electronic medical records, results of laboratories, images, medical reports and so on. For this reason high-level modelling language should also take into account this kind of possibility during the workflow modelling.

In previous works we proposed a methodology for the classification and the management of sensitive data, able to retrieve and associate the proper concepts to the clinical records. It is based on semantic approaches for relevant concepts identification in textual data by means of lexical-statistical techniques [4]. The main idea of that work was to design a reconfigurable framework for documents processing that accepts in inputs a collection of heterogeneous data, including textual and multimedia, belonging to specialist domains and provides semi-automatic procedures for structuring data and extracting information of interest.

This work represents a first step towards a new Model-Driven methodology for the verification, assessment and efficiency of workflows. Long run objectives of the overall methodology reside in the generation of formal models for quantitative properties and correctness evaluation, and in the support for the execution of workflows on the basis of specific data. In doing this, the original contribution of this work resides in the adoption of a modelling formalism based on state machines, specifically tailored for clinical process modelling. In detail, we customize the Dynamic State Machine formalism (DSTM), specifying a proper syntax to decorate transitions (i.e., trigger, condition and action). The obtained formalism is hence very powerful, and allows for a customizable representation of workflows, according to possible additional end-user requirements. The specific syntax, ad-hoc defined, ensures also that the modelled workflows respects the specific set of constraints given by a specific organization.

## 2  A Model-Driven Methodology in Clinical Domain

The proposed methodology is depicted in Fig. 1. It is a Model-Driven methodology specific for E-Health workflows verification and assessment. The core of the methodology resides in the workflow modelling by adopting the Dynamic STate Machine (DSTM) formalism: this formalism provides modellers with very powerful and high level syntactical concepts, with a precise and strong semantics. The adopted formalism represents workflows as hierarchical state-machines, extended with parallel and concurrent execution. The usage of this formalism helps also the model reuse: for example, a specific procedure, such as the treatment of a given disorder, can be easily modelled as a unique state machine, which can be instantiated by an higher-level machines where needed.

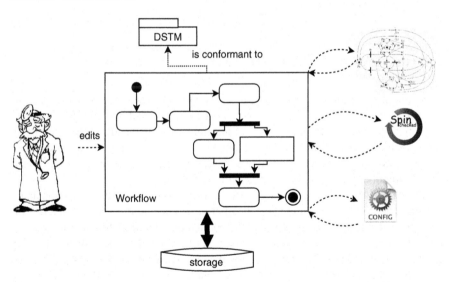

**Fig. 1**  Methodology

Another important advantage offered by the proposed approach is that decoration of transitions can be easily customized by a specific organization, which can define its own specific syntax: in this way the specific organization can customize, as transition triggers the set of possible events, as firing conditions the set of rules (based also on the results of automatic analysis of clinical records), and as actions to be executed during a transition firing, the set of possible actions.

The first step of our methodology is the representation of the workflow by the realization of a DSTM model, which captures the different states and the possible forks of the workflow itself. The conformity to the DSTM formalism allows the highly-customized annotation on the transitions of the machines such as: patient admission and dismission, patient transfer, exam start or completion and so on. Since the semantics of DSTM is defined, the execution of the modelled workflow can be simulated in appropriate simulation environments.

Automatic transformations are then defined in order to generate formal models or configuration files for semantic processing tools. Formal models can be used to assess quantitative or other kind of properties. Different properties can be also evaluated with a model checking tool, such as the SPIN model checker: in fact the DSTM model can be transformed into a Promela specification which can be adopted to prove specific properties of the workflow. Hence the proposed approach can help the usage of formal models in workflow verification, not only allowing for a complete model-based approach, but also a Model-Driven process able to generate automatically formal models without extra-effort in training activity of end-users.

The third pillar of our methodology is constituted by a workflow repository, where existing models of both entire workflows and specific procedures can be stored as state machine; this allows to: trace model changes, reuse assessed procedures, extract and combine historical data.

## 3 Dynamic State Machine (DSTM)

Dynamic State Machine (DSTM) [8] is a new formalism which extends Hierarchical State Machines [2] by adding concepts of fork, join and allowing for dynamic instantiation and concurrent execution of machines. Two main advantages have been individuated: (1) each state machine may be parametric over a finite set of dynamically evaluated parameters, and (2) the same machine may be instantiated many times without explicitly replicating its entire structure.

DSTM was born in the context of railway control system specifications; the main objective of DSTM in [8] is the specification of railway control systems in terms of expected behaviour, in order to enable the application of a Model-Driven methodology which allows for the automatic generation of test cases, as described in [7]. To meet this goal, DSTM4Rail shall be formal and not ambiguous; the system specification is hence transformed, by a set of automatic transformations, into a Promela specification (executed by the SPIN model checker) [9], from which test cases can be generated as counterexamples of ad-hoc defined properties.

DSTM is more powerful than the UML State Machines [18] since it adds mechanisms for dynamic instantiation and recursive execution. For this reason, it could be applied into different application domains. The whole language, in fact, well supports modification and customization of specific syntax of triggers, conditions and actions, used to decorate transitions between nodes. An excerpt of the DSTM meta-model is shown in Fig. 2, where the Ecore diagram is depicted. This Ecore diagram represents the realization of DSTM, formally defined in [8], in the Eclipse Modeling Framework [19]. The main class is Dynamic State Machine *(DSTM)*, which repre-sents the entire specification model. It is characterized by the attribute *max_proc*, which indicates the maximum number of processes active in each instant of time (this attribute avoid the infinite activation of machines inside a DSTM). A DSTM is composed of different *Machines*, *Channels* and *Variables*. *Channels* and *Variables* allow for communication between machines. A single *Machine* is composed of *Ver-tex*es, *Transition*s and may have a set of *Parameter*s.

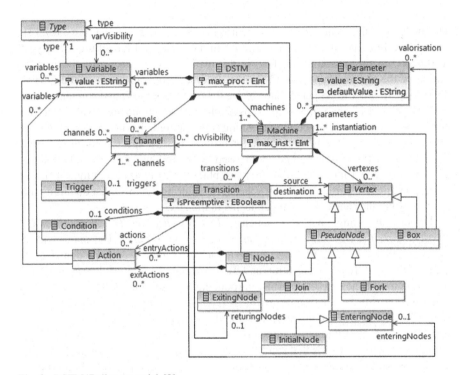

**Fig. 2** DSTM4Rail metamodel [8]

The class *Vertex* is abstract since different kinds of vertexes (with different fea-tures and constraints) may be present in a machine. The vertex types are similar to those contained in the UML State Machine, but with a different semantics for the *Fork* and *Join* concepts. The former splits an incoming transition into more outgo-ing transitions; it allows for instantiating one or more processes either synchronously or asynchronously with the currently executing process. The latter merges outgoing transition from concurrently executing processes; it synchronizes the termination of concurrently executing processes or allows to force the termination when a process is able to perform a *preemptive* exiting transition. The classes *Fork*, *Join* and *EnteringNode* are inherited from the abstract class *PseudoNode* which encompasses different types of transient vertexes in the machine.

The extension point inside DSTM is represented by the specific syntax of trigger, condition and action, in order to customize the decoration of transitions and the entry and exit actions of nodes. In the following we will show a specific syntax for clinical application.

## 4 A Methodology for Semantic Processing

The analysis of the content of questionnaires has been addressed in a past work [3]. This possibility shall be taken into account in the high-level modelling activity of workflows, for this reason the analysis of this methodology is necessary for the definition of the specific syntax used for the decoration of transitions in the DSTM formalism. In order to properly characterize the content of medical documents, such as clinical records, analysis results or psychological questionnaires evaluation, the application of semantic processing techniques is required. To this aim, we adopted the methodology for information extraction proposed in [3]. Specifically, the adopted methodology for text processing is composed by the following stages: (1) breaking up a stream of text into a list of words and phrases, (2) marking up the tokens as corresponding to a particular part of speech; (3) filtering the token list obtaining the most relevant ones and identifying concepts in the text.

The first stage implements in sequence: *Text Tokenization*, *Text Normalization*, *Part-Of-Speech (POS) Tagging* and *Lemmatization* procedures. The main goal of these procedures is the extraction of relevant terms that are used to recognize concepts in the text. Text Tokenization and Text Normalization procedures perform a first grouping of the extracted terms, introducing a partitioning scheme that establishes an equivalence class on terms [6]. *Text Normalization* procedures take variations of the same lexical expression that should be reported in a unique way, such as words that assume different meanings if are written in small or capital letter; compounds and prefixed words that can be (or not be) separated by a hyphen; dates that can be written in different ways; acronyms and abbreviations. This phase is also responsible for the transformation of capital letter that, for example, helps in distinguish a common noun used at beginning of a sentence from a proper name. *Part-Of-Speech (POS) Tagging* consists in the assignment of a grammatical category (noun, verb, adjective, adverb, etc.) to each lexical unit identified within the text collection. Morphological information about the words provides a first semantic distinction among the analysed words. The words can be categorized in: *content words* and *functional words*. Content words represent nouns, verbs, adjectives and adverbs. In general, nouns indicates people, things and places; verbs denote actions, states, conditions and processes; adjectives indicate properties or qualities of the noun they refer to; adverbs, instead, represent modifiers of other classes (place, time, manner, etc.).

Functional words are made of articles, prepositions and conjunctions; they are very common in the text. Automatic POS tagging involves the assignment of the correct category to each word encountered within a text. But, given a sequence of words, each word can be tagged with different categories [5].

The second stage realize the *word-category disambiguation* and *Text Lemmatization*. The disambiguation involves two kinds of problems: (1) finding the POS-tag or all the possible tags for each lexical item; (2) choosing, among all the possible tags, the correct one. Here the vocabulary of the documents of interest is compared with an external lexical resource, whereas the procedure of disambiguation is carried out through the analysis of the words in their contexts. *Text Lemmatization* is performed in order to reduce all the inflected forms to the respective lemma, or citation form, coinciding with the singular male/female form for nouns, the singular male form for adjectives and the infinitive form for verbs. Lemmatization introduces a second partitioning scheme on the set of extracted terms, establishing a new equivalence class on it. All these procedures are language dependent, consisting of several sub-steps, and are implemented by using the state of the art NLP modules. At this point, a list of tokens is obtained from the raw data and the third stage is applied.

In order to identify concepts, not all words are equally useful: some of them are semantically more relevant than others, and among these words there are lexical items weighting more than other. The third stage aims at filtering the token list in order to obtain a reduced list, containing only the relevant tokens. To do that, there are several techniques in literature that "weight" the importance of a term in a document, based on the statistics of occurrence of the term. TF-IDF index (*Term Frequency - Inverse Document Frequency*) [17] is actually the most popular measure used to evaluate terms semantic relevance [1]. Having the list of relevant terms, concepts are detected by relevant token sets that are semantically equivalent (synonyms, arranged in sets named synset). In order to determine the synonym relation among terms, it is possible to use statistic-based techniques of unsupervised learning, as clustering, or external resources like thesaurus (an example is wordnet). At this point it is possible to codify concepts by means of ontology data models (RDF, OWL, etc.).

In this way, we map a document in an ordered vector of concepts, recurring to a standard bag of word representation [17]. We evaluate the TF-IDF for every concept occurring in the document. Once the concepts are associated to the document, we can apply a semantic metric, measuring the semantic distance between two documents. In this way we can state if a document contains topic belonging to a particular field or not. In particular, we state that two documents contain similar concepts recurring to a similarity distance [17], using the normalized scalar product (cosine of angle):

$$sim(V_a, V_b) = \frac{V_a \cdot V_b}{\|V_a\|_2 \cdot \|V_b\|_2}$$

If $sim(V_a, V_b)$ is over a given, empirically defined, threshold, we can state that $V_a$ and $V_b$ contains the same topics.

# 5 Describing Workflows Through DSTM

As stated in Sect. 3, the formalization of workflows with the DSTM formalism can be allowed by defining a proper syntax for decoration of transitions and for entry and exit actions of nodes. The definition of this syntax is a complex activity since it requires the analysis of the specific clinical application. The defined syntax must take into account the possible questionnaires and examinations that are adopted in the specific application, to which the semantic analysis shall be applied. In this Section we show how this specific syntax can be defined; it will be applied in next Section in the modelling of a real world use case, in order to show the expressive power of this representation.

Let us consider a ward of an hospital where both questionnaire and examinations can be submitted to patient in order to assess their disturbs. We want to define a syntax which allows for the following: **trigger** representing the arrival of a new patient, the completion of a specific questionnaire, the completion of a specific examination or the request for interrupt of all treatments performed by a patient; **condition** representing boolean conditions related to the result of the semantic analysis of a questionnaire or to comparison between internal model variables; **action**: numeric variable assignment, patient admission, patient dismission, the request for an examination or the request for a questionnaire.

So let us define: $E$ the set of available examinations in the ward, $Q$ the set of available questionnaires in the ward. Let $\mathcal{T}$ be the set of triggers, $C$ be the set of conditions, $\mathcal{A}$ be the set of actions. Let also $\tau \in \mathcal{T}$ be a trigger, $\gamma \in C$ be a condition and $\alpha \in \mathcal{A}$ be an action.

---

$binarop$ : "$+$" | "$-$" | "$*$" | "$/$"

$comparop$ : "$<$" | "$<=$" | "$>$" | "$>=$" | "$==$" | "$!=$"

$intOrDouble$ : $v_i$ | $v_d$

$constant$ : $k_i$ | $k_d$

$expr$ : $constant$ | $intOrDouble$ | "(" $expr$ ")" | $expr$ $binarop$ $expr$

$\tau$ : "$patient\_arrival$" | "$questionnaire\_completed$"$(q)$ | "$exam\_completed$"$(e)$ |
        | "$treatment\_interrupt$"

$\gamma$ : $intOrDouble$ $comparop$ $constant$ | $intOrDouble$ $comparop$ $intOrDouble$ |
        | "$sim$ {" $v_q$ "," $v_d$ "}" $comparop$ $intOrDouble$

$action$ : $v$ "$=$" $expr$ | "$patient\_admission$" | "$patient\_dismission$" | "$exam\_start(e)$" |
         "$quest\_start(q)$"

$\alpha$ :        $action$ | $\alpha$ "$;$" $action$

---

**Fig. 3** Grammar Rules

Let us define the following basic types: $Int$: the integer type; $Double$: the double type; $Questionnaire$: the type representing a questionnaire; $DisorderKW$: the type representing a vector of keywords for a disturb. Let us also define $\mathcal{V}$ as the variable set and identify $v \in \mathcal{V}$ as a variable; $v$ must be of one of the following kinds: (1) $v_i \in Int$ a variable typed as integer, $k_i \in Int$ an integer constant, $v_d \in Double$ a variable typed as double, $k_d \in Double$ a double constant, $v_q \in Questionnaire$ a variable typed as questionnaire, $v_d \in DisorderKW$ a variable typed as disorder keywords.

We define the following grammar rules (Fig. 3):

where: $e \in E$ is an examination between the set of available examinations $E$ and $q \in Q$ is a questionnaire between the set of available questionnaires $Q$.

# 6 Application of the DSTM Formalism

On the basis of the described syntax, it is possible to model specific workflows. As an example, we want to describe a process for the admission, the diagnosis, the treatment and the discharge of patients with psychological problems. Once admitted at the acceptance, to psychiatric patients is administered a questionnaire. From the responses you can get an early diagnosis of the disorder. The example contains two types of disorder: bipolar disorder, or mood disorder and we defined a disorder vector for each one of them. The semantic procedures, activated at this decision point, analyse the content of the questionnaires in order to map them in a vector of concepts: the evaluation of the semantic distance among the vector of the questionnaire and the defined disorder vectors, we can preliminary address the patient to the proper treatment. Medical staff always supervise the system, which is designed to provide medical support.

In our case of study we define the following vectors of concepts:

- $v\_bip\_concepts$ = {Talkativeness, Inflated ego, Foolishness, Hyperactivity, Asleep, More_social, More_sex_interested};
- $v\_mood\_concepts$ = {Anxiety, Depression,Irritability, Distraction}.

The defined vector of questionnaire are $v\_quest$, $v\_bip\_diag$ which represent the resulting questionnaires, compiled by the patient. We define also three different double values, representing three thresholds of the semantics analysers: $T\_bip$, $T\_mood$, $T\_diagn$. We define also an integer variable $cont$, which counts the number of patients under treatment.

Figure 4 depicts a portion of the complete model; the entire model is not reported for sake of space. The first machine, the *MAIN* one, sets the *count* variable to 0; then it waits an arrival of a patient in its *wait for patient* state. When a new patient arrives, if the number of already accepted patients is not greater than 10, the patients is accepted (the *patient_admission* action is performed) and *count* is incremented by 1. For each new patient, if accepted, the machine *MANAGE_DISORDER* is instantiated. When in finishes its execution, through the join, the *count* variable is decremented by 1.

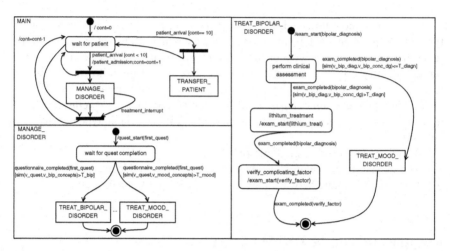

**Fig. 4** Case study

If a patient cannot be accepted, since 10 patients have been already accepted, the *TRANSFER_PATIENT* machine is instantiated in order to transfer the patient to a different ward.

The *MANAGE_DISORDER* machine submit the questionnaire to the patient and waits for its completion in the *wait for quest completion* state. Two different transitions exit from this state, instantiating respectively the *TREAT_BIPOLAR _DISORDER* or the *TREAT_MOOD_DISORDER* if the semantic analysis of the submitted questionnaire gives a result greater than the *T_bip* or *T_mood* thresholds. Let us note that this evaluations has been modelled as conditions of the transitions in the specific syntax indicated before. At last, the machine *TREAT_BIPOLAR _DISORDER* models the treatment of the bipolar disorder as the sequence of followed steps.

# 7 Conclusions

This paper describes a first step towards the application of a Model-Driven methodology, which would address the automatic document analysis inside high-level models. The Dynamic State Machine (DSTM) formalism is reused in order to represent workflow processes as state machine. The choice of this formalism allows also for the definition of a specific syntax for the decoration of transitions, and for the definition of entry and exit actions inside nodes. Specifically, the define syntax takes into account also an approach for semantic processing of medical documents. The expressive power of this approach is showed by the application to a real-world case study, which is easily modelled by adopting the proposed approach. As future work, we plan the implementation of automatic transformations which allows for the definition of formal models, which will be helpful in applying formal verification techniques in the addressed domain.

**Acknowledgments** This work was supported by Smart Health 2.0 project, founded by the Italian Initiative "Smart cities and communities and social innovations PON ricerca e competivita AVVISO N. 84/RIC dee 2/03/2013.

# References

1. Albanese, M., d'Acierno, A., Moscato, V., Persia, F., Picariello, A.: A multimedia semantic recommender system for cultural heritage applications. In: Semantic Computing (ICSC), 2011 Fifth IEEE International Conference on, pp. 403–410. IEEE (2011)
2. Alur, R., Kannan, S., Yannakakis, M.: Communicating hierarchical state machines. In: Automata, Languages and Programming, pp. 169–178. Springer, New York (1999)
3. Amato, F., Casola, V., Mazzeo, A., Romano, S.: A semantic based methodology to classify and protect sensitive data in medical records. In: Information Assurance and Security (IAS), 2010 Sixth International Conference on, pp. 240–246. IEEE (2010)
4. Amato, F., Mazzeo, A., Moscato, V., Picariello, A.: Semantic management of multimedia documents for e-government activity. In: Complex, Intelligent and Software Intensive Systems, 2009. CISIS'09. International Conference on, pp. 1193–1198. IEEE (2009)
5. Amato, F., Mazzeo, A., Moscato, V., Picariello, A.: A system for semantic retrieval and long-term preservation of multimedia documents in the e-government domain. Int. J. Web Grid Serv. **5**(4), 323–338 (2009)
6. Amato, F., Mazzeo, A., Penta, A., Picariello, A.: Building RDF ontologies from semi-structured legal documents. In: Complex, Intelligent and Software Intensive Systems, 2008. CISIS 2008. International Conference on (2008)
7. Barberio, G., et al.: An interoperable testing environment for ERTMS/ETCS control systems. In: Computer Safety, Reliability, and Security, pp. 147–156. Springer, New York (2014)
8. Gentile, U., Nardone, R., et al.: Dynamic state machines for formalizing railway control system specifications. In: Proceedings of the 3rd International Workshop on Formal Techniques for Safety-Critical Systems (FTSCS 2014), Communications in Computer and Information Science (2015)
9. Holzmann, G.J.: The model checker spin. IEEE Trans. Software Eng. **23**(5), 279–295 (1997)
10. Marrone, S., Flammini, F., Mazzocca, N., Nardone, R., Vittorini, V.: Towards model-driven V&V assessment of railway control systems. Int. J. Softw. Tools Technol. Transfer **16**(6), 669–683 (2014)
11. Moscato, F., Amato, F., Amato, A., Aversa, R.: Model-driven engineering of cloud components in metamorp(h)osy. Int. J. Grid Util. Comput. **5**(2), 107–122 (2014)
12. Moscato, F., Aversa, R., Amato, A.: Describing cloud use case in metamorp(h)osy. pp. 793–798 (2012)
13. Moscato, F., Di Martino, B., Aversa, R.: Enabling model driven engineering of cloud services by using mosaic ontology. Scalable Comput. **13**(1), 29–44 (2012)
14. Moscato, F., Venticinque, S., Aversa, R., Di Martino, B.: Formal modeling and verification of real-time multi-agent systems: the remm framework. Studies Comput. Intell. **162**, 187–196 (2008)
15. Pérez, B., Porres, I.: Authoring and verification of clinical guidelines: a model driven approach. J. Biomed. Inform. **43**(4), 520–536 (2010)
16. Marrone, S., Mazzocca, N., Merseguer, J., Nardone, R., Bernardi, S., Flammini, F., Vittorini, V.: Enabling the usage of UML in the verification of railway systems: the DAM-rail approach. Reliab. Eng. Syst. Saf. **120**(0):112–126 (2013)

17. Sivic, J., Zisserman, A.: Efficient visual search of videos cast as text retrieval. IEEE Trans. Pattern Anal. Mach. Intell. **31**(4), 591–606 (2009)
18. OMG Available Specification: Omg unified modeling language OMG UML, superstructure, v2.4.1. object management group (2011)
19. Steinberg, D., Budinsky, F., Merks, E., Paternostro, M.: EMF: Eclipse Modeling Framework. Pearson Education, Boston (2008)
20. van der Aalst, W.MP.: The application of petri nets to workflow management. J. Circuits Syst. comput. **8**(01):21–66 (1998)

# A Monitoring System for the Recognition of Sleeping Disorders in Patients with Cognitive Impairment Disease

Antonio Coronato and Giovanni Paragliola

**Abstract** Alzheimer's disease (AD) is the most common type of dementia; accounts report the impact of AD in a range of 60–80 % with respect all the dementia related pathologies. Patients with Alzheimer's could show early symptoms such as sleep disturbances, well-formed visual hallucinations, and muscle rigidity or other AD movements disorders. In this work we focus on disturbances related to sleep disorders, for this reason, we propose a monitoring system for the recognition of these kind of disorders. In detail, we present: (1) a novel model for the detection of a particular type of sleep disorders, the Periodical Legs Movements (PLM) and (2) a prototype of monitoring system able to recognize the PLM events. Preliminary experiment have been made by monitoring a patient with a early stage of the sleep disorder.

## 1 Introduction

Alzheimer's disease is the most common type of dementia; accounts for an estimated 60–80 % of cases. About half of these cases involve solely Alzheimer's pathology; many have evidence of pathology changes related to other dementia [1].

Patients with AD could show early symptoms such as sleep disturbances, well-formed visual hallucinations, and muscle rigidity or other Alzheimer movement features.

In this work we focus on those types of disturbances related to sleep disorders. The attention on such kind of disorders is growing in the last decades as proof of this the cost of the treatments has been estimated in the range of 30$ to 35$ billion [2].

A. Coronato (✉) · G. Paragliola
National Research Council (CNR) Institute for High-Performance Computing
and Networking (ICAR), Naples, Italy
e-mail: antonio.coronato@na.icar.cnr.it

© Springer International Publishing Switzerland 2015
E. Damiani et al. (eds.), *Intelligent Interactive Multimedia Systems and Services*,
Smart Innovation, Systems and Technologies 40,
DOI 10.1007/978-3-319-19830-9_12

For this reason, technologies such as ubiquitous computing and ambient intelligence are increasing their role on the development of ICT solutions for the monitoring of such kind of disorders.

In detail, in this paper we present: (1) a novel model for the detection of a particular type of sleep disorders, the Periodical Legs Movements (PLM) and (2) a prototype of monitoring system able to recognize PLM events.

The recognition process is based on the evaluation of the *momentum* of the patient, that is the amount of movements during the sleep.

This parameter is calculated by measuring the patient's arms and legs acceleration. For this reason, an accelerometer sensor is placed on the patient's legs in order to acquire the x, y, z acceleration' s value.

The paper is organized as following: in Sect. 2 we provide a overview on the related work about ICT solutions for the detection of sleeping disorders, in Sect. 3 we present the model for the recognition of the PLM, in Sect. 4 we describe the prototype of the monitoring architecture, and in Sect. 4 the preliminary results are shown.

## 2 Related Work

With the term *Sleep disorders*, we describe a very big set of different disorders which happen while the patient sleeps.

In this section we are going to present some ICT solutions able to monitor some of sleeping disorders shown by an AD patient.

Alves de Mesquita et al. [3] present a monitoring system to recognize sleep breathing disorders by means of nasal pressure recording technique.

Occhiuzzi et al. [4] investigate the feasibility of the passive RFID technology for the wireless monitoring of human body movements in some common sleep disorders by means of passive tags equipped with inertial switches.

Poree et al. [5] propose a sleep recording system to perform the monitoring of sleeping disorders; the solution adopts five electrodes: two temporal, two frontal and a reference. This configuration enables to avoid the chin area to enhance the quality of the muscular signal and the hair region for patient convenience. The electroencephalopgram (EEG), eletromyogram (EMG), and elctrooculogram (EOG) signals are separated using the Independent Component Analysis approach.

The analisys of EEG to recognize sleep disorders was adopted from Huy Quan Vu et al. [6] as well. Although the approaches widely used are based on the EEG/EMG signals, other kinds of parameters have been adopted. An example is provided by Flores et al. [7]. In their work the authors use a motion sensor to catch the gross body movements. During the awake, respiratory movements are masked by other motor activities. An automatic pattern recognition system was developed to identify periods of sleep and awake using the piezoelectric generated signal.

Xie et al. [8] adopt machine-learning algorithms to combine the data came from electrocardiogram (ECG) and saturation of peripheral oxygen ($SpO_2$) sensors in order to detect sleep apnea disorders.

Agarwal et al. [9] analyzed the rapid-eye movements in order to discover the involving of sleep disorders. The work adopts Electrooculography (EOG) for measuring the corneo-retinal standing potential that exists between the front and the back of the human eye.

In the best of our knowledge, our solution is the first that adopts an accelerometer sensor to evaluate the patient's momentum and recognize PLMs events.

# 3 Sleep Disorders Model

In this section we are going to describe the sleep disorder model for the recognition of the anomalous movements during the patient's sleep.

PLM are defined as a sequence of four or more leg movements separated by at least 5 (and not more than 90) seconds with a duration between 0.5 and 5 s. [10].

In order to detect these movements we have adopted an approach base on the measure of the momentum of the patient's movements during the sleep. The measurements of the patient's movements have been made by using an accelerometer sensor placed on the his legs.

Figure 1 shows a nite state machine that describes how our solution recognizes the PLM events (the associated transition model is depicted in Fig. 2).

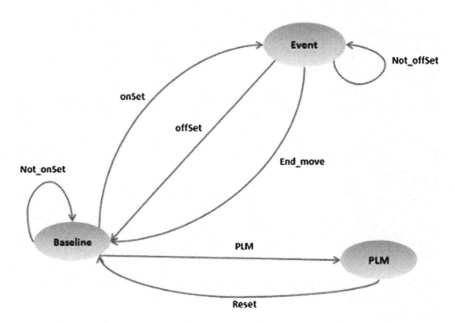

**Fig. 1** Overview of sleep disorders model

The *Baseline* state indicates that none candidate anomalous movements have been detected; the *event* state indicates that and a candidate anomalous movement is in progress; the PLM state indicates that a PLM event has been detected.

It is worth to describe the set of variables and functions that we use in the model.

- $Q$, patient's momentum during the sleeping.
- *QonSet*, if Q limits this values then a candidate anomalous movements starts.
- *Qoffset*, if Q gets under this values then a candidate anomalous movements ends.
- *EC*, a set of at least four candidate anomalous movements.
- $T(*)$, function that returns the time duration of an event *
- $C(*)$, function that returns the number of occurrence of an event *

It is important noting that a single candidate anomalous movement is not a PLM, a PLM is defined as a set of candidate anomalous movements which satisfy a set of temporal constraints [11].

| Name | Starting state | Ending state | Value | Description |
|------|----------------|--------------|-------|-------------|
| Not_onSet | Baseline | Baseline | $Q < Q_{onset}$ | No Candidate Event is detect |
| onSet | Baseline | Event | $Q > Q_{onset}$ And $T(Q > Q_{onset}) < 15s$ | Event Candidate Starts |
| Not_offSet | Event | Event | $Q > Q_{offset}$ | No ending Event Candidate is detected |
| offSet | Event | Baseline | $Q < Q_{offset}$ And $T(Q < Q_{offset}) = > 0.5s$ | Event Candidate Ends |
| End_move | Event | Baseline | $Q < Q_{offset}$ And $T(Q < Q_{offset}) > 15s$ | A no camdicate event ends |
| PLM | Baseline | PLM | $C(\{Ec\}) > = 4$ And * | An PLM events is detected |
| Reset | PLM | Baseline | Not Value | Non specificato |

**Fig. 2** Transitions table of the sleep disorders model

The transitions table of the model is reported in the Fig. 2. It is worth noting that the relations between the variables have been inferred by analyzing a set of medical works about PLMs. [11].

# 4 Proposed Architecture

The Fig. 3 shows the architecture of the system. The system analysis data coming from the accelerometer sensor with the aim to recognize PLM events. The information provided by the analysis process are saved into a internal data repository.

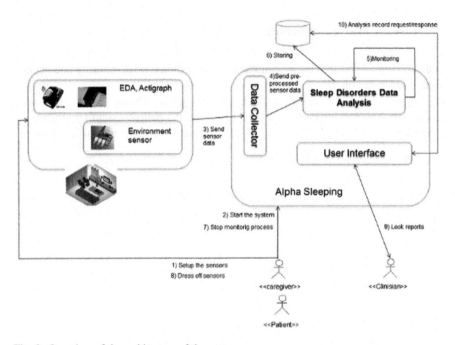

**Fig. 3** Overview of the architecture of the system

The Fig. 3 shows the interaction diagram between intended users and system components. Particularly, it shows the sequence of actions which realizes a collection-analysis-reporting session.

The intended users are: (1) *caregiver*, who is in charge to take care of the patient, he/she sets up the system before the patient goes to bed; (2) *patient*, he/she is the subject of the monitoring process, (3) *clinician*, who retrieves the stored information.

At first point (1), the caregiver has to set up the sensor by placing it on the patient'legs.

From this point at the end of the monitoring (7), the patient wears the sensor. At point 3, the sensor sends the collected data to the Data Collector, that makes them available to the Sleep Disorders Data Analysis (4). At this point, the system works to recognize possible anomalous movements (5) and store both data collected by the sensor (6) and semantic information derived by the analysis process such as reports, tables, etc.

The collection-analysis process can be stopped at any time (7) by the caregiver, this operation is manually done at the time the patient wakes up. Only after action (7), the sensor can be removed from the patient's body (8). Actions (9, 10) can be performed either during the collection-analysis process or after it has been stopped. The clinicians can access to the stored data by using a set of user interface (10).

## 5  Results and Conclusion

In this section we present the preliminary results obtained by monitoring one patient in an early stage of AD.

The patient has been monitored for tree nights, each night the working time was around 8 h. The accelerometer has been placed on the patient's right leg. Figure 4a shows the value of the x,y,z accelerations of the patient's movements during the sleep.

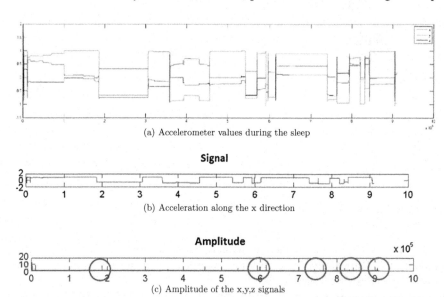

(a) Accelerometer values during the sleep

(b) Acceleration along the x direction

(c) Amplitude of the x,y,z signals

PML event detected, Duration: 28S, Start time: Fri Dec 05 03:05:51 CET 2014
PML event detected, Duration: 40S, Start time: Fri Dec 05 04:06:35 CET 2014
Number of PML events: 2
Frequncy 2 Events/Night

(d) Example of a Stored Report

**Fig. 4** Example of the analyzed signals, **a** accelerometer values during the sleep, **b** acceleration along the x direction, **c** amplitude of the x, y, z signals, **d** example of a stored report

Each signal x, y, z has been analyzed frame-by-frame, where each frame is 1 s length.

For each frame, the momentum is calculated as the amplitude of the x,y,z signals. Figure 4b highlights the acceleration value along the x direction.

An example of the amplitude along the x, y, z signals is shown in the Fig. 4c, in which on the x-axis the time is reported (seconds), on the y-axis the amplitude is shown (dimensionless). The figure shows the trend of the amplitude along the night, the red cycles highlight the frames in which candidate anomalous movements have been detected.

The formula for the calculation of the amplitude is show under, where $N$ is the length of the frame, and $x,y,z$ are the acceleration values (Formula 1).

$$\sum_{0}^{N} \sqrt{(x^2 + y^2 + z^2)} \tag{1}$$

The amplitude signal is submitted to the model for the detection of the PLM. Along the 3 nights the system recognized only two PML events. These information have been stored in a DB for future analysis; other information such as temporal duration of the PML, period of the night in which it happens, frequency of the PLMs, etc. are stores as well, Fig. 4d.

In this paper we present a model for the recognition of periodic legs movements, a type of sleep disorders that affects patients' with Alzheimer's disease. We also describe a first prototype of monitoring system for the detection of such kind of disorders.

The preliminary results show that by using an approach based on the measurement of the patient's momentum during the sleep is possible recognize the occurrence of the PLM events.

# References

1. Alzheimer's association. http://www.alz.org/downloads/Facts_Figures_2014.pdf
2. Hossain, J., Shapiro, C.: The prevalence, cost implications, and management of sleep disorders: an overview. Sleep Breath. 6(2), 85–102 (2002). http://dx.doi.org/10.1007/s11325-002-0085-1
3. Alves de Mesquita, J., de Melo, P.L.: Respiratory monitoring system based on the nasal pressure technique for the analysis of sleep breathing disorders: reduction of static and dynamic errors, and comparisons with thermistors and pneumotachographs. Rev. Sci. Instrum. 75(3), 760–767 (2004)
4. Occhiuzzi, C., Marrocco, G.: The rfid technology for neurosciences: feasibility of limbs' monitoring in sleep diseases. IEEE Trans. Inf. Technol. Biomed. 14(1), 37–43 (2010)
5. Poree, F., Kachenoura, A., Gauvrit, H., Morvan, C., Carrault, G., Senhadji, L.: Blind source separation for ambulatory sleep recording. IEEE Trans. Inf. Technol. Biomed. 10(2), 293–301 (2006)
6. Vu, H.Q., Li, G., Sukhorukova, N., Beliakov, G., Liu, S., Philippe, C., Amiel, H., Ugon, A.: K-complex detection using a hybrid-synergic machine learning method. IEEE Trans. Syst. Man Cybern. Part C: Appl. Rev. 42(6), 1478–1490 (2012)

7. Flores, A., Flores, J., Deshpande, H., Picazo, J., Xie, X., Franken, P., Heller, H., Grahn, D., O'Hara, B.: Pattern recognition of sleep in rodents using piezoelectric signals generated by gross body movements. IEEE Trans. Biomed. Eng. **54**(2), 225–233 (2007)
8. Xie, B., Minn, H.: Real-time sleep apnea detection by classifier combination. IEEE Trans. Inf. Technol. Biomed. **16**(3), 469–477 (2012)
9. Agarwal, R., Takeuchi, T., Laroche, S., Gotman, J.: Detection of rapid-eye movements in sleep studies. IEEE Trans. Inf. Technol. Biomed. Eng. **52**(8), 1390–1396 (2005)
10. Nation sleep foundation. . http://sleepfoundation.org/sleep-disorders-problems/sleep-related-movement-disorders/periodic-limb-movement-disorder
11. Zucconi, M., Ferri, R., Allen, R., Baier, P.C., Bruni, O., Chokroverty, S., Ferini-Strambi, L., Fulda, S., Garcia-Borreguero, D., Hening, W.A., Hirshkowitz, M., Hgl, B., Hornyak, M., King, M., Montagna, P., Parrino, L., Plazzi, G., Terzano, M.G.: The official world association of sleep medicine (wasm) standards for recording and scoring periodic leg movements in sleep (plms) and wakefulness (plmw) developed in collaboration with a task force from the international restless legs syndrome study group (irlssg). Sleep Med **7**(2), 175–183 (2006). http://www.sciencedirect.com/science/article/pii/S1389945706000049

# Mersenne-Walsh Matrices for Image Processing

Nikolay Balonin, Anton Vostrikov and Mikhail Sergeev

**Abstract** This paper presents a modified Paley method for calculation of Mersenne matrices at order values equal to odd prime numbers. Some examples of Mersenne matrix sorting, allowing for calculation of the complete set of functions, are also considered. A comparison of Walsh and Mersenne-Walsh systems of functions in terms of their properties and fields of application is provided. The efficiency of this topic for use in the development of band-pass filters is indicated.

**Keywords** Orthogonal matrices · Quasi-orthogonal matrices · Walsh functions · Hadamard matrices · Modified paley method · Mersenne matrices · Mersenne-Walsh matrices · Image processing

## 1 Introduction

The paper [1] describes the "MMatrix-2" software complex for generating and studying the properties of special orthogonal (quasi-orthogonal) bases, used in image processing [2]. The algorithms of operation of the complex and the theoretical foundations of the quasi-orthogonal matrices of local maximum determinant are discussed in detail in papers [3, 4]. The theory of quasi-orthogonal matrices,

N. Balonin (✉) · A. Vostrikov · M. Sergeev
Department of Computer Systems and Networks, Saint-Petersburg State University
of Aerospace Instrumentation, 67 Bolshaya Morskaya Street, Sankt-Petersburg 190000,
Russian Federation
e-mail: korbendfs@mail.ru

A. Vostrikov
e-mail: vostricov@mail.ru

M. Sergeev
e-mail: mbse@mail.com

M. Sergeev
ITMO University, 49 Kronverksky Ave, Sankt-Petersburg 197101, Russian Federation

© Springer International Publishing Switzerland 2015         141
E. Damiani et al. (eds.), *Intelligent Interactive Multimedia Systems and Services*,
Smart Innovation, Systems and Technologies 40,
DOI 10.1007/978-3-319-19830-9_13

which started to form in works [3, 4, 6], today is not only backed by algorithms and programs, but also by experiences of their applications [2, 5, 7].

For many applications of computational mathematics, coding theory, digital signal processing, etc., an important requirement is simplicity – a finite set of values of ortho-normalized systems functions. The first system of this kind – the Rademacher system [8] – was built in 1922 as a significantly simplified version of the trigonometric system of functions. Although the Rademacher functions (meanders) have only two values $\{1, -1\}$, their disadvantage is that the system is incomplete and therefore is not a basis in the Hilbert space $L_2$. The complete system was first introduced by J. Walsh in 1923 [9]. In contrast to the Rademacher functions, Walsh functions can be divided into even and odd, similarly to the *sin* () and *cos* () functions. The order of their numeration, proposed by Paley [10], proved convenient for calculations. In addition, there is Hadamard ordering, also important when considering this material.

Hadamard matrices are quasi-orthogonal matrices with the elements of two levels $\{1, -1\}$ [11] – they are orthogonal provided their columns are normalized. Since the column vectors form a basis, they generate a complete set of functions. This system is called the Walsh system of functions if the columns are numbered by the number of changes of signs of their elements (frequency analog). Hartmut paper [12] indicates the beginning of the widespread use of Walsh's theoretical results in applied communication tasks. Around the same time, the Hadamard matrices found their immediate application in anti-noise coding of information in radio channels used in automated space missions to Mars [13]. These and other applications have stimulated interest in the generalized theories of orthogonal bases based on these and new systems of functions.

Mersenne-Walsh functions and their generalizations do not belong to well-known functions, making them prospective for use in communications, error control and protective coding, signal processing, and compression and/or masking of images [2, 5, 6].

## 2 Mersenne-Walsh System of Functions

Currently, a huge number of papers are devoted to generalization of Hadamard matrices. An overview of important generalizations of such matrices on the odd orders was first introduced in paper [4]. In paper [14], a class of quasi-orthogonal Hadamard-Mersenne matrices on orders, equal to Mersenne numbers $n = 2^k - 1$, is defined. In papers [15, 16], the hypothesis of their existence for all values of odd orders $n = 4k - 1$, studied in paper [17], is expressed.

In this paper we'll call a Mersenne matrix, generated by columns that are ordered by frequency, a Mersenne-Walsh matrix, and a system of orthogonal functions, generated on the basis of an ordered matrix, we'll call, as distinguished from the classic Walsh functions, a Mersenne-Walsh system of functions.

As an example, columns of a Mersenne matrix of the seventh order are presented as signals in Fig. 1. Orthogonal functions here take two value-levels $\{1, -b\}$, where $|b| < 1$.

Let's list the distinctive features of such a system of orthogonal functions and name the reasons why this basis of signals is of practical interest.

Firstly, the system of Mersenne-Walsh functions is a two-level system, just like the classic system, but it differs from Walsh functions by its lower level $-b$, which is reduced in amplitude, and which approaches $-1$ as the dimensionality of the system increases. In this sense, it is a fairly close approximation of the Walsh system of functions on odd values of the order.

Secondly, the Mersenne-Walsh system of functions is distinguished by a fewer-per-unit number of elements of columns of Mersenne matrices that generate it, which means that it is simpler to calculate than the classic system of Walsh functions.

Finally, the Mersenne Walsh system of functions is higher frequency than the Walsh system of functions, since it does not include the zero frequency function (constant). Therefore, it is preferable for use in the construction of image processing band-pass filters.

**Fig. 1** Columns of an $\mathbf{M}_7$ matrix in the form of signals (meanders)

The first single column and row of normalized Hadamard matrices are unnecessary components, which are useless in band-pass filters, since they correspond to the frequency that is filtered by these filters. Therefore, their presence leads to unnecessary waste of CPU time. However, we would like to note that the simple removal of the canvas of the Hadamard matrix by dropping its first row and column disrupts the orthogonality of columns of the truncated matrix.

## 3  Mersenne-Walsh Matrices

Mersenne-Walsh matrices as well as Hadamard-Walsh matrices can be obtained by sorting the Mersenne [4, 14] and Hadamard [11] matrices, respectively, by the frequency of columns.

The generalization of the methods of calculation of the Hadamard matrices was studied by R. Paley. In 1933 he found the algorithms [18] that significantly increase the set of calculated quasi-orthogonal matrices with levels $\{1, -1\}$.

Let's clarify the concept of quasi-orthogonal matrices, which the matrices we're looking for belong to.

**Definition 1** Quasi-orthogonal matrix $\mathbf{A}_n$ is a square matrix of $n$ order, the values of the elements of each column of which $\leq 1$ (maximal element equals 1), satisfying the condition of column connection expressed as

$$\mathbf{A}_n^{\mathrm{T}}\mathbf{A}_n = \omega\mathbf{I}, \tag{1}$$

where $\mathbf{I}$ is a single diagonal matrix, and $\omega$ is the matrix weight.

Weight $\omega = 1$ is typical for orthogonal matrices, to which quasi-orthogonal matrices, including Hadamard matrices, do not belong because of constraints on the values of their elements. However, these matrices are very close to orthogonal, that are obtainable from $\mathbf{A}_n$ by elementary valuation (normalization) of their rows and columns. After normalization, their maximum element ($m$-norm) [16] for orders $n > 1$ is reduced to $m < 1$.

**Definition 2** $M$-matrices (minimax quasi-orthogonal matrices) are matrices (1), that have a minimum of $m$-norm (global or local [4]) on the class of quasi-orthogonal matrices of the $n$ order.

It is easy to notice that $|\det(A_n)| = \omega^{n/2}$, and $\omega = 1/m^2$.

Hadamard matrices with global determinant maximums have minimal values of $m$-norms, that is they are special cases of $M$-matrices with weight $\omega = n$.

Mersenne matrices, while remaining two-level, differ from Hadamard matrices only in values of their elements: Hadamard matrices of elements have values $\{1, -1\}$, Mersenne matrices $-\{1, -b\}$, where $b = 1/2$ with $n = 3$, and in other cases $b = \frac{q - \sqrt{4q}}{q - 4}$ with $q = n+1$ (the order of constituent Hadamard matrices).

Mersenne matrices can be found using universal search algorithms of the "MMatrix-2" software complex [1]. These universal algorithms and related software solutions are effective in the wide area of research of quasi-orthogonal matrices, relying on their extreme feature – local determinant maximum [4]. They are useful for the identification of systems of orthogonal functions, comparative analysis of their features, building necessary graphs, etc.

Effective algorithms for computation of Mersenne matrices, and then Mersenne-Walsh matrices, are based on methods of the theory of numbers first noticed by Paley [10]. A modified Paley's approach can be used to compute Mersenne matrices.

## 4 Modification of the Paley's Method

Let $n$ be a prime integer, setting the order $n = 4k - 1$ of Mersenne matrix $\mathbf{M}_n$. Then this is a necessary and sufficient condition for the existence of a quasi-orthogonal skew-symmetric cyclic Mersenne matrix of $n$ order with elements, equal to the Legendre symbols $\chi(j - i/n) = \{1, -b\}$, calculated for the differences of pairs of indices $i$, $j$ of their rows and columns [16].

Legendre symbols $\chi(m/n)$ take single values, if $m = j - i$ is a quadratic residue modulo $n$ or 0, and value $-b$, if $m$ is quadratic non-residue modulo $n$. Here $b$ is the absolute value of the negative elements of the Mersenne matrix.

**Example** Let's consider the construction of the Mersenne matrix $\mathbf{M}_7$, associated with finding of Legendre symbols for a set of numbers $\{0, 1, 2, 3, 4, 5, 6\}$, equal to the difference of indices of elements of the first row. Their squares modulo 7 are $\{0, 1, 4, 2, 2, 4, 1\}$, respectively. Numbers $\{1, 2, 4\}$, which are present in both sets, are quadratic residues, and the rest are nonresidues.

Portraits of cyclic Mersenne matrices $\mathbf{M}_3$ and $\mathbf{M}_7$ are shown in Fig. 2. White squares on the portrait are elements with single values, and black are elements $-b$, where $b = 1/2$ with $n = 3$ and $b = 2 - \sqrt{2} \cong 0.5857$ with $n = 7$.

Since the quasi-orthogonal Mersenne matrices contain elements of only two level values $\{1, -b\}$), their multiplication by -1 is not allowed. Consequently, the reducibility to functions, closest in their meaning to Walsh functions, is not obvious for these matrices. However the Mersenne matrices allow for rational ordering of their rows and columns by permutations. For example, Fig. 3 shows the results of ordering of $\mathbf{M}_7$ and $\mathbf{M}_{31}$ matrices.

**Fig. 2** Portraits of cyclic matrices **a** $\mathbf{M}_3$ and **b** $\mathbf{M}_7$    **(a)**                                **(b)**

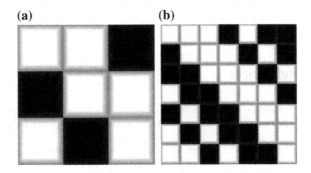

**Fig. 3** Portraits of ordered matrices **a** $M_7$ and **b** $M_{31}$

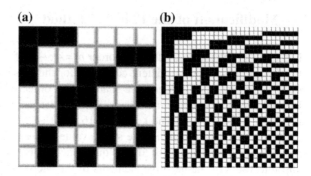

## 5 Conclusion

The newly presented system of Mersenne-Walsh orthogonal functions is intended for use in image processing algorithms, including in the construction of band-pass filters, used in methods of data compression and masking. The algorithms for computing of Legendre symbols, profoundly developed in the number theory, were used to calculate the system of Mersenne-Walsh orthogonal functions.

In general, the value of $b$ is irrational and variable when changing the order $n$ of a matrix. Third parties attempts to unmask an image, protected using Mersenne-Walsh matrices, are more difficult to implement compared to the use of integral Hadamard matrices. At the same time, the resources for storage or calculation of the variable value of $b$ are small, which is important for application of the results of the research in the embedded class systems.

The software complex "MMatrix-2" mentioned above, used in study of Mersenne and Mersenne-Walsh matrices, is built using the implementation of the universal algorithms, which are based on modified methods of Paley and Sylvester. Therefore, it can be used for research of matrices on the orders expressed by simple numbers.

## References

1. Balonin, Y., Vostrikov, A., Sergeev, M.: Software for finding M-matrices. In: Frontiers in Artificial Intelligence and Applications, vol. 262. Smart Digital Futures 2014, pp. 475–480. doi:10.3233/978-1-61499-405-3-475
2. Balonin, N., Sergeev, M.: Expansion of the orthogonal basis in video compression. In: Frontiers in Artificial Intelligence and Applications. Smart Digital Futures 2014, vol. 262, pp. 468–474. doi:10.3233/978-1-61499-405-3-468
3. Sergeev, A.: Generalized Mersenne Matrices and Balonin's Conjecture. Automatic Control and Computer Sciences **48**(4), 214–220 (2014). doi:10.3103/S0146411614040063
4. Balonin, N.A., Sergeev, M.B.: Local maximum determinant matrices. Informatsionno-upravliaiushchie sistemy**1**(68), 2–15 (2014) (In Russian)

5. Balonin, N., Sergeev, M.: Construction of transformation basis for video image masking procedures. In: Frontiers in Artificial Intelligence and Applications. Smart Digital Futures 2014, vol. 262, pp. 462–467. doi:10.3233/978-1-61499-405-3-462
6. Balonin, N.A., Sergeev M.B., Mironovsky L.A.: Calculation of Hadamard-Fermat matrices. Informatsionno-upravliaiushchie sistemy 6(61), 90–93 (2012) (In Russian)
7. Vostrikov, A., Chernyshev, S.: Implementation of novel quasi-orthogonal matrices for simultaneous images compression and protection. In: Frontiers in Artificial Intelligence and Applications. Smart Digital Futures 2014, vol. 262, pp. 451–461. doi:10.3233/978-1-61499-405-3-451
8. Rademacher, H.: Einige Sätze über Reihen von allgemeinen Orthogonalfunktionen. Math. Ann. 87(1–2), 112–138 (1922)
9. Walsh J.L. A closed set of normal orthogonal functions. Am. J. Math. 45, 5–24 (1923)
10. Paley, R.E.A.C.: A remarkable series of orthogonal functions. I, II. *Proc. Lond. Math. Soc.* 34, 241–279 (1932)
11. Hadamard, J.: Résolution d'une question relative aux déterminants. Bulletin des Sciences Mathématiques 17, 240–246 (1893)
12. Harmuth, H.F.: Applications of Walsh functions in communications. IEEE Spectr. 6, 82–91 (1969)
13. Eliahou, S.: La conjecture de Hadamard (I) – Images des Mathématiques. CNRS. 2012. http://images.math.cnrs.fr/La-conjecture-de-Hadamard-I.html. Accessed 05 Jan 2014
14. Balonin, N.A., Sergeev M.B., Mironovsky L.A.: Calculation of Hadamard-Mersenne matrices. Informatsionno-upravliaiushchie sistemy. 5(60), 92–94 (2012) (In Russian)
15. Balonin, N.A., Sergeev, M.B.: Two ways to construct Hadamard-Euler Matrices. Informatsionno-upravliaiushchie sistemy. 1(62), 7–10 (2013) (In Russian)
16. Balonin, N.A., Sergeev, M.B.: On the issue of existence of hadamard and Mersenne matrices. Informatsionno-upravliaiushchie sistemy 5(66), 2–8 (2013) (In Russian)
17. Balonin, N.A.: Existence of Mersenne matrices of 11th and 19th orders. Informatsionno-upravliaiushchie sistemy. 2, 89–90 (In Russian)
18. Paley, R.E.A.C.: On orthogonal matrices. J. Math. Phys. 12, 311–320 (1933)

# Frequency Characteristics for Video Sequences Processing

Andrei Bogoslovsky, Irina Zhigulina, Igor Maslov
and Tatiana Mordovina

**Abstract** A technique of video sequence processing through the energy parameters analysis of video signal is discussed in this paper. The purpose of the analysis here is to determine the object movements in a real-life scene. The new technique is theoretically justified. It is based on the dependence of energy signal at the initial phase. The signal is represented as a spatial cosine segment. The differences of the energy characteristics are introduced to improve the efficiency of video sequences processing. An example of a video sequence processing of the real scene is considered. It is shown that if pixels orderly move inside the object, then the borders and the velocity of the moving object are well determined by using the characteristics difference. However, the pixels groups with the movement differ from the object movement can exist within the object boundaries. In this case, the pixels must be regularly zeroed out, starting from the vision area edges.

**Keywords** Video sequence · Detection · Interframe difference · Energy spectrum · Frequency characteristic · Zeroed pixels

## 1 Introduction

Visual search, a vital task for humans and animals, has also become an important tool for studying many topics central to active vision and cognition. While visual search often seems effortless to humans, trying to recreate human visual search abilities in

A. Bogoslovsky · I. Zhigulina (✉) · I. Maslov · T. Mordovina
Tambov State Technical University, 106 Sovetskaya, Tambov, Russian Federation
e-mail: irazhigulina@gmail.com

A. Bogoslovsky
e-mail: p-digim@mail.ru

I. Maslov
e-mail: p-digim@mail.ru

T. Mordovina
e-mail: p-digim@mail.ru

© Springer International Publishing Switzerland 2015     149
E. Damiani et al. (eds.), *Intelligent Interactive Multimedia Systems and Services*,
Smart Innovation, Systems and Technologies 40,
DOI 10.1007/978-3-319-19830-9_14

machines has represented an incredible challenge for computer scientists and engineers [1]. There are a lot of different algorithms for moving objects detection. The result of processing significantly depends on the operating conditions and problems to be solved. Further progress of video sequences processing inhibits several factors:

- The techniques for processing of moving and static images are significantly different from each other. Processing of static images is based on the integral transformations, while the video sequences processing is mainly based on differential transformations. A search for the uniform techniques of any images treatment is very important. These techniques are likely to be integral.
- Underused properties of the human visual analyzer are: receptive fields [2], "ON-OFF" – centers [3], variable resolution in the retina field [4]. These properties are indirect evidence in favor of integral techniques. For example, the pixels exception (OFF - center) does not give a positive effect by means of the existing techniques.
- The final spatial coordinates (finiteness) of the vision area is not taken into account. However, the movement consideration with respect to the image boundaries can improve the detection characteristics and moving objects identification.

Overcoming these difficulties is possible by analyzing the changes of important universal characteristic – energy of image. The movement will be detected, if the energy analysis between frames for creating and changing "ON-OFF" - centers will be carried out.

This paper is organized as follows: Sect. 2 introduces a short literature review of existing techniques, the proposed techniques are discussed in Sect. 3, and experimental results are represented in Sect. 4 with conclusion in Sect. 5.

## 2   Literature Review

There are several approaches to the motion assessment in video sequences. Block matching algorithms quickly assess the objects displacement between frames [5–7]. They can be used for the evaluation of the unwanted motion in static scenes. Disadvantages of these techniques are following: homogeneous areas influence on the outcome of the motion estimation, and there is a relation to the size of the search block. Block matching algorithms are of the least accurate in comparison with other techniques.

Methods based on the point features allow identifying specific points in the image uniquely [8–10]. They are used for motion estimation in difficult conditions. However, these techniques are sufficiently resource-intensive for a real-time implementation; they require additional hardware and software solutions. The physical equation of transfer processes underlies optical flow techniques [4, 11]. They should be used for the reconstitution of moving objects. The computational complexity of techniques increases sharply with a large motion vector module, and they cease to be effective.

Despite the large number of papers on this subject and as a result of their analysis we could not found any publications, which fully investigated and solved the problem of a universal technique development based on the integral transformation.

# 3   Proposed Methods

Our techniques are based on the dependence of the signal energy upon the initial phase $\varphi_0$ [12, 13]. This signal can be described as a portion of a cosine curve $y = U_0 + U_m \cos(\omega_x x + \varphi_0)$, $-M \leq x \leq M$. In this case the most important thing is a space finite signal, i. e. the vision area edges. Though, the signals have the same spatial frequency $\omega_x$, pulse height $U_m$ and signal constant component $U_0 \geq U_m$, but different initial phases $\varphi_0$, they are not in the same position due to the frame edges.

Video signal can be decomposed into Fourier series [12–14]. Therefore, distribution of its energy for spatial frequencies (energy spectrum), depends on spatial harmonic phases (phase spectrum) [15]. Hence, it is possible to determine the changes of object position even in the images, video signals energy of which does not change. It can be the object of constant brightness and constant shape on a uniform background. Since the movement determination can be factorized [12, 15], i. e. spatial coordinates changes of moving objects can be defined independently, then the line manipulation of video signals will be considered.

## 3.1 Interframe Differences of Energy Spectra

Figure 1 schematically displays two lines with equal numbers of two frames with a signal from a moving object. The contrast of the object, its location and direction of movement is not essential.

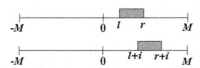

**Fig. 1** Two lines with a moving object

Applicative model will be used as a line model. Then the video signals of the previous frame line and the subsequent frame line can be written as Eq. 1, where $a_t$ is a signal of a background and $b_t$ is a signal of a moving object, $t$ is a frame number, $t = 1$; 2 ($i = 0$ for the first frame, $i \neq 0$ for the second frame).

$$f_t(n) = \sum_{k=-M}^{l+i-1} a_t(k)\ \delta(n-k) + \sum_{k=l+i}^{r+i} b_t(k)\ \delta(n-k) + \sum_{k=r+i+1}^{M} a_t(k)\ \delta(n-k) \quad (1)$$

Spectrum of the line signal takes the form Eq. 2.

$$\dot{F}_t(\omega_x) = \sum_{n=-M}^{M} f_t(n)\, e^{-j\omega_x n} = \left|\dot{F}_{a_t}\right| \cdot e^{-j\,\arg \dot{F}_{a_t}} + \left|\dot{F}_{b_t}\right| \cdot e^{-j\,\arg \dot{F}_{b_t}} \quad (2)$$

Energy spectrum of the line signal can be determined by the Eq. 3.

$$S_t(\omega_x) = \dot{F}_t(\omega_x) \cdot \dot{F}_t^{*}(\omega_x) = S_{a_t} + S_{b_t} + 2\left|\dot{F}_{a_t}\right| \cdot \left|\dot{F}_{b_t}\right| \cdot \cos\left(\arg \dot{F}_{b_t} - \arg \dot{F}_{a_t}\right) \quad (3)$$

Interframe difference of the energy spectra is defined as Eq. 4.

$$\Delta S(\omega_x) = S_2(\omega_x) - S_1(\omega_x) = (S_{a_2} - S_{a_1}) + (S_{b_2} - S_{b_1}) +$$

$$+ 2\left( \left|\dot{F}_{a_2}\right| \cdot \left|\dot{F}_{b_2}\right| \cdot \cos\left(\arg \dot{F}_{b_2} - \arg \dot{F}_{a_2}\right) - \left|\dot{F}_{a_1}\right| \cdot \left|\dot{F}_{b_1}\right| \cdot \cos\left(\arg \dot{F}_{b_1} - \arg \dot{F}_{a_1}\right) \right)$$

$$(4)$$

The first term in Eq. 4 describes the change of the background energy spectrum, and the second term shows the change of the object energy spectrum.

Equation 5 will be executed, if the pulse height background spectrum and the pulse height object spectrum will have been changed slightly between frames.

$$\Delta S(\omega_x) \approx -4\left|\dot{F}_a\right| \cdot \left|\dot{F}_b\right| \cdot \sin\left(\frac{\arg \dot{F}_{b_1} + \arg \dot{F}_{b_2} - \arg \dot{F}_{a_1} - \arg \dot{F}_{a_2}}{2}\right) \times$$

$$\times \sin\left(\frac{\arg \dot{F}_{b_2} - \arg \dot{F}_{b_1} + \arg \dot{F}_{a_1} - \arg \dot{F}_{a_2}}{2}\right) \quad (5)$$

The change of the object location is possible to identify according to the interframe differences of energy spectra. In this case, the analysis of Eq. 5 zeros or the number of sign alternations is interesting.

## 3.2 An Example of Energy Spectra Interframe Difference

Let us consider the simplest case, when the constant brightness object is moving in a uniform background. Then we will obtain Eq. 6.

$$\dot{F}_t(\omega_x) = \sum_{n=-M}^{M} a \cdot e^{-j\omega_x n} + \sum_{n=l+i}^{r+i} (b-a) \cdot e^{-j\omega_x n} \quad (6)$$

Turn to the discrete frequencies $\omega_x = \dfrac{\pi}{M}k$, $k \in [0;\ 2M]$, and we will get Eq. 7.

$$\sum_{n=-M}^{M} a \cdot e^{-j\frac{\pi k n}{M}} = \begin{cases} a \cdot e^{-j\pi k}, & k \neq 0 \\ a \cdot (2M+1), & k = 0 \end{cases} \tag{7}$$

Let us write down the energy spectra differences as Eq. 8.

$$\Delta S(k) = 4a\ (b-a) \cdot \frac{\sin \frac{\pi k(r-l+1)}{2M}}{\sin \frac{\pi k}{2M}} \cdot \sin \frac{\pi k[2M-(r+l+i)]}{2M} \cdot \sin \frac{\pi k i}{2M} \tag{8}$$

The information about the length $(r-l)$ of a moving object, its location $(r+l)$ and its movement $i$ is contained in Eq. 8.

## 3.3 Interframe Difference of Energy Characteristics

The use of the energy spectra causes difficulties even in the simplest cases. Therefore, the difference of energy spectra is expedient to convert. The energy spectrum of a discrete signal is a periodic, even real-valued function of frequency, the $2\pi$ period. The energy spectrum can be decomposed into series over the cosines as Eq. 9.

$$S(\omega_x) = \sum_{p=0}^{2M} \psi(p) \cos p\omega_x = \sum_{p=0}^{2M} \psi(p) \cos \frac{p\pi k}{M} \tag{9}$$

The coefficients $\psi(p)$ in Eq. 9 are values of input sequence $f(n)$ autocorrelation function. It means that Eq. 9 can be converted into Eq. 10.

$$S(\omega_x) = \sum_{p=0}^{2M} \sum_{n=-M}^{M-p} f(n)f(n+p) \cos p\omega_x \tag{10}$$

Thus, Eq. 11 corresponds to each phase number $p$.

$$\psi(p) = \sum_{n=-M}^{M-p} f(n)f(n+p) \tag{11}$$

Presented function will be called the *energy characteristic*.

Equations 10–11 are not recorded accurately. We can obtain Eq. 10 from Eq. 3, but $\psi(p)$ value will doubled for $p \geq 1$. However, this is not essential for the further work, and we shall determine the energy characteristic as Eq. 11 for $p \geq 0$.

The energy characteristic is the amplitude of the corresponding harmonic $\cos p\omega_x$ and completely determines the energy spectrum of the input sequence $f(n)$.

Therefore, we can consider the energy characteristics difference instead of the energy spectra difference.

The dependence of $\Delta\psi(p) = \psi_2(p) - \psi_1(p)$ on the phase number $p$ for a signal which was discussed in Sect. 3.2 is presented in Fig. 2. Characteristic points $\{A, B, C, D, E, F, G, H\}$ revealing the movement and its parameters are marked in the chart. The frame differences of energy characteristics are more visualizing in comparison with the differences of the energy spectra. Another advantage is the operation with the values of autocorrelation function. They are also used in 2D digital filtration (Wiener-Hopf theory, for example) [13, 14].

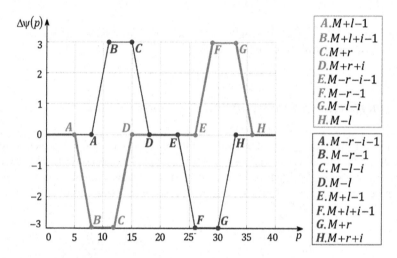

**Fig. 2** Energy characteristics differences

## 4 Experimental Results

The interframe differences of frequency characteristics for the real-life video sequence processing were used. Figure 3 displays a frame of the video sequence. The object (woman) is moving forward and leftward. Its image contains as large spots of brightness, as well as small parts (especially in the middle and in the upper part). Their movement may contain differences from the general object movement. Therefore, we will process the line № 180 (upper part) and the line № 266 (middle part). These lines are marked by white color in Fig. 1. The length of the lines is equal to $2M + 1 = 641$ $(M = 320)$.

The interframe differences are illustrated in Figs. 4 and 5. There is a movement of many small parts near the line № 266 (Fig. 5) while near the line № 180 the movement is more ordered (Fig. 4).

**Fig. 3** Video sequence frame

**Fig. 4** Interframe difference
(near the line № 180)

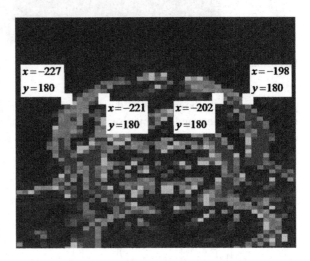

## 4.1 Detection of Moving Object and Its Boundaries

Figures 6 and 7 show the differences of the energy characteristics for the line № 180 and № 266 respectively. Boundaries of specific areas [15] are marked there.

Distant-pulse decay is defined in Fig. 6. This point has the abscissa $M - l + i = 546$. Therefore, we can find the left boundary of the moving object in the second frame: $l - i = -226$. Also we can define the beginning of the far pulse front: $M - r - 1 = 519$. Hence, we obtain the right edge of the moving object in the first frame: $r = -200$. Object displacement also can be measured.

The frequency differences are symmetric with respect points $(M + 0.5)$. Therefore, near-impulse can be found. Impulses can be also noticed in the inter-impulse area and in the area of near phases. This fact is caused by the presence of moving

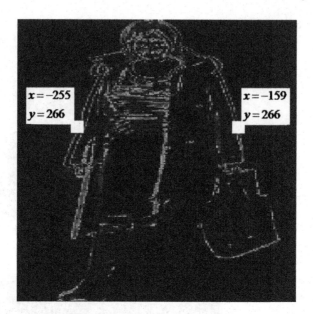

**Fig. 5** Image interframe difference

pixels between the boundaries of the object (Fig. 5). There are other distortions (at the tops of the near and far impulses in the area of distant impulse phase). But the displacement of the moving object can be determined with sufficient accuracy.

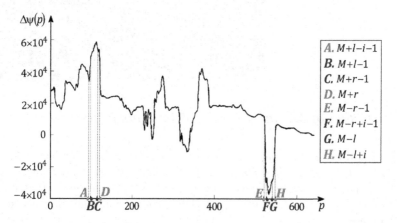

**Fig. 6** Energy characteristics differences (line № 180)

Interframe frequency differences are significantly distorted because of the large number of changing pixels between frames within the boundaries of the moving

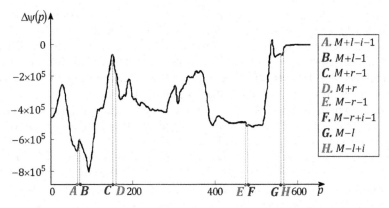

**Fig. 7** Energy characteristics differences (line № 266)

object (Fig. 7). In this case, the boundaries of the moving object (especially the right boundary) and its displacement are difficult to determine.

## 4.2 Improvement of Detection Characteristics

The detection of moving object can be improved by increasing of the absolute values for interframe differences of energy spectrum [12, 15]. This result can be achieved by nullification of pixels with the same number in both frames. The differences changes should be insignificant if zeroed pixels fall on parts corresponding to the background or to the object in the neighbor frames. The interframe frequency characteristics are changed substantially, if zeroed pixels correspond to the background in one frame and they correspond to the object in other frame.

The differences of the energy characteristics for the line № 266 are shown in Fig. 8. They are obtained at regular nullification of one pixel. Cases of getting zeroed pixels of both frames on the background are indicated by black color. There are small changes in limited areas of graphs. (See a blue circle in Fig. 8).

**Fig. 8** Energy characteristics differences if a pixel is regularly nullificated (colour figure online)

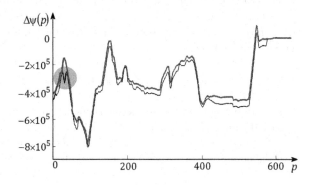

The energy characteristics difference (red line in Fig. 8) varies considerably across the whole range as soon as a zeroed pixel has the coordinate $l - i = -255$. In this case, the left edge of a moving object in the second frame is clearly defined.

The object displacement can be found by this way: measure the difference of the frequency characteristics at any phase, sequentially increasing the number of zeroed pixels (start from pixel $l - i$ with further magnifying).

Figure 9 represents the interframe difference dependence of energy characteristics on the amount $q$ of successively zeroed pixels; the phase $p$ is equal to 100. Dependence growth stops when $q = 5$, that is why $i = 5$.

Then we need to make line $l$ segmentation. To find the right edge of the moving object, it is advisable to perform nullification of truncate lines on the right and then the right edge of the object will be defined. After that it is necessary to classify the movement within its borders.

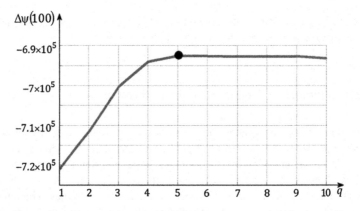

**Fig. 9** Interframe difference dependence on the amount of successively zeroed pixels

## 4.3 Double Differences of Frequency Characteristics

More qualitative discovering of a moving object is possible, if we reduce the impact of the background pixels. For this purpose *double differences of frequency characteristics* can be accomplished [12, 15]. They can be written as Eq. 12 where function $\psi_t(p)$ is an energy line characteristic, $t$ is a frame number, $N$ and $N + 1$ are zeroed pixels numbers.

$$\Delta \psi''(p) = (\psi_1(p) - \psi_2(p))_N - (\psi_1(p) - \psi_2(p))_{N+1} \tag{12}$$

The double differences of energy characteristics for line № 266 are shown in Fig. 10. If zeroed pixels are more to the left of pixels $l - i$, then the lines are black. If one of zeroed pixel has the coordinate $l - i$, then the lines are red. If zeroed pixels have the coordinates $l - i$ and $l - i + 1$, then the lines are blue.

**Fig. 10** Double differences of energy characteristics for line № 266 (colour figure online)

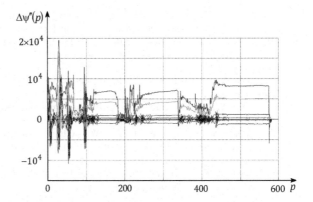

Overview of Fig. 10 leads to the following results:

- There will be a sharp change of double values differences when zeroed pixels get on the object.
- In this case there is a drift and an oscillation of double energy differences.
- Rare oscillations correspond to the movement of small details in the image. Monotone areas generally correspond to the far impulse, partially to the near impulse and to the interpulse area.

# 5 Conclusion

Proposed energy characteristics differences are integral. Every value is formed by a lot of input video images values but not only by one value. This allows to change detection characteristics through the manipulation of video signals values, for example, through a partial nullification. For example, the "browse" mode of a moving object is possible, i.e. we can receive information about its details by changing the values, which are even not corresponded to the object. The discussed feature allows implementing algorithms inherent to a human visual system, for example, treatment with the help of « ON-OFF » centers. In addition, the proposed techniques are sufficiently flexible and allow various modifications.

**Acknowledgments** The work of the authors was supported by the Russian Fund of Basic Researches project No. 15-01-08043.

# References

1. Eckstein, M.P.: Visual search: a retrospective. J. Vis. **11**(5), 1–36 (2011)
2. Bradley, C., Abrams, J., Geisler, W.S.: Retina-V1 model of delectability across the visual field. J. Vis. **14**(22), 1–22 (2014)

3. Olmedo-Paya, A., Martinez-Alvarez, A., Cuenca-Asensi, S., Ferrandez-Visente, J.M., Fernandez, E.: Modeling the effect of fixational eye movements in natural scenes. In: Ferrandez Visente, J.M., Sánchez, J.R.A., de la Paz López, F., Moreo, F.J.T. (eds.) Natural and Artificial Models in Computation and Biology-1. LNCS, vol. 7930, pp. 332–341. Springer, Heidelberg (2013)
4. Jimenez, E.V.C., Navarro, D.Z.: Intelligent active vision systems for robots. Cuvillier, Göttingen (2007)
5. Cai, J., Pan, W.: On fast and accurate block-based motion estimation algorithms using particle swam optimization. Inf. Sci. **197**, 53–64 (2012)
6. Ho, H., Klepko, R., Nam, N., Demin, W.: A high performance hardware architecture for multi-frame hierarchical motion estimation. IEEE Trans. Consum. Electron. **57**(2), 794–801 (2011)
7. Srinivasarao, B.K.N., Chakrabarti, I.: A parallel architectural implementation of the fast three step search algorithm for block motion estimation. In: 5th International Multi-Conference on Systems, Signals and Devices, pp. 1–6. IEEE Press, Amman, USA (2008)
8. Lucas, B.D., Kanade, T.: An iterative image registration technique with an application to stereo vision. In: 7th International Joint Conference on Artificial Intelligence, vol. 2, pp. 674–679. Morgan Kaufmann Publishers Inc., San Francisco, CA, USA (1981)
9. Lowe, D.G.: Distinctive image features from scale-invariant keypoints. Int. J. Comput. Vision **60**(2), 91–110 (2004)
10. Bay, H., Ess, A., Tuytelaars, T., Van Gool, L.: Speeded-up robust features (SURF). Comput. Vis. Image Underst. **110**(3), 346–359 (2008)
11. Reducindo, I., Arce-Santana, E., Campos-Delgado, D.U., Vigueras-Gomez, F.: Non-rigid multimodal image registration based on local variability measures and optical flow. In: Annual International Conference of the IEEE Engineering in Medicine and Biological Society, pp. 1133–1136. IEEE Press, San Diego, CA, USA. (2012)
12. Bogoslovsky, A.: Multidimensional signal processing. Radiotec, Moscow (in Russian) (2013)
13. Gonzalez, R.C., Woods, R.E.: Digital Image Processing, 2nd edn. Prentice Hall, Upper Saddle River (2002)
14. Jähne, B.: Digital Image Processing, 6th edn. Springer, Berlin (2005)
15. Bogoslovsky, A., Zhigulina, I.: A way of energy analysis for image and video sequence processing. In: Favorskaya, M.N., Jain, L.C. (eds.) Computer Vision in Control Systems-1. ISRL, vol. 73, pp. 183–210. Springer, Switzerland (2015)

# Expansion of the Quasi-Orthogonal Basis to Mask Images

Anton Vostrikov and Mikhail Sergeev

**Abstract** The article discusses a significant expansion of the basis of two-level quasi-orthogonal Hadamard-Mersenne matrices, used for compression and masking of images and video stream frames. The article presents the dependences of the levels of matrix elements on their order, and the algorithm for calculation through chains that include Hadamard-Euler matrices. It is demonstrated that the sizes of Hadamard and Hadamard-Mersenne matrix families are the same. The procedure of masking and compression of images using quasi-orthogonal matrices, the orders of which correspond to the sequence of Mersenne numbers, is analyzed.

**Keywords** Orthogonal matrices · Quasi-orthogonal matrices · M-matrices · Hadamard matrices · Hadamard-Mersenne matrices · Hadamard-Euler matrices · Mersenne numbers · Image masking · Image compression

## 1 Introduction

Protection of information from unauthorized access and spoofing is of great importance in today's world, especially the protection of video data in the public domain [1–3]. A variety of perfect protection systems that are successfully used in practice have been already developed. But most of those systems cannot be directly used to protect digital video in real-time systems because they are based on encryption algorithms and require significant computational power.

A. Vostrikov (✉) · M. Sergeev
Department of Computer Systems and Networks, Saint-Petersburg State University
of Aerospace Instrumentation, 67 Bolshaya Morskaya Street, 190000 Sankt-Petersburg,
Russian Federation
e-mail: vostricov@mail.ru

M. Sergeev
e-mail: mbse@mail.com

M. Sergeev
ITMO University, 49 Kronverksky Ave, 197101 Sankt-Petersburg, Russian Federation

© Springer International Publishing Switzerland 2015
E. Damiani et al. (eds.), *Intelligent Interactive Multimedia Systems and Services*,
Smart Innovation, Systems and Technologies 40,
DOI 10.1007/978-3-319-19830-9_15

161

The monograph [4] discusses the matrix methods of video coding based on bases of orthogonal transformations, which may be used, particularly but not exclusively, to protect video frames against unauthorized viewing.

For this purpose, the authors previously proposed a relatively simple image masking procedure [5, 6], based on the use of quasi-orthogonal bases, that can be performed in real time. The study of the theory of their construction and their properties has intensified significantly in recent years [7, 8] and will be discussed in this paper.

## 2  Basic Definitions

**Definition 1** Matrix masking is a computational procedure of matrix transformation of images, breaking them down to the form, visually perceived as noise.

**Definition 2** Matrix unmasking is a computational procedure of inverse transformation with the use of an inverse matrix that restores the original image from the masked one.

The task of organizing the process of matrix masking/unmasking of images (individual frames of a video stream) lies in the simplicity and symmetry of transformations applied to images. Obviously, if the system is symmetric, the inverse matrix is required for inverse transformation of an image.

**Definition 3** Levels are the values of the entries of a matrix.

The definition of matrix levels was first introduced in paper [9]. It allowed for the interpretation of mathematical objects with flat and volumetric graphic portraits of matrices [10], which, in turn, allows for the identification of previously unknown patterns and relationships of quasi-orthogonal matrices.

Hadamard matrices [11–13], for example, are two-level matrices, while symmetric conference matrices [14, 15] and weighing matrices [16, 17] are three-level matrices.

**Definition 4** A real square matrix $\mathbf{X}_n$ of order $n$ is called *quasi-orthogonal* if it satisfies $\mathbf{X}_n^T\mathbf{X}_n = \omega\,\mathbf{I}$, where $\omega \leq n$ is a constant real number, $\mathbf{I} = \mathrm{diag}\{1,1,\dots,1\}$.

In this paper, we consider only quasi-orthogonal matrices with real elements, in which at least one element in each row and column is 1. Hadamard matrices are a well-known example of such matrices with elements placed in the unit circle. Symmetric conference matrices, in particular, are also an important class of quasi-orthogonal matrices.

Level $a = 1$ is pre-determined for all quasi-orthogonal matrices.

We will denote the levels of two level matrices as $a$, $b$; for positive $0 \leq b \leq a$, $a = 1$.

The concept of Hadamard-Mersenne matrices was first introduced in paper [9]. These matrices complement the set of Hadamard matrices in the class of orthogonal matrices of odd orders $n = 2^k - 1$ ($k$ – integer) equal to the numbers of Mersenne sequence.

**Definition 5** A Hadamard-Mersenne matrix Mn [9, 18] is a quasi-orthogonal matrix of the n = 2k−1 order, where k is an integer with two level functions a = 1, −b. A Mersenne matrix, in the general case, of n = 4t−1 order, is a Quasi-orthogonal matrix with the same level function b as the Hadamard-Mersenne matrix: $b = \frac{t}{t-\sqrt{t}}$.

The invariant of Mersenne matrices is a difference (equal to 1) between the number of positive and negative elements in each column (row), so the weight $\omega(n) = \frac{n+1}{2} + \frac{n-1}{2}b^2$.

The important thing is that, according to Balonin's Conjecture [18, 19], the sets of Hadamard and Mersenne matrices are equal in their sizes. Their orders $n = 4t−1$ cover Mersenne numbers $n = 2^k−1$, where $k$ is an integer for a so-called elementary set of Mersenne matrices (or *pure* Mersenne matrices).

The number of quasi-orthogonal Mersenne matrices is not less than the number of integer Hadamard matrices, but the values of the coefficients of two-level Mersenne matrices are real numbers, which allows for better protection of video. They are denoted as follows: $\mathbf{M}(4t − 1; \ b = \frac{t}{t+\sqrt{t}})$.

A lot of Hadamard-Euler matrices are related to Hadamard-Mersenne matrices.

**Defenition 6** A Hadamard-Euler matrix En [20] is a quasi-orthogonal matrix of order n = 2m, m = 2k−1, where k is an integer with level functions a = 1, −b constructed with two blocks A, B of the Hadamard-Mersenne type, from which it is constructed.

An Euler matrix, in the general case, of order n = 4t−2, is a quasi-orthogonal matrix with the same level function b, as a Hadamard-Euler matrix: $b = \frac{t}{t-\sqrt{2t}}$.

The number of quasi-orthogonal Euler matrices, in turn, is also not less than the number of integer Hadamard matrices. It is also a set of matrices with real numerical coefficients.

They are denoted as $\mathbf{E}(4t − 2; \ b = \frac{t}{t+\sqrt{2t}})$.

The current state of digital signal processors, characterized by an increase in productivity and structural orientation for performance of convolution operations in real number format, allows for effective use of described and more complex bases (multilevel M-matrices) [21, 22].

## 3 The Hadamard-Mersenne Matrices and Their Calculation

The sequence of Mersenne numbers, given by the formula $n = 2^k−1$, starts with numbers 1, 3, 5, 15, 31, ... and belongs to a subset of numbers of the form 4t−1. The two-level Hadamard-Mersenne matrix of the third order has the form [9]:

$$\mathbf{M}_3 = \begin{pmatrix} a & -b & a \\ -b & a & a \\ a & a & -b \end{pmatrix}$$

To iteratively obtain two-level orthogonal Hadamard-Mersenne matrices of subsequent orders from previous ones using the Sylvester formula [11], a four-level Hadamard- Euler matrix should be built on the first iteration step:

$$\mathbf{E}_n = \begin{pmatrix} \mathbf{M}_{n/2} & \mathbf{M}_{n/2} \\ \mathbf{M}_{n/2} & -\mathbf{M}_{n/2} \end{pmatrix},$$

where $\mathbf{M}_{n/2}$ is a two-level Hadamard-Mersenne matrix of the half odd order, which has two levels $\{a = 1, -b\}$ [20]. Under this transformation, the number of levels doubles due to the inversion of the two-level Hadamard-Mersenne matrix.

On the second iteration step, the Hadamard-Euler matrix is recalculated to a Hadamard-Mersenne matrix by addition of a row and a column to it (bordering):

$$\mathbf{M}_{n+1} = \begin{pmatrix} -\lambda & e^{\mathrm{T}} \\ e & \mathbf{E}_{2n}^* \end{pmatrix},$$

where $\lambda = -a$ – eigen value, and $e$ – eigenvector of "conjugated" matrix $\mathbf{E}_{2n}^* = \begin{pmatrix} \mathbf{M}_{n/2} & \mathbf{M}_{n/2} \\ \mathbf{M}_{n/2} & \mathbf{M}_{n/2}^* \end{pmatrix}$, $\mathbf{M}_{n/2}^*$ is obtained from the Mersenne matrix of the corresponding order by interchanging of elements $a = 1$ and $-b$, in which the first half of coefficients of the eigenvector, different from $a$, are $-b$ elements.

The formulas above make it evident that such a sequence of actions during iteration allows for a return to the two-level version of the Hadamard-Mersenne matrix, the order of which corresponds to the following number in the Mersenne sequence.

The values of matrix elements (levels) are the roots of characteristic algebraic equations. Their parameters for levels, corresponding to the condition of orthogonality of the columns of the Hadamard-Mersenne matrix, are given in paper [9].

## 4   On Balonin's Conjecture

Unlike Hadamard matrices, Hadamard-Mersenne matrices exist on odd orders. Odd order matrices are prime because their order not only begins with 1, but also with all other prime integers.

In this analogy lies a deeper interpretation. Recall that the Hadamard matrix is a square two-level $\mathbf{H}_n$ matrix of $n$ order, consisting of numbers $\{1, -1\}$, the columns of which are orthogonal $\mathbf{H}_n^{\mathrm{T}}\mathbf{H}_n = n\mathbf{I}$.

Unlike Euler-Hadamard matrices, from the elementary Hadamard-Mersenne matrix, equal to one according to the Sylvester Rule, we can only obtain a Hadamard matrix of the second order, and further iterations continue with it. The orders of such matrices are spaced too far apart and the distance between the orders of adjacent matrices increases intensely. However, Hadamard gave examples of two-level orthogonal matrices with unit values of moduli of elements of intermediate orders $n = 12$ and $n = 20$. According to his conjecture, there are Hadamard matries of orders 1, 2 and $4k$, where $k$ is an integer.

Balonin's conjecture [18, 19] consists of the assertion of the existence of Mersenne matrices of orders $4k-1$. Matrices of orders $n = 11$ and $n = 19$ [18] with the signs of Hadamard-Mersenne matrices were found in support of equality of Hadamard and Hadamard-Mersenne matrix sets, referred to simply as Mersenne matrices for convenience.

With the apparent symmetry of both conjectures, there are, however, certain differences.

Hadamard matrices are based on square three-level Belevich matrices $\mathbf{C}_n$ of $n$ order, consisting of numbers $\{1, 0, -1\}$, the columns of which are orthogonal $\mathbf{C}_n^\mathrm{T} \mathbf{C}_n = (n-1)\mathbf{I}$, and the zero elements are concentrated on the diagonal.

The existence of the Belevich matrices is determined by the Euler criterion (and its interpretations, the case of non-prime numbers) of decomposability of the $n-1$ multiplier to the sum of two squares of numbers.

When the Belevich matrices do not exist, which is true for orders 22, 34, 58, etc., the Hadamard-Euler matrices exist. The Hadamard-Euler matrix of $n = 22$ order can be obtained, for example, based on the above mentioned Hadamard-Mersenne matrix of $n = 11$ order. Thus, the gaps in the set of three-level orthogonal Belevich matrices get filled, which has a significant theoretical and practical value.

Obviously, the existence of Hadamard-Mersenne matrices on $4k-1$ orders significantly expands the basis of orthogonal matrices for use in the above mentioned applied tasks of video-information protection.

## 5  Image Masking Algorithm

The typical processing path in the procedures of the matrix image transformation [4–6] consists of several steps:

- Discrete Fourier transform to obtain the spectral decomposition.
- Use of a low-pass filter to eliminate the high-frequency part of the spectrum.
- Huffman statistical processing to eliminate redundancy.

All these steps are present in the masking algorithm [5]. However, it is essential that the discrete Fourier transform matrix is replaced by the original Hadamard-Mersenne matrix of quasi-orthogonal basis (Sylvester sequence of matrices). The low-pass filter is replaced with multiplication by the quantization matrix, coordinated with the selected Hadamard-Mersenne matrix.

Paper [8], which discusses the results of the image masking using the $\mathbf{M}_{15}$ matrix, gives an example of the use of Hadamard-Mersenne matrices.

In addition to the discreteness of levels of matrix elements, as in the Hadamard matrix, an equally important role is played by the originality of the basis that provides the secrecy of transformed data. This makes the matrices with rational and irrational values of levels convenient for use in the procedure of image masking on digital devices.

Unlike traditional bases, in the task of masking the great importance belongs to qualities, apart from those mentioned above, acquired from the extreme properties of basis sets [23–26]. For example, Hadamard matrices and Mersenne matrices that are close to them are optimal in terms of neutralizing the effects of point interference when transmitting via communication channels.

The masking matrix order-increasing recursive procedures, adapted to the size of the image being masked, are also very important in the context of our paper.

Another argument to rationalize the use of bases constructed on the sequence of numbers [23] is that the algorithm of their construction is fractal, and matrices in the specific configuration of the algorithm have increased sensitivity to changes in processor capacity and the initial data.

The method of image transformation considered allows, first of all, for keeping the possibility in principle of compression of masked information, for example, by adaptation of the filtering procedure to the structural specificities of the basis. Secondly, an unusual matrix and masking key in the form of a vector of permutation of rows and columns, unknown to third parties, provide reliable concealment of an image or a sequence of frames in a video stream from unauthorized use and spoofing.

An instrument used for search for new quasi-orthogonal matrices is the [22, 23, 27] software, which can be used to simplify the task of video masking, when the orthogonal transformation matrix is not calculated in advance, but is the result of the work of the algorithm. Only the configuration of parameters for its calculation are sent as a key via an open communication channel.

# 6   Conclusion

In the search for original orthogonal matrices of odd orders close to Hadamard matrices in terms of their properties, for use in the image masking algorithms, a preferred class of two-level matrices, called Hadamard-Mersenne matrices, has been selected. The orders of these matrices are equal to Mersenne numbers of the form $2^k - 1$ and their number is greatly expanded by Balonin's Conjecture. The elements of Hadamard-Mersenne matrices, with the increase of the value of the integer argument k, tend toward the values $\{1, -1\}$ like the Hadamard matrices. The authors monitor the basis matrices on the Internet: http://mathscinet.ru/ catalogue.

The practical application of matrices considered in this paper is expedient to increase the degree of noise immunity and protection during transmission of information, since the matrix methods imply an effective implementation in modern microprocessor structures oriented to digital information processing.

# References

1. Bezzateev, S.V., Litvinov, M.Y., Troyanovskij, B.K., Filatov, G.P.: The choice of the transformation algorithm that ensures a structural change of videoinformation. Informatsionno-upravliaiushchie sistemy **6**, 2–6 (2006) (In Russian)
2. Vostrikov, A.A., Chernyshev, S.A.: On distortion assessment of images masking with m-matrices. Sci. Tech. J. Inf. Technol. Mech. Opt. **5**, 99–103 (2013) (In Russian)
3. Erosh, I.L., Filatov, G.P., Sergeev, A.M.: Protection of images during transfer via communication channels. Informatsionno-upravliaiushchie sistemy **5**, 20–22 (2007) (In Russian)
4. Mironovsky, L.A., Slaev, V.A.: Strip-Method for Image and Signal Transformation. DeGruyter, Berlin (2011). 163 p
5. Balonin, N., Sergeev, M.: Construction of transformation basis for video image masking procedures. Frontiers in Artificial Intelligence and Applications, Volume 262: Smart Digital Futures, pp. 462–467 (2014). doi:10.3233/978-1-61499-405-3-462
6. Vostrikov, A., Chernyshev, S.: Implementation of novel Quasi-Orthogonal matrices for simultaneous images compression and protection. Frontiers in Artificial Intelligence and Applications, Volume 262: Smart Digital Futures, pp. 451–461 (2014). doi:10.3233/978-1-61499-405-3-451
7. Balonin, N.A., Sergeev, M.B.: Local maximum determinant matrices. Informatsionno-upravliaiushchie sistemy **1**, 2–15 (2014) (In Russian)
8. Balonin, N., Sergeev, M.: Expansion of the orthogonal basis in video compression. Frontiers in Artificial Intelligence and Applications, Volume 262: Smart Digital Futures, pp. 468–474 (2014). doi:10.3233/978-1-61499-405-3-468
9. Balonin, N.A., Sergeev, M.B., Mironovsky, L.A.: Calculation of Hadamard-Mersenne matrices. Informatsionno-upravliaiushchie sistemy **5**(60), 92–94 (2012) (In Russian)
10. Balonin, Yu. N., Sergeev, M.B.: The algorithm and program for searching and studying of M-matrices. Sci. Tech. J. Inf. Technol. Mech. Opt. **3**, 82–86 (2013) (In Russian)
11. Hadamard, J.: Résolution d'une question relative aux déterminants. Bulletin des Sciences Mathématiques **17**, 240–246 (1893)
12. Paley, R.E.A.C.: On orthogonal matrices. J. Math. Phys. **12**, 311–320 (1933)
13. Seberry, J., Yamada, M.: Hadamard matrices, sequences, and block designs. In: J. H. Dinitz, D. R. Stinson (eds.) Contemporary Design Theory: A Collection of Surveys, John Wiley and Sons, Inc., pp. 431–560 (1992)
14. Belevitch, V.: Theorem of 2n-terminal networks with application to conference telephony. Electr. Commun. **26**, 231–244 (1950)
15. Balonin, N.A., Seberry, J.: A review and new symmetric conference matrices. Informatsionno-upravliaiushchie sistemy **4**(71), 2–7 (2014)
16. Wallis (Seberry), J.: Orthogonal (0, 1, −1) matrices. Proceedings of First Australian Conference on Combinatorial Mathematics, TUNRA, Newcastle, pp. 61–84 (1972)
17. Balonin, N. A., Seberry, J.: Remarks on extremal and maximum determinant matrices with real entries ≤1. Informatsionno-upravliaiushchie sistemy **5**(71), 2–4 (2014)
18. Balonin,N.A.: Existence of Mersenne matrices of 11th and 19th Orders. Informatsionno-upravliaiushchie sistemy **2**, 89–90 (2013) (In Russian)

19. Sergeev, A.: Generalized Mersenne matrices and Balonin's conjecture. Automat. Control Comput. Sci. **48**(4), 214–220 (2014). doi:10.3103/S0146411614040063
20. Balonin, N.A., Sergeev, M.B.: Two ways to construct Hadamard-Euler matrices. Informatsionno-upravliaiushchie sistemy **1**(62), 7–10 (2013) (In Russian)
21. Balonin, N.A., Sergeev, M.B.: M-matrices. Informatsionno-upravliaiushchie sistemy **1**, 14–21 (2011) (In Russian)
22. Balonin, Yu. N.: Program complex "MMatrix-1" and searched M-matrices. Vestnik komp'iutornykh i informatsionnykh tekhnologii **10**, 58–63 (2013) (In Russian)
23. Balonin, Yu., Sergeev, M., Vostrikov, A.: Software for finding M-matrices. Frontiers in Artificial Intelligence and Applications, Volume 262: Smart Digital Futures, pp. 475–480 (2014). doi:10.3233/978-1-61499-405-3-475
24. Balonin, N.A., Sergeev, M.B. Mironovsky L.A.: Calculation of Hadamard-Fermat matrices. Informatsionno-upravliaiushchie sistemy **6**(61), 90–93 (2012) (In Russian)
25. Balonin, N.A., Sergeev, M.B.: Matrix of golden ratio G10. Informatsionno-upravliaiushchie sistemy **6**(67), 2–5 (2013) (In Russian)
26. Balonin, Yu. N., Sergeev, M.B.: M-matrix of 22nd Order. Informatsionno-upravliaiushchie sistemy **5**(54), 87–90 (2011) (In Russian)
27. Balonin, N.A., Sergeev, M.B.: On the issue of existence of Hadamard and Mersenne matrices. Informatsionno-upravliaiushchie sistemy **5**(66), 2–8 (2013) (In Russian)

# Performance Analysis of Prediction Methods for Lossless Image Compression

Nickolay Egorov, Dmitriy Novikov and Marat Gilmutdinov

**Abstract** Performance analysis of several state-of-the-art prediction approaches is performed for lossless image compression. To provide this analysis special models of edges are presented: bound-oriented and gradient-oriented approaches. Several heuristic assumptions are proposed for considered intra- and inter-component predictors using determined edge models. Numerical evaluation using image test sets with various statistical features confirms obtained heuristic assumptions.

**Keywords** Lossless image compression · Intra- and inter-component prediction · Edge modeling · Bound-oriented prediction · Gradient-oriented prediction

## 1 Introduction

This paper is dedicated to performance analysis of prediction schemes used in lossless image compression. The selection of predictors is limited by state-of-the-art predictors used in compression standards and popular compression schemes. This limitation depends on complexity requirements. Several lossless compression schemes, e.g. PAQ [1], GraLic [2], have excellent compression ratio. Unfortunately referenced compressors have very high complexity unsuitable for real-time application requirements.

The main purpose of prediction is removal of inter-symbol correlation (dependency). Unfortunately de-correlation depends on input data statistical properties that cannot be estimated a priori, especially for images. Another key image feature which

N. Egorov · D. Novikov · M. Gilmutdinov (✉)
Saint Petersburg State University of Aerospace Instrumentation,
Bolshaya Morskaya, 67, 190000 Saint Petersburg, Russia
e-mail: mgilmutdinov@gmail.com

N. Egorov
e-mail: negorov.91@gmail.com

D. Novikov
e-mail: dnovikov.suai@gmail.com

© Springer International Publishing Switzerland 2015                                      169
E. Damiani et al. (eds.), *Intelligent Interactive Multimedia Systems and Services*,
Smart Innovation, Systems and Technologies 40,
DOI 10.1007/978-3-319-19830-9_16

impacts prediction performance is different statistics for various image regions, so image statistical properties are not stationary in general. Most predictors are not effective in regions where statistic changes extremely. Typical example of fast statistical properties changing is object's edge. The ability of predictors to adapt to edges affects prediction performance significantly. For this reason local-adaptive methods are more preferable, because they have fast adaptation rate, even though their complexity is higher.

Besides complexity limitation the following requirements should be represented to provide high prediction performance: minimum standard deviation and maximum number of zero values of prediction errors. One of the additional goals of this paper is performance improvement of lossless image compression scheme with Binary Layers Scanning (BLS) introduced in previous work [3]. One of the key features of BLS is adaptation to correlation of prediction errors, so-called *residual correlation*. As demonstrated in [3] this algorithm of prediction errors encoding outperforms conventional lossless image compression standards, e.g. JPEG-LS [4, 5] and JPEG-2000 [6]. To estimate the impact on compression performance, the combination of each analyzed predictor and BLS codec is considered.

The paper is organized as follows. Section 2 contains predictors classification and heuristic simple edge classification-based analysis of several predictors. Intra- and inter-component prediction approaches are considered. Comparison of prediction performance is presented in Sect. 3. This comparison uses image test sets and numerical criteria based on statistical properties of prediction errors which are also described in Sect. 3.

## 2 State-of-the-art Prediction Methods Analysis

### 2.1 General Statements

At this moment a lot of pixel prediction methods were developed. They use a large spectrum of approaches from simple neighboring pixels averaging [7] to methods that use solution of a large number of equations [8]. Prediction methods can be classified by:

- number of passes: single- or multi-pass,
- number of components used: intra- or inter-component;
- adaptation type: global optimization based, local-adaptive or non-adaptive;
- scan order: box-out, raster or wipe scans [9];
- data specificity accounting: universal or specific-oriented, e.g. predictors for medical images.

Main objects of interest are single-pass local-adaptive prediction methods which use raster scan order. These methods are low-complexity and have high performance for specific image classes.

As far as prediction removes spatial correlation between neighboring pixels, it should use previously processed pixels. A set of all pixel positions considered in prediction algorithm is called *context*. Previously processed neighboring pixel in context has special assignment according to cardinal direction. An example of context assigment is depicted in Fig. 1. Note, that only top and positions in the same row can be used in context for the currently processed component. Pixel positions in previously processed components do not have any restrictions.

## 2.2 Edge Modeling

Most local-adaptive predictors use approaches for detection of intensity discontinuities [10], which are always presented in images especially at object borders. These discontinuities are usually part of image textures. Intensity discontinuities can be divided into two basic classes: *bounds* and *gradients*. Examples of models from these classes are depicted in Fig. 2.

**Fig. 1**  Cardinal direction assignment to neighboring pixels (raster scan)

| NWW | NW | N | NE |
|-----|-----|-----|-----|
| WW | W | X | E |
|  |  | S |  |

**Fig. 2**  Models of intensity discontinuities with corresponding intensity profiles: (a) – bound; (b) – gradient

a)                                    b)

*Edge pixel* is a pixel located in bound or gradient region. All co-located edge pixels are united into *edge*. In this paper *forward edge* and *backward edge* definitions are used to determine direction of diagonal edges. Forward edge direction is visually similar to main diagonal of matrix, while backward edge direction is visually similar to reverse diagonal of matrix. These definitions will be used for further prediction methods analysis.

## 2.3 The Spatial Prediction Methods

**Median Edge Detection (MED).** MED is the most popular method used in series of JPEG-LS standards [4, 5]. It provides high compression performance with low computational complexity. Prediction result $\hat{x}$ for current pixel is calculated using the following equation:

$$\hat{x} = \begin{cases} min(n, w), & \text{if } nw \geqslant max(n, w); \\ max(n, w), & \text{if } nw \leqslant min(n, w); \\ n + w - nw, & \text{otherwise}. \end{cases} \tag{1}$$

The main idea of MED is estimation of edge direction near the predicted pixel $x$. This algorithm considers three edge cases. In the first and second conditions of Eq. (1) horizontal and vertical border edge direction cases are verified. These border edges are approximated by bound model described in Sect. (2.2). Thus either $n$ or $w$ values are used as prediction in this model. The last condition of (1) is used in forward egde case. In such case the method uses gradient edge model. Thus the predicted value can differ from $n$ and $w$ neighbor values.

The main disadvantages of MED are *coarse approximation* of vertical and horizontal borders (first and second conditions of (1)) and *insensitivity* to backward edges. Figure 3 demonstrates the selection of condition for edge pixel processing. Edge pixels are white-highlighted for third condition of MED applying. Otherwise edge pixels are black-highlighted corresponding to backward edges.

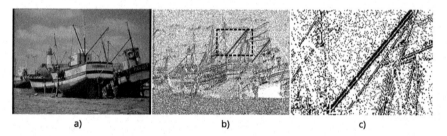

a)                          b)                          c)

**Fig. 3** Edge pixel processing by MED algorithm: (a) – original image, (b) – processed image, (c) – image fragment. *White* – third condition of (1), *black* – other conditions of (1)

**Active Level Classification Model Predictor (ALCM).** The prediction method described in [11] calculates sum of context pixels $p_k$ weighted by $b_k$, where $k$ is an index of pixel in context. ALCM algorithm is designed to decrease the prediction error for the next predictions assuming high *spatial correlation* of neighboring pixels. According to this assumption ALCM uses approximation of *Gradient Descent* algorithm to perform the following optimization:

$$E = \min_{\{b_k\}} \left| x - \sum_k b_k p_k \right|, \tag{2}$$

where the set of $b_k$ is obtained during previous pixel processing. Choise of pixel context has great influence on ALCM prediction performance. Author of [12] proposed to use context that consists of six pixels: $nw$, $n$, $ne$, $w$, $ww$, $nww$ (see Fig. 1). Such small amount of pixels is explained by extreme decreasing of pixel correlation function. Thus only closest pixels are considered to provide local-adaptive property.

Figure 4 depicts performance comparison of MED and ALCM algorithms using pixels grayscale pattern. Visual comparison demonstrates that ALCM outperforms MED in still regions and on gradient type edges, but MED is more effective on bound type edges due to fast adaptation rate.

There is one popular prediction method — Gradient Adjusted Prediction (GAP) from CALIC codec [13]. It uses gradient-oriented approach as well as ALCM but with predetermined sets of $b_k$ coefficients. Since GAP is a specific case of ALCM, its consideration is out of this paper scope.

a)                              b)                              c)

**Fig. 4** Prediction performance comparison: ALCM vs. MED. (a) – processed image, (b) – fragment of original image, (c) – fragment of processed image. Greyscale pattern for processed image: $white |x - \hat{x}_{ALCM}| < |x - \hat{x}_{MED}|$, $black |x - \hat{x}_{ALCM}| > |x - \hat{x}_{MED}|$, $grey |x - \hat{x}_{ALCM}| = |x - \hat{x}_{MED}|$

## 2.4 Inter-Component Predictions

**Conventional Color Transforms**. These predictors are conventionally used in image compression standards, e.g. JPEG-LS (Part 2) [5] and JPEG-2000 (lossless mode) [6]. They use typical assumptions about correlation between red, green and blue components. These predictors are usually non-adaptive, so they are out of scope of this paper.

**Simple Inter-color Lossless Image Coder (SICLIC)**. SICLIC [14] combines both intra- and inter-component predictions. Spatial prediction is identical to the MED prediction from [4]. Inter-color prediction is based on MED idea, but it uses differences (gradients) $D_k$ between pixels from current and previously processed components as depicted in Fig. 5. SICLIC selects the best predictor from intra- and inter-component parts according to prediction error magnitude.

**Fig. 5** SICLIC prediction algorithm [14]

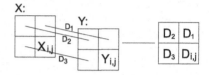

The main advantage of SICLIC codec is that it uses two types of predictors. Since some images are better compressed using the spatial predictor, while other are better compressed using the inter-color predictor, it is important to use both. In regions with high inter-color correlation SICLIC shows good prediction performance on bound edges as well as gradient edges. The main disadvantage is low prediction performance in regions with low correlation between components because MED shows low prediction accuracy in still regions and on gradient type edges.

# 3 Experimental Results

## 3.1 Experiment Setup

To estimate compression performance all prediction errors are encoded using JPEG-LS part 2 [5], JPEG-2000 [6] and BLS [3] codecs. For color images JPEG-LS part 2 codec uses reversible color transform which is similar to transform described in JPEG-2000 (lossless compression mode). This transform is used as inter-component prediction. JPEG-2000 codec is used in lossless compression mode.

BLS codec is used in combination with various intra- and inter-component predictors analyzed in this paper [3]. MED or ALCM are used as intra-component predictors for the first (green) image component processing. SICLIC-based algorithm or ALCM approach are used as inter-component predictors to provide second and third image component processing. IC-ALCM (Inter-Component ALCM) acronym is used to distinguish intra- and inter-component predictors based on ALCM. The following pixel positions in the previously processed image components are used as context for IC-ALCM: $n, w, x, e, s$.

Most image test sets contain color images. The next subsections contains: description of image test sets, numerical criteria used in comparison and comparison summary.

## 3.2 Images Test Sets for Prediction Performance Comparison

The experiments are performed using image test sets which have various parameters: one or three color components images; 8 or 10 bits per pixel in component. Table 1 presents basic properties of the following image sets:

- *Lossless Photo Compression Benchmark* (LPCB) [15]: contains large set of high resolution images. This set consists of three image types: natural images, Earth surface pictures and pictures of space objects;
- *Squeeze Chart* (SC) [16] test set: contains high resolution photorealistic images;
- *Computing Tomography* (CT) [17]: contains 10bpp grayscale image sequence of body detailed scans; they are used for testing predictors on data with specific statistical properties;
- *Kodak* [18] image test set: popular test set of photorealistic images;
- *Rawzor corpus* [19]: high resolution image test set presented by Rawzor Inc.;
- *State-of-the-art image test set*: conventionally used images, e.g. lena, baboon, airplane etc.

**Table 1**  Test sets description

| Corpus name | Files number | Size, bytes | Maximum Resolution | Number of components | BPP |
|---|---|---|---|---|---|
| LPCB | 107 | 3 456 571 880 | $6800 \times 6800$ | 3 | 3*8 |
| SC | 7 | 242 932 451 | $5184 \times 3456$ | 3 | 3*8 |
| CT | 175 | 91 753 200 | $512 \times 512$ | 1 | 10 |
| Kodak | 24 | 28 311 912 | $768 \times 512$ | 3 | 3*8 |
| Rawzor | 14 | 470 611 702 | $7216 \times 5412$ | 3 | 3*8 |
| State-of-the-art | 7 | 6 927 657 | $787 \times 576$ | 3 | 3*8 |

## 3.3 Numerical Criteria for Prediction Methods Comparison

The following numerical criteria are used for prediction performance comparison:

- *SD*-criterion: *Standard Deviation* of prediction errors;
- *Z*-criterion: number of *Zeros* among prediction errors;
- Bits Per Pixel (BPP): estimation of compression performance for corresponding lossless compression scheme;
- *G*-criterion: compression *Gain* (in percents) for $X1$ compression scheme compared to $X2$ (for $i$-th file):

$$G_i(X1, X2) = \frac{S_i^{X2} - S_i^{X1}}{S_i^{X2}} \cdot 100\%,  \quad (3)$$

where $S_i^X$ – size of $i$-th file compressed by algorithm $X$. Negative $G$-criterion value means loss of $X1$ algorithm.

## 3.4 Comparison Summary

Compression results for different prediction methods using BLS compression system are listed in Table 2. This table also shows the compression results for state-of-the-art codecs with similar complexity: JPEG-LS and JPEG-2000. Combination of ALCM prediction for the first color component and SICLIC-based approach for two other components outperforms all state-of-the-art prediction algorithms in all test sets except LPCB and Rawzor corpus which consist of high resolution images. ALCM & SICLIC combination is more efficient for images that are characterized by domination of gradient type edges with high spatial correlation and extremely changing inter-component statistic. ALCM & IC-ALCM combination outperforms other prediction methods on images with high inter-component correlation especially for high resolution images in LPCB and Rawzor corpus.

To improve numerical analysis, pairwise comparisons of considered prediction algorithms are performed for particular images. These results are listed in Table 3. The best result of the first prediction algorithm is presented in top line and the best result of the second method is presented in bottom line of each row. *SD* and *Z* criteria are presented by two values separated by symbol "/".

**Table 2** Compression performance for different prediction algorithms using BLS system (BPP)

| Set name | MED & SICLIC | ALCM & SICLIC | ALCM & IC-ALCM | JPEG-LS | JPEG-2K |
|---|---|---|---|---|---|
| LPCB | 7.91 | 7.87 | **7.84** | 8.45 | 8.65 |
| SC | 10.18 | **10.07** | 10.22 | 10.38 | 10.52 |
| CT | 3.25 | **3.18** | 3.18 | 3.3 | 3.54 |
| Kodak | 8.59 | **8.55** | 9.35 | 8.78 | 8.9 |
| Rawzor | 9.92 | 9.84 | **9.71** | 10.32 | 10.62 |
| State-of-the-art | 12.46 | 12.41 | 12.43 | **12.3** | 13.13 |

Obtained results listed in Tables 2, 3 confirm the following heuristic assumptions from prediction methods analysis:

- Domination of the gradient type edges and smoothly varying statistic are specific for high resolution images. Usage of gradient-oriented ALCM prediction is preferable for this type of images (PIA13815.ppm)
- Domination of the bound type edges are specific for artificial images. MED prediction outperforms ALCM predictor for this specific type of images by proposed G-, SD- and Z-criteria due to ALCM slow weights adaptation speed (PIA13779.ppm and artificial.ppm)
- MED prediction also is more preferable than ALCM for noisy images and pictures with extreme statistic changes. Abrupt statistic changes is specific for low-definition images (baboon.ppm)

- Extended bit depth (10 BPP or more) and medical scanner features are the reasons of gradient edges domination for computer tomography images. Gradient-oriented prediction is more preferable for this type of images
- SICLIC-based inter-component approach is robust to extreme changes of inter-component statistics and more effective on images with significant number of color objects (sony.ppm, artificial.ppm and PIA13779.ppm)
- IC-ALCM prediction is more accurate than SICLIC-based approach in case of high inter-component correlation (PIA13803.ppm, spider-web.ppm and hubble.ppm)

**Table 3** Pairwise prediction algorithms comparison (best gain results are selected)

| Compared predictions: | Best gain on file | G, % | SD | Z, $10^5$ |
|---|---|---|---|---|
| | | LPCB | | |
| MED vs ALCM | PIA13779.ppm | 20.05 | 5.79 / 5.7 | 0.86 / 0.36 |
| | PIA13815.ppm | 5.55 | 9.04 / 13.36 | 19.38 / 17.4 |
| SICLIC vs IC-ALCM | PIA13779.ppm | 53.46 | 3.18 / 16.16 | 52.83 / 48.68 |
| | PIA13803.ppm | 18.66 | 3.38 / 6.14 | 8.68 / 3.37 |
| | | SC | | |
| SICLIC vs IC-ALCM | sony.ppm | 13.74 | 2.45 / 15.77 | 45.47 / 39.9 |
| | hubble.ppm | 3.85 | 8.55 / 9.99 | 8.94 / 7.35 |
| | | Rawzor | | |
| SICLIC vs IC-ALCM | artificial.ppm | 26.37 | 2.48 / 3.99 | 57.85 / 53.25 |
| | spider_web.ppm | 5.64 | 0.82 / 0.86 | 68.28 / 61.59 |
| | | State-of-the-art | | |
| MED vs ALCM | baboon.ppm | 0.82 | 25.72 / 22.33 | 0.33 / 0.1 |
| | pepper.ppm | 1.28 | 9.3 / 10 | 0.3 / 0.26 |
| SICLIC vs IC-ALCM | barbara.ppm | 4.02 | 4.18 / 4.93 | 1.39 / 0.82 |
| | airplane.ppm | 2.89 | 3.8 / 3.86 | 0.51 / 0.44 |

## 4 Conclusion

Experimental results confirm the main heuristic assumption about edge feature impact on prediction performance. Trend of image resolution increasing leads to rise of correlation between neighboring pixels and number of edges decreasing. Therefore ALCM intra-component prediction looks more preferable. It is also true for medical imaging. Unfortunately slow adaptation rate and insensibility to low correlation between image components make ALCM unsuitable for images with extreme statistic changes. Thus ALCM intra-component prediction is the optimal prediction scheme among considered approaches. Inter-component prediction should be selected basing on image specific: SICLIC-based prediction for images with extremely changing inter-component correlation or IC-ALCM in other case. Anyway, increasing the adaptation rate in edge regions is an obvious way to improve the ALCM in the future work.

# References

1. Mahoney, M.: The ZPAQ Open standard format for highly compressed data. http://mattmahoney.net/dc/zpaq.html
2. Ratushnyak, A.: GraLIC - new lossless image compressor. http://encode.ru/threads/595-GraLIC-new-lossless-image-compressor
3. Gilmutdinov, M., Egorov, N., Novikov D.: Lossless image compression scheme with binary layers scanning. In: 2014 XIV International Symposium on Problems of Redundancy in Information and Control Systems (2014)
4. Information Technology Lossless and Near-Lossless Compression of Continuous-Tone Still Images: Baseline. ISO/IEC, ISO/IEC 14495–1:1999 (1999)
5. Information technology - Lossless and near-lossless compression of continuous-tone still images: Extensions. ISO/IEC 14495–2 (2003)
6. Information Technology JPEG 2000 image coding system: Core coding system, ISO/IEC 15444–1:2004 (2009)
7. Xiaojun Q., Tyler, J.M.: Linear-prediction-based multi-resolution approach for lossless image compression. In: Proceedings of SCI/ISAS on Image, Acoustic, Speech and Signal Processing, vol. IV, pp. 217–222, Orlando, Florida (2003)
8. Li, X., Orahard,M.T.: Orchard, edge-directed prediction for lossless compression of natural images. IEEE Trans. Image Process. **10**(6), 813–817 (2001)
9. Richardson, I.: The H.264 Advanced Video Compression Standard. Wiley, Chichester (2010)
10. Gonzalez, R.C., Woods, R.E.: Digital Image Processing, 3rd edn. Prentice Hall, Upper Saddle River (2007)
11. Speck, D.: Proposal for next generation lossless compression of continous-tone still pictures: Activity level classification model (ALCM). ISO Working Document ISO/IEC JTC1/SC29/WG1 N198 (1995)
12. Speck, D.: Fast robust adaptation of predictor weights from min/max neighboring pixels for minimum conditional entropy. 1995 Conference Record of the Twenty-Ninth Asilomar Conference on Signals, Systems and Computers, vol. 1, pp. 234–238 (1995)
13. Wu, X, Memon, N.: CALIC-a context based adaptive lossless image codec. IEEE Acoust, Speech, Signal Process **4**, 1892–1893 1996)
14. Barequet, R., Feder, M.: SICLIC: a simple inter-color lossless image coder. In: Proceedings of Data Compression Conference, DCC '99, pp. 501, 510 (1999)
15. Lossless Photo Compression Benchmark by Alexander Ratushnyak. http://www.imagecompression.info/gralic/LPCB.html
16. Squeeze Chart Benchmark. http://www.squeezechart.com/bitmap.html
17. Dataset_small_CT_175_Image. http://public.cancerimagingarchive.net/ncia/login.jsf
18. Brocchi, S.: Bidimensional pictures: reconstruction, expression and encoding. Universita degli studi di Firenze. Dottorato di ricerca in ingegneria informatica e dell'automazione Ciclo XXII, Settore disciplinare INF/01 - Informatica (2009)
19. High-resolution images benchmark presented by Rawzor - Lossless compression software for camera raw images. http://www.imagecompression.info/test_images/

# Non-stationary Correlation Noise Modeling for Transform Domain Wyner-Ziv Video Coding

Anton Veselov, Boris Filippov, Victor Yastrebov and Marat Gilmutdinov

**Abstract** In this paper the problem of correlation noise modeling (CNM) for Transform Domain Wyner-Ziv (TDWZ) video coding is considered. The CNM algorithm from DISCOVER codec is analyzed. An extended set of assumptions about correlation noise in transform domain is considered and a new CNM algorithm is proposed. The proposed algorithm is a generalization of the reference one for the case of non-stationary noise. The evaluation results show that the proposed method demonstrates higher performance than the state-of-the-art approach on test video sequences with complex motion.

**Keywords** Distributed video coding · Wyner-ziv video coding · Virtual dependency channel · Correlation noise · Transform domain

## 1 Introduction

Distributed Video Coding (DVC), also known as Wyner-Ziv Video Coding (WZVC), is a modern video compression paradigm applied in energy-constrained applications, e.g. mobile video streaming. The theoretical background of DVC was proposed in

This work has been financially supported by the Russian Ministry of Education and Science within a framework of the basic task to the university in 2015 (project number 2452).

A. Veselov (✉) · B. Filippov · V. Yastrebov · M. Gilmutdinov
Saint Petersburg State University of Aerospace Instrumentation,
Bolshaya Morskaya St. 67, 190000 Saint Petersburg, Russia
e-mail: anton.veselov@gmail.com http://www.suai.ru

B. Filippov
e-mail: filippovboris@gmail.com

V. Yastrebov
e-mail: victor.yastrebov@vu.spb.ru

M. Gilmutdinov
e-mail: mgilmutdinov@gmail.com

© Springer International Publishing Switzerland 2015
E. Damiani et al. (eds.), *Intelligent Interactive Multimedia Systems and Services*,
Smart Innovation, Systems and Technologies 40,
DOI 10.1007/978-3-319-19830-9_17

179

works of Slepian and Wolf [1], Wyner and Ziv [2] in 1970-s as distributed source coding theory. But this approach was not required in practice till the beginning of 2000-s when the rise of energy-constrained mobile applications predetermined the further development of DVC-based systems. Several compression schemes based on DVC were implemented and thoroughly analyzed. The most popular DVC codec is DISCOVER project [3] which utilizes Stanford DVC architecture [4]. Two parts of Stanford-based DVC systems have significant impact on compression performance: side information generation conventionally based on temporal interpolation and correlation noise modeling. The latter is used at the decoder to estimate the mismatch between side information available at decoder and original video frames. This paper is dedicated to a new method of accurate correlation noise estimation. Proposed method is based on assumption about non-stationarity of mismatching between side-information and original frame. According to this assumption, pixels positions which have interpolation errors (correlation noise) with similar statistical properties are concentrated in the same regions. This assumption WAS originally investigated in works of Westerlaken et al. [5]. In [6] authors proposed new correlation noise model in terms of Hidden Random Markov Field. This model is also based on the same assumption about non-stationarity of correlation noise. This paper extends the model proposed in [6] for Transform Domain Wyner-Ziv (TDWZ) coding. The performance of proposed correlation noise modeling method is compared with the approach used in DISCOVER which is based on works of Pereira and Brites [7]. To apply numerical evaluation both methods are embedded in TDWZ video coding system similar to DISCOVER.

This paper is organized as follows. The background and proposed correlation noise modeling method are described in Sect. 2. Numerical evaluation of the proposed and state-of-the-art methods is provided in Sect. 3.

## 2 Correlation Noise Modeling

### 2.1 Correlation Noise

Correlation noise is a key idea of practical DVC schemes. It provides alternative interpretation of video encoding procedure connecting it with the methods of data transmission over a noisy channel. Two sources $U_X$ and $U_Y$ produce messages which are the observations of the correlated random variables $X \in \mathcal{X}$ and $Y \in \mathcal{Y}$ correspondingly. These variables are drawn from the bivariate distribution

$p(X, Y) \neq p(X)p(Y)$. Let $x \in \mathcal{X}^n$ and $y \in \mathcal{Y}^n$ be the correlated information sequences observed at the output of the sources. It should be noted that

$$p(x, y) = \prod_{i=1}^{n} Pr[X = x_i, Y = y_i].$$

Suppose that the information sequence $y$ is known to the decoder and it is required to reconstruct $x$. The idea of reconstruction procedure is as follows. Suppose that sequence $x$ has been transmitted over a noisy channel and the corrupted sequence is $y$, i.e.

$$y = x + n,$$

where $n$ represents noise. It means that the reconstruction procedure can apply error correction techniques to eliminate the noise component $n$ from $y$ thus reconstructing $x$.

This alternative view of distributed source coding is known as Virtual Dependency Channel (VDC) based model. Noise $n$ which "arises" in VDC is called *correlation noise*.

Now consider the idea of correlation noise in practical TDWZ video coding based on Stanford architecture. First let us introduce the notation which will be further used in this paper. Denote the previously decoded frames used for Wyner-Ziv (WZ) frame approximation by $F^{(p)}$ and $F^{(f)}$. WZ frame available at encoder is denoted by $F^{(i)}$, WZ frame approximation generated by decoder is denoted by $F^{(a)}$.

Decoder by analogy with encoder applies block spectral transform and combines the calculated spectral coefficients into bands. Each band contains the spectral coefficients located at a specific positions of the transformed blocks. If block size is $4 \times 4$ pixels then there are 16 bands of spectral coefficients. Then decoder quantizes each band and forms a vector of quantized spectral coefficients $\hat{\mathbf{q}}^{(b)} = (\hat{q}_1^{(b)}, \hat{q}_2^{(b)}, \ldots, \hat{q}_N^{(b)})$, where $b$ is the band index and $N$ is the number of spectral coefficients in a band. It should be noted that the encoder forms a vector of quantized spectral coefficients $\mathbf{q}^{(b)} = (q_1^{(b)}, q_2^{(b)}, \ldots, q_N^{(b)})$ by applying the same sequence of operations to original intermediate frame $F^{(i)}$. Since frame $F^{(a)}$ is the approximation of $F^{(i)}$, $\mathbf{q}^{(b)}$ and $\hat{\mathbf{q}}^{(b)}$ can be considered as the correlated messages of the two sources. Thus $\hat{\mathbf{q}}^{(b)}$ is the side information of the decoder and correlation noise $\mathbf{n}^{(b)} = (n_1^{(b)}, n_2^{(b)}, \ldots, n_N^{(b)})$ for each band $b$ is defined as

$$\mathbf{n}^{(b)} = \mathbf{q}^{(b)} - \hat{\mathbf{q}}^{(b)}.$$

Also it should be noted that decoder can not calculate $\mathbf{n}^{(b)}$ explicitly because it has no information about original frame $F^{(i)}$. Nevertheless decoder should estimate the statistical parameters of $\mathbf{q}^{(b)}$ by analyzing the available information. This estimation procedure is known as Correlation Noise Modeling (CNM).

Practical approaches to correlation noise modeling are considered in the following sections of this paper.

## 2.2 State-of-the-Art Correlation Noise Modeling Techniques

In earliest work on DVC correlation noise was modeled as an additive stationary zero-mean random process [8]. Such an approach is known as *stationary virtual dependency channel* modeling. Within the given context term "stationarity" means that the statistical characteristics of noise do not change within a band. This simple model can be easily implemented in practice however it was shown to be inefficient in many cases because of interpolation effects known as motion occlusions and holes. As a result *non-stationary virtual dependency channel* modeling techniques were further developed for correlation noise estimation algorithms.

Consider a set of state-of-the-art assumptions about correlation noise.

**Assumption 1** Decoder can estimate $\tilde{\mathbf{n}}^{(b)} = (\tilde{n}_1^{(b)}, \tilde{n}_2^{(b)}, \ldots, \tilde{n}_N^{(b)})$, s.t. $\tilde{n}_i^{(b)} = n_i^{(b)} + e_i^{(b)}$, where $e_i^{(b)}$ is a zero-mean random variable.

**Assumption 2** Noise is a sequence of independent identically distributed (IID) random variables which probability density function can be approximated as a zero-mean Laplacian Mixture Model:

$$p(x) = \sum_{k=1}^{K} w_k \frac{\alpha_k}{2} e^{-\alpha_k|x|},$$

where $w_k$ is the weight of mixture component $k$ and $\alpha_k$ is the Laplacian distribution parameter for component $k$.

We will call $\tilde{\mathbf{n}}^{(b)}$ *correlation noise approximation*. As a rule correlation noise approximation is obtained from spectral coefficients of motion compensated residuals of inter-frame approximation.

Taking into account the above assumptions the goal of correlation noise modeling is to estimate Laplacian distribution parameter $\alpha_i$ for each $\hat{q}_i$. Such an approach allows modeling non-stationarity by adapting the number of components $K$ in mixture. However the problem of determining the relation between mixture components and quantized spectral coefficients is an open issue. This problem was considered in work [5]. The authors used a synthetic test sequences with manual partition of regions and demonstrated the bit rate savings up to 30 % for $K = 2$.

State-of-the-art correlation noise modeling algorithm used in DISCOVER codec was proposed in [7]. In this algorithm each coefficient in band is classified into inlier or outlier class. Inliers are those coefficients whose absolute value is close to the average magnitude value in band. Outliers are the remaining coefficients in band. Then Laplace parameter is calculated for each coefficient using the following rule:

$$\hat{\alpha}_i^{(b)} = \begin{cases} \sqrt{\frac{2}{\sigma_b^2}}, & \text{if inlier coefficient} \\ \sqrt{\frac{2}{d_i^2}}, & \text{otherwise} \end{cases},$$

where $\sigma_b^2$ is a band magnitude variance estimation, $d_i$ is the distance between average magnitude in band and magnitude of spectral coefficient in number $i$.

The described approach manages the problem of non-stationary VDC modeling by adapting the Laplacian parameter to band and coefficient levels. However this model has several disadvantages. The main disadvantage is that the model does not consider possible errors of correlation noise approximation when large values of errors are estimated for the correctly approximated regions and vice versa. One of the approaches which can be used to process such regions is based on spatial correlation analysis of spectral coefficients. This approach is presented in the next section of the paper.

## 2.3 Proposed Correlation Noise Modeling Algorithm

In this section a description of the new CNM algorithm is given. The algorithm is based on a generalized model proposed in [6]. The model provides a mathematical framework for noise parameter estimation in Pixel Domain Wyner-Ziv (PDWZ) video coding. In this section the model extension to transform domain is considered and a heuristic CNM algorithm is proposed.

Consider a set of new assumptions about correlation noise in transform domain. This set extends the previously defined assumptions by saving Assumption 1 and replacing Assumption 2 with the following new assumptions.

**Assumption 2** Statistical characteristics of correlation noise in band $b$ can be described using a set $\mathcal{D}^{(b)}$ of $K$ various distributions

$$\mathcal{D}^{(b)} = \{D_1^{(b)}(\theta_1^{(b)}), D_2^{(b)}(\theta_2^{(b)}), ..., D_K^{(b)}(\theta_K^{(b)})\},$$

where $D_i^{(b)}(\theta_i^{(b)})$ represents the probability distribution law and parameters for distribution number $i$. Distributions in $\mathcal{D}^{(b)}$ can have identical laws but the distribution parameters should be different. In the course of correlation noise estimation each position in $\tilde{\mathbf{n}}^{(b)}$ should be associated with one of distributions from $\mathcal{D}^{(b)}$.

**Assumption 3** Distribution laws for different spectral coefficients within a band are not independent, i.e.

$$P(\tilde{n}_i^{(b)} \sim D_k^{(b)}(\theta_k^{(b)}), \tilde{n}_j^{(b)} \sim D_l^{(b)}(\theta_l^{(b)})) \neq P(\tilde{n}_i^{(b)} \sim D_k^{(b)}(\theta_k^{(b)}))P(\tilde{n}_j^{(b)} \sim D_l^{(b)}(\theta_l^{(b)})).$$

These assumptions allow using the same mathematical foundation as in [6] to model correlation noise in transform domain. It means that the correlation noise in each band can be described using Hidden Markov Random Field (HMRF) with hidden nodes denoting the association of distributions in $\mathcal{D}^{(b)}$ to the quantized spectral coefficients and the observed nodes corresponding to the estimated noise. Then the correlation noise estimation can be reformulated as a labeling problem within a HMRF framework. This task can be solved using standard energy minimization techniques [9], however to apply it in practice the potential functions describing the joint likelihood of neighboring nodes should be specified which is a difficult problem. In this context a simpler approach to model correlation noise is considered. It is proposed to apply estimation in three steps.

1. Preliminary independent labeling of nodes given the observed noise.
2. Spatial filtering of labels to take into account the spatial dependence of nodes.
3. Noise parameters estimation for each coefficient.

At the first step each quantized spectral coefficient is assigned with a label denoting the distribution index in $D^{(b)}$. We propose an iterative procedure to simultaneously determine both labeling and corresponding set of distributions $D^{(b)}$.

The algorithm starts with initial set of distributions $D^{(b)}$ and iteratively calculates labeling and updates $D^{(b)}$ until either labeling stays unchanged between iterations or maximal number of iterations is achieved. Then spatial filtering is applied to smooth the preliminary labeling.

**Algorithm** Proposed correlation noise modeling algorithm for band $b$.
**Input**:

– noise approximation $\tilde{\mathbf{n}}^{(b)}$;
– initial set of distributions $D^{(b)}$;
– maximal number of iterations in preliminary labeling $N_1$;
– number of median filtering iterations $N_2$.

**Output**: estimate of noise parameters for each coefficient.

1. Calculate preliminary labeling $\mathbf{l} = (l_1, l_2, \ldots, l_N)$.

    (a) Calculate labeling for current set $D^{(b)}$:

    $$l_i = \arg \max_{j=\{1,2,\cdots,N\}} p(\tilde{n}_i^{(b)} | D_j^{(b)}(\theta_j^{(b)})),$$

    where $p(\tilde{n}_i^{(b)} | D_j^{(b)}(\theta_j^{(b)}))$ is the likelihood of the observed random variable $\tilde{n}_i$ given distribution $D_j^{(b)}(\theta_j^{(b)})$.

    (b) Update $D^{(b)}$. For each $D_j^{(b)}(\theta_j^{(b)}) \in D^{(b)}$ estimate parameters $\theta_j^{(b)}$ using those random variables which have been labeled with label $j$ in the previous step.

    (c) If number of iterations exceeds $N_1$ or labeling varies insignificantly go to step 2. Otherwise go to step 1a.

2. Iterative median filtering of $\mathbf{l}$.
3. Estimate parameters for each pixel. If Laplacian model is assumed then the approach from the state of-the-art correlation noise modeling can be used at this step:

$$\hat{\alpha}_i^{(b)} = \begin{cases} \sqrt{\dfrac{2}{(\sigma_{l_i}^{(b)})^2}}, & \text{if inlier coefficient} \\[3mm] \sqrt{\dfrac{2}{d_i^2}}, & \text{otherwise} \end{cases},$$

where $(\sigma_{l_i}^{(b)})^2$ is the variance estimate of coefficients labeled by $l_i$ (region $l_i$); $d_i$ is the distance between magnitude of coefficient $i$ and average magnitude in region $l_i$.

To summarize the description let us provide the relation of the proposed algorithm to the state-of-the-art approach. The developed algorithm can be considered as a generalization of the state-of-the-art scheme. The algorithms are identical when the initial set $\mathcal{D}^{(b)} = \{D_1^{(b)}(\theta_1^{(b)})\}$ and $D_1^{(b)}(\theta_1^{(b)})$ represents Laplacian distribution. But when $\mathcal{D}^{(b)}$ contains more than one distribution the proposed correlation noise modeling scheme is able to adapt to the spatial characteristics of noise more accurately: noise intensity estimate is reduced for regions characterized by high approximation quality and increased for low approximation quality regions.

## 3 Experimental Results

In order to evaluate the proposed virtual channel model the DVC framework was implemented. This framework is a DVC system similar to DISCOVER codec [3] with following features. Error-correcting module is based on LPDCA code with accumulated syndrome width equal to 1584 for QCIF and 6336 for CIF input video resolution (according to latest DISOVER release). Quantization matrices were chosen according to [10]. The implemented temporal prediction method is described in [11], GOP is set to 2.

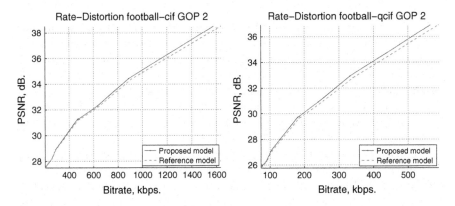

**Fig. 1** PSNR-bitrate for 'football' video sequence

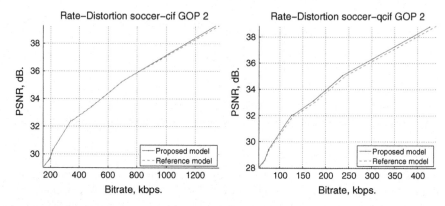

**Fig. 2**  PSNR-bitrate for "soccer" video sequence

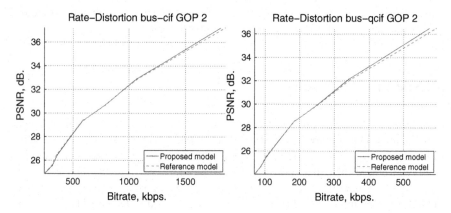

**Fig. 3**  PSNR-bitrate for "bus" video sequence

In a first set of experiments test video sequences were encoded using implemented framework with reference channel model [7]. For a second experiments set reference model of virtual channel was replaced with proposed channel model. In general, proposed model is compatible with arbitrary distribution but due to DISCOVER specifics only Laplace Mixture Model with $K = 2$ was considered.

Following video sequences were processed during the experiments: foreman, soccer, football, coastguard and bus [12]. Each video sequence was represented in two resolutions: CIF ($352 \times 288$) and QCIF ($176 \times 144$).

The BD-RATE and BD-PSNR values [13] for the processed video sequences are presented in Table 1. Rate-distortion curves are shown in Figs. 1, 2, 3, 4 and 5.

In order to analyze the experimental results let us introduce the "motion complexity" measure which reflects the speed and direction of objects movements in video sequence. The faster and more randomly objects are moving the higher is "motion

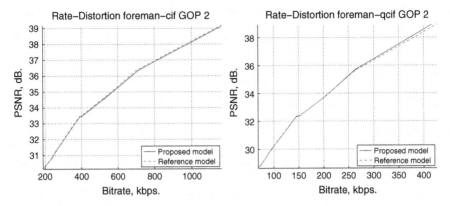

**Fig. 4** PSNR-bitrate for "foreman" video sequence

**Table 1** BD-PSNR and BD-RATE for the processed videos

| Video sequence | BD-PSNR, dB | BD-RATE, % | Motion complexity |
|---|---|---|---|
| foreman_cif | −0.058 | 1.156 | 2 |
| foreman_qcif | 0.038 | −0.631 | 2 |
| football_cif | 0.141 | −2.732 | 5 |
| football_qcif | 0.214 | −4.111 | 5 |
| soccer_cif | 0.016 | −0.282 | 4 |
| soccer_qcif | 0.167 | −3.257 | 4 |
| bus_cif | -0.004 | −0.012 | 3 |
| bus_qcif | 0.047 | −0.931 | 3 |
| coastquard_cif | −0.075 | 1.502 | 1 |
| coastquard_qcif | −0.071 | 1.42 | 1 |

complexity" measure. "Motion complexity" value for each sequence is provided in Table 1.

Experimental results show that the proposed virtual channel model has better performance for video sequences with higher motion complexity (bus, soccer, football). The explanation of this fact is following. Since the quality of inter-frame prediction is high for low "motion complexity" sequences the number of errors in virtual channel is low. The statistics of errors do not vary throughout the approximated frame and the conventional correlation noise modeling techniques demonstrate good performance. In case of complex motion the approximation errors are caused by moving objects – the statistics of such errors are different throughout the frame, though it does not change inside an area of moving object. Proposed virtual channel model considers the possible clusterization of errors therefore it shows better performance for video sequences with complex motion. Therefore it is reasonable to analyze the complexity of motion in video sequence before encoding and choose suitable CNM algorithm.

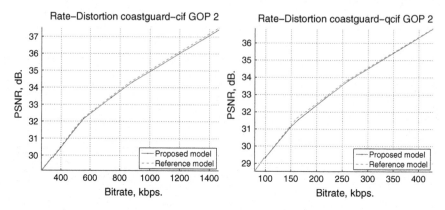

**Fig. 5** PSNR-bitrate for "coastguard" video sequence

## 4 Conclusion

In this paper the problem of non-stationary correlation noise modeling for Transform Domain Wyner-Ziv video coding was considered. The importance of non-stationarity is proven and an extended set of assumptions about noise was proposed. Then a new CNM algorithm was considered. The algorithm generalizes the state-of-the-art approach to the case of non-stationary characteristics of noise by applying noise segmentation into regions and performing per-region estimation of noise parameters. The efficiency of the algorithm is demonstrated on video sequences with complex motion which are characterized by non-stationary noise characteristics.

## References

1. Slepian, D., Wolf, J.K.: Noiseless coding of correlated information sources. IEEE Tran. Inf. Theory **19**(4), 471–480 (1973)
2. Wyner, A.D., Ziv, J.: The rate-distortion function for source coding with side information at the decoder. IEEE Tran. Inf. Theory **22**(1), 1–10 (1976)
3. Artigas, X., Ascenso, J., Dalai, M. et al.: The DISCOVER codec: architecture, techniques and evaluation. In: Picture Coding Symposium, vol. 17, pp. 1103–1120, Lisbon, Portugal (2007)
4. Aaron, A., Rane, S., Setton, E., Girod, B.: Transform-domain wyner-ziv codec for video. In: Proceedings Visual Communications and Image Processing (2004)
5. Meyer, P., Westerlaken, R., Gunnewiek, R., Lagendijk, R.: Distributed Source Coding of Video with Non-Stationary Side-Information. In: Proceedings of Visual Communications and Image Processing, pp. 857–866 (2005)
6. Veselov, A., Gilmutdinov, M.: A generalized correlation noise model for pixel domain Wyner-Ziv video coding. In: 6th International Congress on Ultra Modern Telecommunications and Control Systems (2014)
7. Brites, C., Pereira, F.: Correlation noise modeling for efficient pixel and transform domain wyner-ziv video coding. IEEE Trans. Circuits Syst. Video Technol. **18**(9), 1177–1190 (2008)

8. Ishwar, P., Prabhakaran, V., Ramchandran, K.: Towards a theory for video coding using distributed compression principles. Proc. Int. Conf. Image Process. **3**, 687–690 (2003)
9. Szeliski, R., Zabih, R., Scharstein, D., Veksler, O., Kolmogorov, V., Agarwala, A., Tappen, M., Rother, C.: A comparative study of energy minimization methods for Markov random fields with smoothness-based priors. IEEE Trans. Pattern Anal. Mach. Intell. **30**(6), 1068–1080 (2008)
10. Dragotti, P.L., Gastpar, M.: Distributed Source Coding: Theory, Algorithms and Applications. Academic Press, San Diego (2009)
11. Veselov, A., Gilmutdinov, M.: Iterative hierarchical true motion estimation for temporal frame interpolation. In: 2014 IEEE 16th International Workshop on Multimedia Signal Processing (MMSP), pp. 1–6 (Sept 2014)
12. Xiph.org video test media [derf's collection]. http://media.xiph.org/video/derf/
13. G. Bjontegaard: Calculation of average PSNR differences between RD-curves (VCEG-M33)

# Fusion of Airborne LiDAR and Digital Photography Data for Tree Crowns Segmentation and Measurement

**Margarita Favorskaya, Anastasia Tkacheva, Igor M. Danilin
and Evgeny M. Medvedev**

**Abstract** During airborne laser scanning, different types of information are available including Light Detection And Ranging (LiDAR) data as a cloud of 3D points, aerial digital photography data (hyperspectral or color visual images), and additional information about parameters of shooting. Difficulties of large image stitching due to parallax effects lead to distortions between ground truth 3D LiDAR coordinates and 2D visual coordinates of the same point. Our contribution is to develop a method for fusion of LiDAR and visual information for accurate segmentation of individual tree crowns in order to receive biomass measurements. The shearlet theory was used to improve boundaries and texture of airborne images. Also in this paper, a higher-order active contour model is applied for area evaluation of tree crowns in a plane. The received area measurements are promising and coincide with expert estimations providing accuracy 92–96 %. The modeling results are good for non-Lambert space of forest.

**Keywords** Lidar · Data fusion · Tree crown measurement · Higher-order active contour · Shearlets · Forest inventory

M. Favorskaya (✉) · A. Tkacheva · I.M. Danilin
Siberian State Aerospace University, Institute of Informatics and Telecommunications,
31 Krasnoyarsky Rabochy, 660014 Krasnoyarsk, Russian Federation
e-mail: favorskaya@sibsau.ru

A. Tkacheva
e-mail: tkacheva@sibsau.ru

I.M. Danilin
Institute of Forest Russian Academy of Sciences, Siberian Branch, Akademgorodok,
50/28, Krasnoyarsk, 660036 Krasnoyarsk, Russian Federation
e-mail: danilin@ksc.krasn.ru

E.M. Medvedev
Department of Geography and Human Environment, Tel Aviv University,
Ramat Aviv, 39040, Tel Aviv 69978, Israel
e-mail: evgeny.medvedev.1963@gmail.com

© Springer International Publishing Switzerland 2015
E. Damiani et al. (eds.), *Intelligent Interactive Multimedia Systems and Services*,
Smart Innovation, Systems and Technologies 40,
DOI 10.1007/978-3-319-19830-9_18

191

# 1 Introduction

Airborne LiDAR and aerial digital photography (in following, visual) data have
been used extensively in forest inventory from an average forest stand scale to
individual tree level [1]. The modern LiDAR technology provides the higher pulse
rates and the increased LiDAR posting densities that permits to classify points
cloud into three main classes: Earth's surface, trunks, and crowns (branches in
winter or branches and foliage in summer). In general, the task of forest modeling
includes the stages:

- 2D LiDAR and 2D visual data processing in order to segment in plane individual trees and evaluate indirectly crown biomass.
- Forest modeling on a point of view from the Earth's surface by use 3D LiDAR
  and 2D visual data as a software tool of assistance for forest appraisers [2].
- Improvement of 3D visualization of individual tree using L-system theory [3],
  space colonization algorithm [4], irregular graph interpolation [5], among others, with goal of individual trees inventory.
- Improvement of 3D visualization of landscape (forest) by use environment
  approximate models in time based on life-saving resources (water, light, min-
  erals, and location balances, also weather and climate conditions) with goal of
  forest inventory.

Each of mentioned above stages is a complicated sub-task. This paper is devoted
to first stage implementation in order to receive accurate crown measurements by
use advanced methods of computer vision. The airborne LiDAR shooting is concerned to indirect measurements, while the ground LiDAR shooting provides valid
data of direct measurements. Usually the last type of shooting is used as verification
of obtained results because it is impossible to make a hand-held laser shooting of
each tree.

The LiDAR data are captured by system capable of scanning in three directions
—forward, downward, and backward and providing 200–500 thousand
pulses/measurements per second. In spite of 30 % beams loss through a forest
scanning, the highest density of laser scanning pulses achieves 1 pulse per 5–7 cm
on the Earth's surface. Thus, the measurement accuracy of vegetation morpho-
structural elements such as trunks and crowns of trees and other ground objects is
about ±5–10 cm. A tree crown is a semi-transparent object for laser scanning that
provides two or three back signal responses. 3D LiDAR coordinates of points are
the ground truth coordinates. Using a density of the LiDAR points, one can judge
about type of known object. Therefore, the LiDAR data help to identify, not recognize natural or man-caused object. A recognition task is solved using airborne
digital visual data, which can be multispectral monochrome/color or as digital color
stereo photographs stitching as a single aerial large-sizes image.

The rest of the paper is organized as follows. Section 2 reviews related works.
The details of fusion LiDAR and visual data are presented in Sect. 3, while Sect. 4
describes the proposed method of accurate segmentation of tree crowns based on

preliminary shearlet transform and high-order active contours model. Some measurement results are added in Sect. 5. Finally, conclusion is proposed in Sect. 6.

## 2 Related Work

The scopes of airborne shooting are very wide, ranging from GIS maps design and use in virtual worlds of computer games to "serious games" application – forest inventory. One can present typical task, when urban area reconstruction from dense aerial LiDAR point clouds was investigated [6]. Zhou and Neumann designed some classifying criteria based on the 2.5D characteristic of building structures (smooth surface of walls and rooftop). The criteria connected with energy terms and classified aerial LiDAR point clouds into three classes – buildings, trees, and ground. In the case of mountain, this method may fail due to a ground surface will not be flat. For forest inventory, the classification of aerial LiDAR point clouds is also necessary. This procedure is based on two evident assumptions. First, all LiDAR responses can be separated in height at the boundary between ground and above-ground objects. Second, a removal procedure of noisy responses using the local characteristics of the surface (tree crown, building, or ground) is executed.

Consider state-of-the-art in forest inventory. In literature, two main approaches of forest inventory are based on laser-scanning data prevail. They are called Area-Based Approach (ABA) and Individual Tree Detection (IDT). In the ABA, regression, nearest neighbor, and random forests models are used for forest inventory [7]. Thus, the laser height metrics for basal area-weighted mean diameter (Dg), height (Hg), trunk number, basal area (BA), and mean volume (V) are used in the parametric regression model while in the k-Most Similar Neighbor (k-MSN) only V parameter is estimated. The IDT approach has been widely studied in scientific literature but related to the practical use it meets difficulties with accuracy detection of tree and more expansive modeling [8]. The main advantage of the ITD is a receiving of the true trunk distribution series. In most ITD methods, a maximum in the Canopy Height Model (CHM) is used for trees segmentation and crown edge detection. As mentioned in [9], Terrestrial Laser Scanning (TLS), Airborne Laser Scanning (ALS), and Vehicle-based Laser Scanning (VLS) can be used for biomass estimation and dimensions measurement at the IDT level. For 3D crown modeling by the TLS, Fernández-Sarría et al. [9] proposed to estimate four parameters such as the total tree height, crown base height, trunk height, and crown diameter extracted from the TLS point clouds and then to calculate a crown volume by four operators modeling in MATLAB:

- Design a global convex hull of each crown from the point cloud using Delaunay triangulation.
- Design a convex hull by slices 5 cm in height in each crown from the points belonging to slices using Delaunay triangulation.

- Volume calculation by sections for every 10 cm of height and the area of each
  section. The total volume is obtained by adding the surface of each section
  multiplied by 10 cm.
- Transformation of the point cloud into small units of volume using a grid in 3D
  space (voxel).

The received voxels indicate the internal presence or absence of points and are
represented by matrix with their coordinates. Then the total volume is obtained by
multiplying a number of voxels with points inside by a volume of voxel. The
structure of a tree crown is depicted by locations of such voxels.

Use of the IDT and the ABA advantages resulted in Individual Tree Crown
(ITC) method [10] and a semi-ITC method [11]. Lindberg et al. [10] calibrated the
ITD predicted tree lists using area-based information to obtain unbiased estimates at
area level. Breidenbach et al. [11] calculated the segmentation errors of volumes of
segments for non-individual trees.

## 3   Fusion of LiDAR and Visual Data

As mentioned in [9], the apparent volumes of the tree crowns $V$ can be estimated
from crown diameter $dc$ and crown height $hc$ by applying three simple geometric
shapes provided by Eq. 1, where $k$ is a coefficient defining a geometrical shape,
$k = 2$ for cone, $k = 3$ for paraboloid, and $k = 2 \cdot dc/hc$ for hemisphere.

$$V = k \cdot \frac{\pi \cdot dc^2 \cdot hc}{24} \tag{1}$$

Estimations using Eq. 1 are very approximate. Realistic trees possess crowns that
cannot be approximate by simple geometric shapes. Thus, the accurate crown
segmentation is critical in forest inventory. In this paper, a method for crown
segmentation in a plane based on a fusion LiDAR and visual data with following
accurate contour segmentation is developed.

First, a global point cloud ought to be classified into Regions Of Interest (ROI)
—the individual trees using visual data and utilizing the well-known methods of
texture and fractal analysis [12]. Let a point cloud of a single detected ROI be
represented by a set $ROI_j = \{(x_1, y_1, z_1, l_1), (x_2, y_2, z_2, l_2), \ldots, (x_i, y_i, z_i, l_i),$
$\ldots, (x_k, y_k, z_k, l_k)\}$, $j = 1, \ldots, N_{ROI}$. The LiDAR points are the reflections of the
laser beam at positions $(x_i, y_i, z_i)$ with intensity value of the returned signal $l_i$. (Not
all LiDAR systems provide additional effective parameter for classifying crown and
trunk points – the pulse width.) This disadvantage can be removed by application of
visual data. However, due to incorrectness of complicated stitching of aerial photographs the essential bias exists between laser and visual coordinates. For this
purpose, a procedure for accurate matching of laser and visual coordinates in local
ROI based on a genetic algorithm was developed. During several generations, a

population representing visual coordinates in the proposed center of tree crown is moved to the region with higher values of laser coordinates in the same local ROI. Let us notice that usually a forest appraiser utilizes a single highest point for a crown top detection. Our procedure uses three highest points as minimum to detect a crown top coordinates.

Then the laser points ought to be separated into ground, trunk, and crown sets in vertical direction. Ground points have a small height, not more 1 m in Digital Terrain Model (DTM). The remaining laser points are divided into trunk and crown points. The trunk detection is predictable due to typical shapes of trunks while crown includes branches and foliage. In the case of individual tree, the task is solved by dispersion calculation of laser points in $m$ layers, in which a virtual tree height is split by 0.5 m. Layers with large values of dispersion are considered to a trunk. Also a heuristic rule exists according to which 0.15 % of tree points correspond to a trunk. Such procedure can be enforced by hierarchal clustering, RANdom SAmple Consensus (RANSAC)-algorithm, and 3D reconstruction of a trunk. Difficult situation occurs, when several trees are located close each other, and the task cannot be solved without tree model corresponding to the concrete type of tree. Experiments show that the proposed procedure is well for grown trees separation, the underwood is rejected.

## 4 Accurate Segmentation of Trees Crown in Image

Boundaries of tree crowns can be approximate by geometric shapes; however, the accuracy of such approximation is not high. For segmentation of arbitrary contours, a framework of active contour models is used reasonably. Such models are based on a specific energy function moving and deforming an initial contour, which ought to be minimal, when the contour is delineating the object of interest. Different energy functions have been proposed; however, two main groups are distinguished in the active contour framework: first group represents by a parameterized curve called as snakes, when a contour generally converges towards edges or ridges in an image, and second group based on region properties such as intensity (geometric active contours) apply the energy functions, where a contour is represented using geodesic level sets. In this research, first approach was implemented.

For low resolution images, one can recommend a contour and texture improvement based on shearlet transform (Sect. 4.1). Boundaries of tree crowns are extracted by higher order active contour model (Sect. 4.2). Crown measurements in 2D plane and 3D space are discussed in Sect. 4.3.

## 4.1 Shearlet Transform

The necessity of a noise-robust edge detector leads to complicated methods based on isotropic and anisotropic Gaussian kernels, Gabor wavelets, and also contourlet, tetrolet, and shearlet transforms [13], among others. Each wavelet transform fulfills its own task. In our case, it is required to preserve edges and textures in low resolution images, when the noise suppression has become more relevant.

The shearlet transform $SH_\psi f(\cdot)$ of function $f(\cdot)$ is determined by Eq. 2, where $a$, $s$, and $t$ are the scale, shear, and translation parameters in Euclid coordinates, respectively, $t \in R^2$.

$$SH_\psi f(a, s, t) = \langle f(a, s, t), \psi_{a, s, t} \rangle \tag{2}$$

In the case of continuous 2D space, shearlets are defined by Eq. 3, where $\mathbf{M}_{a,s}$ is a transform matrix, $\psi$ is a generating function, $x$ is a coordinate

$$\psi_{a, s, t} = |\det \mathbf{M}_{a, s}|^{-\frac{1}{2}} \psi(\mathbf{M}_{a, s}^{-1} x - t), \tag{3}$$

where matrix $\mathbf{A}_s$ is an anisotropic dilation, matrix $\mathbf{B}_s$ is a shearing matrix

$$\mathbf{M}_{a, s} = \begin{pmatrix} a & \sqrt{a}\, s \\ 0 & \sqrt{a} \end{pmatrix} = \mathbf{B}_s \mathbf{A}_s \quad \mathbf{A}_s = \begin{pmatrix} a & 0 \\ 0 & \sqrt{a} \end{pmatrix} \quad \mathbf{B}_s = \begin{pmatrix} 1 & s \\ 0 & 1 \end{pmatrix}.$$

The discrete form of the shearlets is obtained by following replacements: $a = 2^{-j}$, $s = -l$, where $j, l \in Z$, and $t \in R^2$ is replaced by points $k \in Z^2$. Then Eq. 3 is transformed to discrete form (Eq. 4).

$$\psi_{j, l, k} = |\det \mathbf{A}_0|^{\frac{j}{2}} \psi(\mathbf{B}_0^l \mathbf{A}_0^j x - k) \quad \mathbf{A}_0 = \begin{pmatrix} 4 & 0 \\ 0 & 2 \end{pmatrix} \quad \mathbf{B}_0 = \begin{pmatrix} 1 & 1 \\ 0 & 1 \end{pmatrix} \tag{4}$$

As a result, the images processed by shearlet transform are more promising for crown contour extraction of individual trees.

## 4.2 Generic, Higher Order and Probabilistic Active Contour Models

Parametric active contour models (snakes) are based on a compromise balance between the smoothness parameters of contour (internal energy) and object parameters in an image (external energy). Many types of internal energy functions have been proposed depending on the application. Kass et al. were the first, who introduced contour, deformable curve, and its energies based on variational approach [14]. The conventional energy function $E_{snake}(\mathbf{r})$ includes internal $E_{int}[\mathbf{r}(\cdot)]$ and external $E_{ext}[\mathbf{r}(\cdot)]$ energy of contour. This function moves in spatial domain until reaches the minimum of functions defined by Eq. 5, where $\mathbf{r}(s):[0, 1]$ is a parameterized planar curve, $r(s) = (r_x(s), r_y(s))$, $s$ is a curve parameter.

$$E_{snake}[\mathbf{r}(\cdot)] = E_{int}[\mathbf{r}(\cdot)] + E_{ext}[\mathbf{r}(\cdot)] \tag{5}$$

Equation 5 can be rewritten by Eq. 6, where $I$ is an image, $\alpha$, $\beta$, and $\lambda$ are real positive constants ($\alpha$ and $\beta$ impose an elasticity and rigidity of the curve, respectively, and $\lambda$ means an attraction force of contour towards the object).

$$E_{snake}[\mathbf{r}(\cdot)] = \alpha \int_0^1 \left|\frac{dr(s)}{ds}\right|^2 ds + \beta \int_0^1 \left|\frac{d^2r(s)}{ds^2}\right|^2 ds - \gamma \int_0^1 |\nabla I(r(s))|\, ds \tag{6}$$

The first two terms in Eq. 6 define a smoothness of contour (internal energy) and the third term attracts a contour towards an object in an image (external energy). For a given set of constants $\alpha$, $\beta$, and $\lambda$, a curve $C$ ought to minimize function $E(\cdot)$. The image energy tends to the negative magnitude of image gradient. Thus, a snake is attracted to the regions with low image energy, i.e. edges. The optimization is ended prematurely after one-two hundred iterations.

The generic internal energy functions regularize an active contour by updating a contour point's location based on first and second spatial derivatives in neighboring points. In a discrete case, it means that two previous and two consecutive contour points are used to update the location of active contour point. The higher order energy functions generalize this concept, when all contour points within a predefined distance are applied to update a contour point. Rochery et al. proposed a higher order energy function under assumption that contour points in each other's vicinity should have a similar normal to the curve [15]. This energy function has a view of Eq. 7, where $t(\cdot)$ is a normal to active contour, $h(\cdot)$ are weighs depending on the Euclidean distance between the contour points.

$$E_{int}[\mathbf{r}(\cdot)] = -\int_0^1 \int_0^1 \langle \mathbf{t}(s), \mathbf{t}(u) \rangle\, h(\|\mathbf{r}(s) - \mathbf{r}(u)\|)\, ds\, du \tag{7}$$

Both generic and higher order models regularize an active contour using local features, i.e. only a set of contour points. Another approach is based on a global feature extraction, i.e. all contour points. Among these methods, one can mention probabilistic active contour models, which can segment the object of interest in cluttered and noise background. Probabilistic method does not search optimal weighting parameters but find contour based on prior knowledge about the probability of edges. The idea of this method is to find the most likely to be the true border of an object features especially in noisy or blurred images containing multiple objects [16]. Let $H(\mathbf{r}(\cdot))$ be a hypothesis that a contour $\mathbf{r}(\cdot)$ delineates an object of interest. Equation 8 provides maximum of probability $p(\cdot)$, where $f(\cdot, \cdot)$ is a set of image features $f(x, y)$, e.g. an edge strength in a set of pixels, $\mathbf{r}*(\cdot)$ is an optimal contour.

$$\mathbf{r}^*(\cdot) = \arg \max_{\mathbf{r}(\cdot)} p(H(\mathbf{r}(\cdot))|f(\cdot, \cdot)) \tag{8}$$

Then Eq. 8 can be rewritten according the Bayes' rule in a logarithmic view:

$$\begin{aligned}
\log \mathbf{r}^*(\cdot) &= \arg \max_{\mathbf{r}(\cdot)} \log \frac{p(f(\cdot, \cdot)|H(\mathbf{r}(\cdot))) \, p(H(\mathbf{r}(\cdot)))}{p(f(\cdot, \cdot))} \\
&= \arg \max_{\mathbf{r}(\cdot)} \left( \log \frac{p(f(\cdot, \cdot)|H(\mathbf{r}(\cdot)))}{p(f(\cdot, \cdot))} + \log \frac{p(H(\mathbf{r}(\cdot)))}{p(f(\cdot, \cdot))} \right),
\end{aligned} \tag{9}$$

where internal and external probabilities are defined as follows:

$$p_{int}(\mathbf{r}(\cdot)) = \log \frac{p(H(\mathbf{r}(\cdot)))}{p(f(\cdot, \cdot))} \qquad p_{ext}(\mathbf{r}(\cdot)) = \frac{p(f(\cdot, \cdot)|H(\mathbf{r}(\cdot)))}{p(f(\cdot, \cdot))}.$$

In spite of probability nature of landscape images, a method of probabilistic active contours requires a priori information about data and their distributions that is not possible always in large amount. Therefore, a higher order active contour model is a reasonable approach.

## 4.3 Crown Measurements

The direct measurement of components such as a crown or a trunk is not a straightforward procedure because of objective circumstances, e.g. the individual tree crown segmentation in the forest image is not perfect in a reason of overlapping leaves of neighboring trees. The lower part of tree might be covered by undergrowth vegetation. Data of airborne shooting are difficult matched with data of terrestrial laser scanner directly, without tree modeling. Also a terrestrial laser shooting provides the own measurement errors. The most reasonable approach is based on Density of High Points (DHP) methods and estimations of crown distribution by layers [17].

One can estimate an area of each layer by two ways: as a sum of flat segments with vertices in the center of gravity as it is show in Fig. 1b and as a sum of "crown" pixels in a binary image of individual tree. More complex measurement occurs in 3D space because of missing data in vertical direction. The promising decision is to design 3D blobs with centers in laser points, which will fill up the inter-space between conditional layers of tree crown. The structure of such 3D blobs is determined by a type of tree and is based on preliminary observations.

## 5   Experimental Results

The airborne LiDAR survey was executed from the airplane AN-2 equipped by laser scanner RIEGL Q560 with digital airborne device IGI DigiCAM, which includes digital camera Hesselblad H39/mp and GPS receiver Novatel OEM 4/5. The planning and tracing of route was implemented by satellite images of near infrared range with resolution 50 cm per pixel.

3D visualization of laser data with an example of segmented tree crown is presented in Fig. 1. An image of tree crown is divided into 72 segments with 5° intervals. Sometimes in a single line three edge points appear; in this case only farthest point was considered. Then a total area was summarized by areas of separate segments.

a                                                          b

**Fig. 1** 3D visualization of laser data: **a** tree stand with imposed boundaries of measurement area and inventory circles with constant radius, **b** tree crown with imposed lines

One can see a part of forest aerial image improving by fusion of visual and laser data as well as segmented tree crowns in Fig. 2. The shapes of tree crowns are differed from geometric shapes such as circles or ellipses. However, such rough approximation is used de facto in most measurement procedures. Our approach helps to receive the reasonable measurements of tree crowns in a plane.

Table 1 includes estimations, which were obtained by approximation of geometric shape (circle or ellipse) and higher order active contour model relative expert segmentation of individual tree crowns presented in Fig. 2b (the order of visual projections of trees is from top to bottom and from left to right). As one can see, the area measurements of tree crowns are promising in the last case and coincide with expert estimation providing accuracy 92–96 %. The received results are good for forest airborne modeling because the forest is a non-Lambert space, for which the conventional methods of visual processing are difficult to apply.

a                                        b

**Fig. 2** Visual forest image: **a** part of forest aerial image, **b** set of segmented tree crowns

**Table 1** Accuracy of area measurements of tree crowns

| Crown approximation | Tree 1 | Tree 2 | Tree 3 | Tree 4 | Tree 5 | Tree 6 | Tree 7 | Tree 8 | Tree 9 | Tree 10 |
|---|---|---|---|---|---|---|---|---|---|---|
| By circle or ellipse, % | 82.67 | 75.38 | 81.95 | 78.09 | 77.89 | 78.86 | 73.08 | 80.98 | 76.31 | 83.67 |
| By active contour, % | 91.93 | 92.94 | 95.61 | 93.52 | 96.05 | 94.39 | 93.42 | 95.83 | 96.20 | 94.02 |

# 6 Conclusion

The task of forest modeling based on a fusion of LiDAR and visual data is non-solved task because of natural variability and complicated conditions of airborne shooting. In this paper, all stages of forest modeling are mentioned, and the stage devoting to accurate segmentation and crown measurement is developed. First, a procedure for fusion of LiDAR and visual data was proposed. Second, method of accurate segmentation of tree crowns based on shearlet transform and high-order active contours model was represented. Finally, area measurements of tree crowns were obtained, which have coincided well with expert estimation providing accuracy 92–96 %.

# References

1. Roberts, S.D., Dean, T.J., Evans, D.L., McCombs, J.W., Harrington, R.L., Glass, P.A.: Estimating individual tree leaf area in loblolly pine plantations using LiDAR-derived measurements of height and crown dimensions. For. Ecol. Manage. **213**(1–3), 54–70 (2005)
2. Favorskaya, M.N., Zotin, A.G., Danilin, I.M., Smolentcseva, S.N.: Realistic 3D-modeling of forest growth with natural effect. In: Phillips-Wren, G., Jain, L.C., Nakamatsu, K., Howlett, R. J. (eds.) Advances in Intelligent Decision Technologies. SIST, vol. 4, pp. 191–199. Springer, Heidelberg (2010)
3. Lindenmayer, A.: Mathematical models for cellular interaction in development J. Theor. Biol. 18(3), 300–315 (1968)
4. Favorskaya, M., Tkacheva, A.: Rendering of wind effects in 3D landscape scenes. Procedia Comput. Sci. **22**, 1229–1238 (2013)
5. Xu, L., Mould, D.: A procedural method for irregular tree models. Comput. Graph. **36**(8), 1036–1047 (2012)
6. Zhou, Q.Y., Neumann, U.: Complete residential urban area reconstruction from dense aerial LiDAR point clouds. Graph. Models **75**(3), 118–125 (2013)
7. Junttila, V., Finley, A.O., Bradford, J.B., Kauranne, T.: Strategies for minimizing sample size for use in airborne LiDAR-based forest inventory. For. Ecol. Manage. **292**, 75–85 (2013)
8. Yu, X., Hyyppä, J., Vastaranta, M., Holopainen, M.: Predicting individual tree attributes from airborne laser point clouds based on random forest technique. ISPRS J. Photogrammetry Remote Sens. **66**(1), 28–37 (2011)
9. Fernández-Sarría, A., Velázquez-Martí, B., Sajdak, M., Martínez, L., Estornell, J.: Residual biomass calculation from individual tree architecture using terrestrial laser scanner and ground-level measurements. Comput. Electron. Agric. **93**, 90–97 (2013)
10. Lindberg, E., Holmgren, J., Olofsson, K., Wallerman, J., Olsson, H.: Estimation of tree lists from airborne laser scanning by combining single-tree and area-based methods. Int. J. Remote Sens. **31**(5), 1175–1192 (2010)
11. Breidenbach, J., Næsset, E., Lien, V., Gobakken, T., Solberg, S.: Prediction of species specific forest inventory attributes using a nonparametric semi-individual tree crown approach based on fused airborne laser scanning and multispectral data. Remote Sens. Environ. **114**(4), 911–924 (2010)
12. Favorskaya, M., Zotin, A., Chunina, A.: Procedural modeling of broad-leaved trees under weather conditions in 3D virtual reality. In: Tsihrintzis, G.A., Virvou, M., Jain, L.C., Howlett, R.J. (eds.) Intelligent Interactive Multimedia Systems and Services, vol. 11, pp. 51–59. Springer, Heidelberg (2011)
13. Lim, W.Q.: The discrete shearlet transform: a new directional transform and compactly supported shearlet frames. IEEE Trans. Image Process. **19**(5), 1166–1180 (2010)
14. Kass, M., Witkin, A., Terzopoulos, D.: Snakes: Active contour models. Int. J. Comput. Vision **1**, 321–331 (1988)
15. Rochery, M., Jermyn, I.H., Zerubia, J.: Higher order active contours. Int. J. Comput. Vision **69**(1), 27–42 (2006)
16. De Vylder, J., Ochoa, D., Philips, W., Chaerle, L., Van Der Straeten, D.: Leaf segmentation and tracking using probabilistic parametric active contours. In: Gagalowicz, A., Philips, W. (eds.) Mirage, LNCS, vol. 6930, pp. 75–85. Springer-Verlag, Berlin Heidelberg (2011)
17. Jakubowski, M.K., Guo, Q., Kelly, M.: Tradeoffs between lidar pulse density and forest measurement accuracy. Remote Sens. Environ. **130**, 245–253 (2013)

# Digital Gray-Scale Watermarking Based on Biometrics

**Margarita Favorskaya and Eugenia Oreshkina**

**Abstract** In this paper, a technical solution of digital gray-scale watermarking for law enforcement and copyright protection for digital media is developed. The biometrically generated keys provide higher level of security, when a key is created using biometrical information, in our case—the fingerprint images. First, feature points and, second, fast binary descriptors are applied for secret keys creation. Also a map of feature points is used to randomize the embedded watermark. A blind scheme of gray-scale watermark transform is implemented, when the secret keys ought to be transferred through a secret channel without necessity to transfer whole host image. Two types of transforms such as Discrete Cosine Transform (DCT) and Discrete Wavelet Transform (DWT) are investigated for embedding of watermark in a host image. Numerical estimations of attacked image restoration show a superiority of the DWT for geometric attacks including shifts and rotations.

**Keywords** Digital watermarking · Biometrics · Discrete wavelet transform · Discrete cosine transform · Biometrically generated key

## 1 Introduction

Digital watermarking is a suitable solution for copyright protection and authentication of media resources. The earliest works can be concerned to 1990–2000. Since that time, a digital watermarking develops from simple and evident encryption methods in the spatial domain to complicated ones with multi-level protection in the frequency domain. Methods from both domains have developed

M. Favorskaya (✉) · E. Oreshkina
Institute of Informatics and Telecommunications, Siberian State Aerospace University,
31 Krasnoyarsky Rabochy, 660014 Krasnoyarsk, Russian Federation
e-mail: favorskaya@sibsau.ru

E. Oreshkina
e-mail: oreshkinaei@gmail.com

© Springer International Publishing Switzerland 2015
E. Damiani et al. (eds.), *Intelligent Interactive Multimedia Systems and Services*,
Smart Innovation, Systems and Technologies 40,
DOI 10.1007/978-3-319-19830-9_19

203

persistency. Thus, the basic Least Significant Bits (LSB) method was transformed in LSB matching and LSB matching revised methods. The chaos-based cryptographic algorithms are concerned to spatial methods and have some advantages in comparison the pixel-based methods providing higher security and computational speed.

One of main problems in digital watermarking techniques is to design the schemes, which cannot fail under ambiguity attacks. In terms of robustness, the watermarking approaches can be classified as robust watermarking approach or fragile one. The robust image watermarking systems are able to extract the watermark from geometrically modified image. At the same time, a fragile image watermarking system ought to be sensitive for tampering of digital image only. Most of watermarking systems are oriented on the robust watermarking. If a host image does not used in watermark extraction process, then a watermarking scheme is called blind, otherwise, it is called non-blind watermarking scheme. The digital watermarking ought to be invisible (except the special cases of transparent watermarks) and robust to the main algorithms of signal processing. The attacks of digital watermarks can be categorized as mentioned below:

- Attacks against embedding message are directed on removal or distortion of digital watermarks by host image manipulations. Such methods do not try to detect or extract a watermark.
- Attacks against stego-detector make hard or impossible the right work of detector. A watermark is remained in a host image. However, a possibility of its extraction is lost. The affine transformations are concerned to such type of attacks.
- Attacks against protocols transmitting of digital watermarked image are connected with creation of forgery watermarks, forgery stego-containers, inversed watermarks, additional watermarks, etc.
- Attacks against watermark are used to extract a watermark from stego-message without distortions of stego-container.

The development of watermarking techniques is connected with all steps of embedding and extracting processes. Our contribution is to create biometrically generated secrete keys based on unique location of feature points in fingerprint images. Then the changed watermark is embedded in a host image by the DCT or the DWT. Such technique is available for embedding of gray-scale watermarks or textual messages with a high-level security.

The rest of this paper is organized as follows. Section 2 introduces a short literature review of existing watermarking techniques. A procedure of biometrically generated keys is described in Sect. 3. The watermarking scheme is discussed in Sect. 4. Numerical results are situated in Sect. 5. The paper is finalized by conclusion (Sect. 6).

## 2   Literature Review

Three criteria such as robustness to attacks, quality of host image with embedded watermark, and capacity of embedded watermark have influence on final result. The strengthening of one criterion leads to a weakening of two others. Many methods use the robustness as the main criterion and are based on various frequency transforms. The most popular methods from frequency domain are built on Fourier transform [1], cosine transform [2], wavelet transform [3], fractional Fourier transform [4], dual-tree complex wavelet transform [5], etc. Methods from frequency domain are classified in two categories: block-based and image-based. The DFT, the DCT, and the Singular Value Decomposition (SVD) are concerned to block-based methods while the DWT modifications transform a whole image.

The implementation of Fourier domain method based on the DCT was proposed by Cox et al. in [1]. By opinion Cox et al. each coefficient in any frequency domain has a perceptual capacity, i.e. an ability of information adding without or with minimal impact to a perceptual fidelity of signal. The extraction is based on the inverse transform and the knowledge of the original signal. It is better to hide a spread spectrum watermark in higher energy ranges in order to receive low visual distortion. The block-based methods consider the spatial correlation inside only a single 2D pixel block and do not consider the correlation between pixels from the neighboring blocks. This leads to undesirable blocking artifacts at the boundaries in the reconstructed images. Also some transforms, the DCT in particularly, cannot adapt to source data, and sharp transitions are the usual artifacts in such cases.

The transformation of whole image as a unit is relevant to human perception and provides higher flexibility. Wavelet function can be chosen freely in localized frequency domain. The DWT does not require to divide an initial image in non-overlapping 2D blocks that avoids blocking artifacts and provides higher compression ratios. A blind watermarking scheme based on multi-level DWT was proposed by Wang et al. [6], when a watermark was added to the middle frequency sub-bands of a host image in order to minimize a value of perceptual error and provide a high robustness against filtering. First, real numbered watermark was weighted by a suitable coefficient. Second, the selected sub-bands were processed by the weighted real valued watermark during extraction. Among improved modifications of the DWT, one can point a wavelet tree-based blind watermarking scheme for copyright protection [7], when each watermark bit is embedded by quantizing a single wavelet coefficient out of a set of coefficients corresponding to a particular spatial region. The use of so called wavelet trees, grouping the coefficients with mutual differences, is a usual practice with the main benefit of the robustness against the popular filtering operation and low quality JPEG/JPEG2000 compression.

Barni et al. [8] proposed a modified wavelet-based watermarking method, which is based on the improved Pixel-Wise (PW) watermarking scheme called as a Selective PW (SPW). The proposed algorithm assigned specific locations to be candidates for holding watermark bits that improved the watermarked image quality and the adaptive size of a watermark. Later the PW and the SPW methods were extended for color images implementation. In recent years, the hybrid transform approaches have been developed, for example, based on the DWT and the SVD as a factorization and approximation technique from linear algebra in order to reduce any complex matrix into smaller and invertible matrices [9].

Some interesting approaches to generate a robust secrete key one can find in chaos-based algorithms. A digital image encryption algorithm based on a composition of two chaotic logistic maps was proposed in [10]. A large external secret key for image encryption, comprising of two chaotic logistic maps, destructs the relationship between the cipher and the original image. During encryption phase, the pixels are encrypted iteratively using the cipher module for mixing the current encryption parameters with previously encrypted information. A hybrid model for image encryption using cellular automata and chaotic signal is developed in [11]. An 8-bits mask was used in order to change the gray level pixels of an image. Each bit value of the mask is selected from one of the 256 standard rules of cellular automata. The standard rules are determined by a chaotic signal.

## 3 Biometrically Generated Key

The known techniques for key generation are based on public and secret keys transmission through non-secret and secret channels, respectively. The idea of biometrically generated key attracts by two reasons: a unique key creation without random or pseudo-random numeric generation based on Bayesian or other rules and an authentication of person by own biometric information. The watermarking technique uses two types of keys called as initial seed keys and a transform key. Values of initial seed keys $KS_1$ and $KS_2$ are extracted from biometric images. A transform key $KT$ as a set of binary values is formed by the DCT or the DWT embedding procedures.

For current task, one can recommend the famous technique for detection of feature points, especially, a detection of binary feature points such as Fast REtinA Keypoint (FREAK), Binary Robust Independent Elementary Features (BRIEF), Oriented fast and Rotated BRIEF (ORB), and Binary Robust Invariant Scalable Keypoints (BRISK).Their short description is situated in Table 1.

In current research, the binary fingerprint images are used, thus our choice is connected with fast BRIEF and BRISK descriptors. The BRIEF descriptor classifies the image patches on the basis of a relatively small number of pair-wise comparisons [12]. The test $\tau$ on patches $p(x)$ and $p(y)$ with sizes $S \times S$ elements is defined by Eq. 1, where $p(\cdot)$ is a pixel intensity of a smoothed patch $p(\cdot)$ in surrounding.

$$\tau(p; x, y) = \begin{cases} 1 & \text{if } p(x) < p(y) \\ 0 & \text{otherwise} \end{cases} \tag{1}$$

**Table 1** Feature point detectors

| Descriptor | Short description |
|---|---|
| FREAK | The FREAK descriptor is based on a cascade of binary strings and is computed by efficiently comparing image intensities over a retinal sampling pattern |
| BRIEF | The BRIEF descriptor supports the conception that whole computation can be shortcut by directly computing binary strings from image patches |
| ORB | The ORB descriptor is a fast robust local feature detector for object recognition or 3D reconstruction, which is based on the BRIEF and Features from Accelerated Segment Test (FAST) keypoint detectors |
| BRISK | The BRISK descriptor uses a circular pattern, where points are equally spaced on circles concentric similar to DAISY (efficient dense descriptor applied to wide-baseline stereo) |

A set of $n_d$ $(x, y)$ location pairs defines uniquely a set of binary tests. As a result the BRIEF descriptor is a $n_d$ dimensional bit-string (Eq. 2), where $n_d = \{128, 256, 512\}$.

$$f_{n_d}(p) = \sum_{1 \le i \le n_d} 2^{i-1} \tau(p; x, y) \qquad (2)$$

The BRISK sampling patterns are circles [13]. For rotation and scale invariance, the BRISK applies the sampling patterns rotated by angle $\alpha = \arctan(g_y, g_x)$, where $g_y$ and $g_x$ are gradient characteristics of long-distance pairs of patches around a keypoint $k$. A bit-vector descriptor $d_k$ is assembled by performing all short distance intensity comparisons of point pairs $\left(p_i^\alpha, p_j^\alpha\right) \in S$, and each bit $b$ corresponds to Eq. 3, where $I(p_i^\alpha, \sigma_i)$ and $I(p_j^\alpha, \sigma_j)$ are the smoothed intensity values at the points, $\sigma_i$, $\sigma_j$ are standard deviations.

$$b = \begin{cases} 1 & \text{if } I\left(p_j^\alpha, \sigma_j\right) > I\left(p_i^\alpha, \sigma_i\right) \\ 0 & \text{otherwise} \end{cases} \qquad \forall \left(p_i^\alpha, p_j^\alpha\right) \in S \qquad (3)$$

Two BRIEF (BRISK) descriptors matching is a simple computation of Hamming distance due to their bit representation.

Before a key generation, it is required to capture two biometric images by fingerprint scanner. The procedure is based on two images of any finger. Due to possible affine transformation, these two images $I_{B1}$ and $I_{B2}$ will not be identical pixel by pixel. Based on the obtained feature points with their following matching, one can calculate values of two secret keys for a watermark embedding using the following steps:

- Apply a procedure of BRIEF (BRISK) descriptors matching in two biometric images.
- Build the robust matching point vectors $\bar{v}_1$ and $\bar{v}_2$ along the position $(X_1, Y_1)$ and $(X_2, Y_2)$ pointed manually in two images $I_{B1}$ and $I_{B2}$, respectively.

- Calculate the average errors of the elements of vectors $\bar{v}_1$ and $\bar{v}_2$ using Eq. 4.

$$k_1 = \frac{1}{l_1} \sum_{i=1}^{l_1} |\bar{v}_{1i} - \text{mean}(\bar{v}_1)| \qquad k_2 = \frac{1}{l_2} \sum_{i=1}^{l_2} |\bar{v}_{2i} - \text{mean}(\bar{v}_2)| \qquad (4)$$

- Normalize values $k_1$ and $k_2$ so that values would be in the range of middle frequency sub-band and form the initial seed keys $KS_1$ and $KS_2$.

## 4   Proposed Watermarking Scheme

The proposed scheme of embedding, extraction, and evaluation processes of gray-scale watermark is presented in Fig. 1. Before embedding, a watermark is inverse changed in positions with coordinates of feature points extracted from fingerprint image, which sizes are normalized under the watermark sizes. During extraction, pixels with such coordinates in the obtained watermark image are filled by 0 or 1 according to surrounding values. In this research, a blind scheme was implemented. Such approach does not require private watermarking scheme. Under

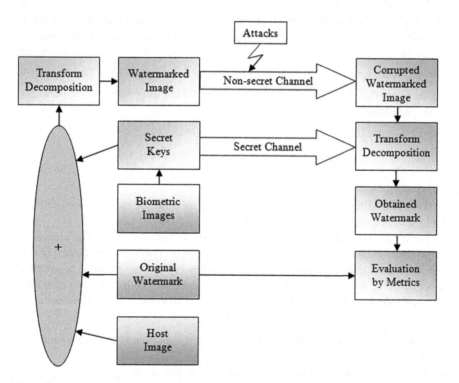

**Fig. 1** Scheme of embedding, extraction, and evaluation processes

attacks, a watermarked image can transform to corrupted watermarked image, and some digital filters can improve a visibility of corrupted watermarked image.

The transform decomposition involves a core transformation and an inverse core transformation. Two cores—the DCT (Sect. 4.1) and the DWT (Sect. 4.2) were investigated. Some evaluation metrics are situated in Sect. 4.3.

## 4.1 Discrete Cosine Transform

The DCT implementation for watermarking embedding was developed by Koch and Zhao [14]. The host image is divided in blocks $8 \times 8$ pixels. Then 2D DCT is applied to each block, as a result the pixel matrices with integer values from spatial domain are transformed in the DCT coefficient matrices with size $8 \times 8$ elements in a frequency domain. The 2D DCT has a view of Eq. 5, where $f(x, y)$ is a discrete image function with sizes $n \times m$, $(x, y)$ are the coordinates, $F(u, v)$ is an element of coefficient matrix in position $(u, v)$, $u \in [0, i - 1]$, $v \in [0, j - 1]$, $a(u)$ and $a(v)$ are coefficients with values: if $u = 0$, then $a(u) = 1/\sqrt{2}$, if $u > 0$, then $a(u) = 1$, if $v = 0$, then $a(v) = 1/\sqrt{2}$, and if $v > 0$, then $a(v) = 1$.

$$F(u,v) = \sqrt{\frac{2}{m}} a(v) \left( \sum_{y=0}^{m-1} \left( \sqrt{\frac{2}{n}} a(u) \sum_{x=0}^{n-1} f(x,y) \cos\left(\frac{\pi(2x+1)u}{2n}\right) \right) \cos\left(\frac{\pi(2y+1)v}{2m}\right) \right)$$

(5)

After embedding the bits of a watermark (digits, text, binary or gray-scale images) into frequency matrices, the inverse DCT, first, along rows and, second, along columns is implemented as this is show in Eqs. 6 and 7.

$$f(u,y) = \sqrt{\frac{2}{m}} \sum_{v=1}^{m-1} a(v) F(u,v) \cos\left[\frac{\pi(2y+1)v}{2m}\right]$$

(6)

$$f(x,y) = \sqrt{\frac{2}{n}} \sum_{u=1}^{n-1} a(u) f(u,y) \cos\left[\frac{\pi(2x+1)u}{2n}\right]$$

(7)

Our implementation is close to scheme proposed in [14]. However, values of initial seed keys $KS_1$ and $KS_2$, which are defined the DCT coefficients for embedding, are determined from biometric images.

## 4.2 Discrete Wavelet Transform

The DWT is another popular often used transformation for watermark embedding. Separable 2D wavelet transform can be performed as two 1D DWT along OX and

OY axis. 2D signal is processed by low-pass and high-pass filtering, thus an image area of low resolution LL is divided into horizontal HL, vertical LH, and diagonal HH components. A watermark is embedded into sub-bands HL, LH, and HH in order to minimize the influence on a quantity of host image.

Let $I$ be a host image with sizes $M \times N$ and $w$ be a watermark as a binary sequence by length $L$. Let coefficient $Y(i, j)$ from corresponding sub-band be a random value, dispersion of which is estimated from surrounding values. An initial choice of coefficient $Y(i, j)$ is determined by seed key $KS$. Let us assume that value of coefficient $Y(i, j)$ and values of its surrounding are random values, and all these values are placed in vector $u_{ij}$ with sizes $p \times 1$. Thus, value of coefficient is recalculated using average absolute values of surrounding and is denoted as $Z(i, j)$ in Eq. 8, where $\mathbf{d}$ is a vector of weighted coefficients.

$$Z(i, j) = d_{ij}^{\mathrm{T}} w_{ij} \tag{8}$$

Vector $\mathbf{d}$ minimizes least square error provided by Eq. 9, where $\mathbf{W}$ is a matrix with sizes $M^2 \times p$, each row of which includes elements $w_{i,j}^{\mathrm{T}}$, $\mathbf{Y}$ is a vector with sizes $M^2 \times 1$ including all coefficients of specified sub-band.

$$d_{LS} = \arg \min \sum_{i,j} \left( |Y(i,j)| - d_{ij}^{\mathrm{T}} w_{ij} \right)^2 = \left( \mathbf{W}^{\mathrm{T}} \mathbf{W} \right)^{-1} \mathbf{W}^{\mathrm{T}} |\mathbf{Y}| \tag{9}$$

Dispersion of coefficient $Y(i, j)$ is estimated using neighboring coefficients relatively surrounding $Z(i, j)$. For this purpose, $Q$ top and $Q$ bottom coefficients relatively $Z(i_0, j_0)$ are considered, with total number $(2Q + 1)$ coefficients. Their dispersion is calculated by Eq. 10, where $B_{i_0, j_0}$ is a set coefficients $\{Y(i, j)\}$ into surrounding.

$$\hat{\sigma}_X^2(i_0, j_0) = \max \left( \frac{1}{2Q+1} \sum_{(k,\, l) \in B_{i_0, j_0}} Y(k, l)^2 - \sigma_n^2, \ 0 \right) \tag{10}$$

Threshold value $T_B(i_0, j_0)$ for $Y(i, j)$ is defined by Eq. 11, where $\sigma_n^2$ is a dispersion of coefficients $Y(i, j)$ in HH sub-band of the DWT.

$$T_B(i_0, j_0) = \frac{\sigma_n^2}{\hat{\sigma}_X^2(i_0, j_0)} \tag{11}$$

Equations 8–11 are applied for calculation of each coefficient $Y(i, j)$ improving threshold values adaptively.

## 4.3 Evaluation

For evaluation of watermark embedding methods, some metrics are used, among which are the following ones:

- Peak Signal to Noise Ratio (PSNR) is a widely used objective estimation of quantity of watermarked image defined by Eqs. 12 and 13, where Max = 255 is a maximum value of intensity, $MSE$ is a mean square error, $M$ and $N$ are sizes of host image, $x$ and $y$ are current coordinates, $\widehat{I}$ is a watermarked image, $I$ is a host image.

$$PSNR = 10\log_{10}\left(\frac{\text{Max}^2}{MSE}\right) = 20\log_{10}\left(\frac{\text{Max}}{\sqrt{MSE}}\right) \tag{12}$$

$$MSE = \frac{1}{M \cdot N}\sum_{x=1}^{M}\sum_{y=1}^{N}\left(\widehat{I}(x,y) - I(x,y)\right)^2 \tag{13}$$

- Mean Structural Similarity Index Measure (MSSIM) show a relative changing of structural information in dependence of luminance and contrast of watermarked image [15] and is calculated by Eq. 14, where empirical constants are determined as $C_1 = (0.01 \cdot MAX)^2$ and $C_2 = (0.02 \cdot MAX)$, MAX = 255.

$$MSSIM\left(\widehat{I}, I\right) = \frac{(2\mu_{\widehat{I}}\mu_I + C_1)\,(2\sigma_{\widehat{I}I} + C_2)}{(\mu_{\widehat{I}}^2 + \mu_I^2 + C_1)\,(\sigma_{\widehat{I}}^2 + \sigma_I^2 + C_2)} \tag{14}$$

- Normalized Cross Correlation (NCC) defines a similarity measure of extracted watermark image relatively original watermark image.
- Visual Signal to Noise Ratio (VSNR) is interpreted as a measure of visual distortions in dependence of luminance and viewing angle [16].

During experiments, two first metrics were applied as the main ones.

## 5 Experimental Results

Software tool "Watermarking", v. 1.02 has been designed in Visual Studio 2008 environment and includes a set of program modules executing the LSB, the DCT, and the DWT. Also the software tool contains Interface Module, Evaluation Module, Journal Module, and the required libraries.

Tables 2 and 3 present the received estimations of watermark extraction without and with geometric attacks. Shift and rotation distortions were tested by PSNR and MSSIM metrics. The DCT under rotation distortions did not provide good results, thus the DCT was used only for estimation of shift distortions.

**Table 2** PSNR estimations (dB)

| Image | Without any distortion | Shift, pixels (DCT) | | | Rotation, ° (DWT) | | |
|---|---|---|---|---|---|---|---|
| | | 3 | 5 | 8 | ±5 | ±10 | ±15 |
| Without biometric keys | | | | | | | |
| Baboon.bmp | 42.044 | 28.311 | 27.035 | 22.513 | 31.478 | 27.605 | 24.030 |
| Boat.bmp | 42.192 | 29.004 | 25.997 | 21.641 | 31.940 | 26.098 | 24.231 |
| Lena.bmp | 39.518 | 28.492 | 28.160 | 21.448 | 29.655 | 26.077 | 21.729 |
| Peppers.bmp | 41.115 | 25.761 | 24.892 | 20.093 | 27.148 | 25.888 | 20.919 |
| With biometric keys | | | | | | | |
| Baboon.bmp | 44.564 | 31.825 | 30.080 | 24.707 | 33.367 | 29.054 | 25.314 |
| Boat.bmp | 43.991 | 30.989 | 28.503 | 22.009 | 32.601 | 27.815 | 24.992 |
| Lena.bmp | 42.207 | 28.882 | 27.412 | 22.758 | 28.985 | 26.900 | 22.045 |
| Peppers.bmp | 43.238 | 27.377 | 25.897 | 20.344 | 30.727 | 27.281 | 21.656 |

**Table 3** MSSIM estimations (%)

| Image | Without any distortion | Shift, pixels (DCT) | | | Rotation, ° (DWT) | | |
|---|---|---|---|---|---|---|---|
| | | 3 | 5 | 8 | ±5 | ±10 | ±15 |
| Without biometric keys | | | | | | | |
| Baboon.bmp | 0.995 | 0.965 | 0.929 | 0.903 | 0.972 | 0.933 | 0.902 |
| Boat.bmp | 0.995 | 0.967 | 0.924 | 0.887 | 0.971 | 0.941 | 0.896 |
| Lena.bmp | 0.988 | 0.961 | 0.934 | 0.898 | 0.978 | 0.927 | 0.889 |
| Peppers.bmp | 0.997 | 0.970 | 0.937 | 0.921 | 0.980 | 0.939 | 0.910 |
| With biometric keys | | | | | | | |
| Baboon.bmp | 0.998 | 0.969 | 0.933 | 0.918 | 0.978 | 0.950 | 0.899 |
| Boat.bmp | 0.997 | 0.972 | 0.929 | 0.898 | 0.983 | 0.948 | 0.899 |
| Lena.bmp | 0.992 | 0.960 | 0.938 | 0.910 | 0.981 | 0.944 | 0.891 |
| Peppers.bmp | 0.998 | 0.979 | 0.946 | 0.904 | 0.990 | 0.959 | 0.903 |

The obtained results show a non-linear dependence from increased shift and rotation parameters. As one can see, any affine distortion leads to decreased results of a watermark extraction. The use of biometrics cannot provide high robust against geometric attacks (up to 0.5–2.5 dB by the PSNR metric evaluation). However, the defense against stego-container attacks is promising. The MSSIM metric provides enough close estimation results because the structural changes in host and water-marked images are non-significant. Therefore, the PSNR metric is preferable.

# 6 Conclusion

In this research, the watermarking methods from frequency domain were considered and implemented in software tool "Watermarking", v. 1.02. A blind scheme of gray-scale watermark transform is realized with two cores—the DCT and the DWT. The use of DWT shows better results in the case of geometric attacks. Numerical results indicated the increase of PSNR estimations up to 0.5–2.5 dB during geometric attacks, when biometrics is used. For evaluation, the PSNR metric is preferable against the MSSIM metric because the structural changes in host and watermarked images are non-significant. In future, the influence of biometrically generated keys against stego-detector and stego-container attacks will be investigated in details.

# References

1. Cox, I., Kilian, J., Leighton, F., Shamoon, T.: Secure spread spectrum watermarking for multimedia. IEEE Trans. Image Proc. **6**(12), 1673–1687 (1997)
2. Barni, M., Bartiloni, F., Cappellini, V., Piva, A.: A DCT domain system for robust image watermarking. Sig. Process. **66**(3), 357–372 (1998)
3. Dawei, Z., Guanrong, C., Wenbo, L.: A chaos-based robust wavelet-domain watermarking algorithm. Chaos, Solitons Fractals **22**(1), 47–54 (2004)
4. Delong, C.: Dual digital watermarking algorithm for image based on fractional Fourier transform. In: 2nd Pacific-Asia Conference on Web Mining and Web-Based Application, pp. 51–54. IEEE Press, Wuhan (2009)
5. Lee, J.J., Kim, W., Lee, N.Y., Ki, G.Y.: A new incremental watermarking based on dual-tree complex wavelet transform. J. Supercomputing **33**, 133–140 (2005)
6. Wang, Y., Doherty, J.F., Dyck, R.E.: A wavelet-based watermarking algorithm for ownership verification of digital images. IEEE Trans. Image Proc. **11**(2), 77–88 (2002)
7. Wang, S.H., Lin, Y.P.: Wavelet tree quantization for copyright protection watermarking. IEEE Trans. Image Proc. **13**(2), 154–165 (2004)
8. Barni, M., Bartolini, F., Pive, A.: Improved wavelet-based watermarking through pixel-wise masking. IEEE Trans. Image Proc. **5**(10), 783–791 (2001)
9. Bhatnagar, G., Raman, B.: A new reference watermarking scheme based on DWT–SVD. Comput. Stand. Interfaces **31**(5), 1002–1013 (2009)
10. Ismail, I., Amin, M., Diab, H.: A digital image encryption algorithm based a composition of two chaotic logistic maps. Int. J. Netw. Secur. **11**(1), 1–10 (2010)
11. Fateri, S., Enayatifar, R.: A new method for image encryption via standard rules of CA and logistic map function. Int. J. Phys. Sci. **6**(12), 2921–2926 (2011)
12. Calonder, M., Lepetit, V., Strecha, C., Fua, P.: BRIEF: binary robust independent elementary features. In: Daniilidis, K., Marago, P., Paragios, N. (eds.) Computer Vision—ECCV 2010, Part IV, LNCS, vol. 6314, pp. 778–792. Springer, Heidelberg (2010)
13. Leutenegger, S., Chli, M., Siegwart, R.Y.: BRISK: binary robust invariant scalable keypoints. In: IEEE International Conference on Computer Vision, pp. 2548–2555. IEEE Press, Barcelona (2011)

14. Koch, E., Zhao, J.: Towards robust and hidden image copyright labeling. In: IEEE Workshop on Nonlinear Signal and Image Processing. pp. 123–132. IEEE Press, Halkidiki (1995)
15. Wang, Z., Lu, L., Bovik, A.C.: Video quality assessment based on structural distortion measurement. Sig. Process. Image Commun. **19**(2), 121–132 (2004)
16. Chandler, D.M., Hemami, S.S.: VSNR: a wavelet-based visual signal-to-noise ratio for natural images. IEEE Trans. Image Proc. **16**(9), 2284–2298 (2007)

# New Non-intrusive Speech Quality Assessment Algorithm for Wireless Networks

Akmal Akmalkhodzhaev and Alexander Kozlov

**Abstract** In the article a new non-intrusive speech quality assessment algorithm for packet switched wireless systems is proposed. The new method is based on PESQ and E-model algorithms and provides quite good accuracy. To estimate performance of the algorithm a model of the 3GPP LTE system is considered. Simulations show that correlation of the results received from the proposed algorithm and PESQ for anchor modes of AMR-NB and AMR-WB speech codecs is 97 and 94 %, respectively.

**Keywords** PESQ · E-model · AMR-NB · AMR-WB · Non-intrusive speech quality assessment

## 1 Introduction

The methods of speech quality assessment can be divided into two main groups: objective and subjective. Subjective methods are based on comparison of the original speech signal and the degraded one. Analysis is done by the listeners, who listen to both signals and make decision on the quality using some defined scale. Usually ACR (Absolute Category Rating) scale is used that takes values between 1 and 5, where 5 stand for excellent quality, 4 - good, 3 - fair, 2 - poor and 1 - bad. The average value of the listeners decision is the required rate, that is called MOS (Mean Option Score). But subjective methods are time-consuming, costly and can not be used for real-time applications. That's why objective methods were developed and currently are actively investigated.

Objective methods use analytical tools based on formulas and algorithms, that try to rate speech quality using some physical measurements and can be used in practical applications. Two types of objective methods are considered: intrusive and

A. Akmalkhodzhaev (✉) · A. Kozlov
Saint-Petersburg State University of Aerospace Instrumentations,
Bol'shaja Morskaja St. 67, Saint-Petersburg, Russian
e-mail: akmal.ilh@gmail.com

© Springer International Publishing Switzerland 2015
E. Damiani et al. (eds.), *Intelligent Interactive Multimedia Systems and Services*,
Smart Innovation, Systems and Technologies 40,
DOI 10.1007/978-3-319-19830-9_20

215

non-intrusive. Schematic representation of both methods is shown in Fig. 1. Intrusive methods for quality rating require both original (reference) and degraded speech signals. This fact limits usage of such algorithms, for example, in wireless networks, since they can not be used on the receiver side where the original signal is not present. In this case non-intrusive algorithms are used.

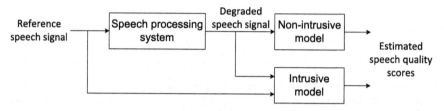

**Fig. 1** Intrusive and non-intrusive speech quality estimation models

Non-intrusive methods rate the quality using information that is available on the receiver side. One of the widely used techniques is parametric algorithms that do not require knowledge of the reference or degraded signals, but the factors that lead to the speech signal degradation. But the accuracy of non-intrusive algorithms is less than the accuracy of the intrusive ones. Thus improvement of their performance is an important and challenging task, especially for cellular systems.

## 2 Speech Quality Assessment Methods

### 2.1 PESQ

PESQ (Perceptual Evaluation of Speech Quality) is the most popular and accurate intrusive estimation method nowadays. PESQ was standardized as ITU-T Rec. P.862 in 2001 for narrowband codecs [1]. It's extension for wideband codecs was published in 2005 as ITU-T Rec. P.862.2 [2].

First PESQ passes the original and the degraded signals through the perceptual transformation model, that transforms signals into perceptually relevant domain: temporal, spectral, loudness or psychoacoustic. After that transformed signals are passed through the judgment module that scores the quality. The scores counted by PESQ lie in the range between −0.5 and 4.5, but they can be mapped to MOS values and are referred as MOS-LQO (Listening Quality Objective) values [3]. Simulations provided in [4] and [5] show that PESQ has very good correlation with subjective methods, and it is the reason why it is widely used for speech quality estimation. This fact explains why PESQ is often used as a reference algorithm for different investigations, and it can be used for accuracy estimation of non-intrusive methods.

For accuracy estimation usually two metrics are used: correlation ($r$) and RMSE (Root Mean Square Error). Correlation value of MOS-LQO and other estimates is obtained using Pearson's correlation formula [6]

$$r = \frac{\sum (x_i - \bar{x})(y_i - \bar{y})}{\sqrt{\sum (x_i - \bar{x})^2 \sum (y_i - \bar{y})^2}} \quad , \tag{1}$$

where $x_i$ and $y_i$ are MOS values obtained from PESQ algorithm and the other considered scheme. RMSE is used for prediction error estimation and is counted using following formula

$$\text{RMSE} = \sqrt{\frac{\sum_{i=1}^{n} (x_i - y_i)^2}{n}} \quad , \tag{2}$$

where $n$ is the number of experiments.

## 2.2 E-Model

One of the widely used techniques for non-intrusive estimations are parametric algorithms. ITU has standardized such computation quality measurement method in ITU-T Rec. G.107 [7] that is called E-Model. Originally it was developed as a "transmission planning tool for assessing the combined effects of variations in several transmission parameters that affect conversational quality of 3.1 kHz handset telephony" [7]. But it is also widely used for non-intrusive speech quality assessment, and many algorithms are based on E-model or on the same approach. In 2011 ITU-T Rec. G.107.1 was published that contains extension for wideband codecs [8].

E-model is based on the principle that "psychological factors on the psychological scale are additive" [9]. This means that each factor that leads to speech quality degradation can be considered independently. A score, estimated by E-model, is called R-factor and depends on 20 parameters that stand for network delay, packet loss, jitter buffer overflow, coding distortion, jitter buffer delay and echo cancellation. The expression for R-factor estimation looks as follows [7]:

$$R = R_0 - I_s - I_d - I_{e-eff} + A \; . \tag{3}$$

$R_0$ stands for basic signal-to-noise ratio, that groups room and circuit noise; $I_s$ is the combination of impairments that occur with the voice signal more or less simultaneously; factor $I_d$ represents impairments caused by delays; $I_{e-eff}$ gathers impairments caused by low bit-rate codec, it also includes impairments due to jitter and packet loss. A is an advantage factor that allows to compensate impairment factors due to user benefits from other types of access to the user. Advantage factor values for different conditions are also specified in [7]. For example for mobility by cellular networks in a building A is upper limited by value 5.

For narrowband codecs R-factor takes values from 0 to 100, while for wideband codecs ranges from 0 to 129. When all the impairment factors are supposed to take default values, then $R_0 = 93.2$ and $R_0 = 129$ for narrowband and wideband codecs, respectively. R-factor can be converted to MOS value using following expression [7]:

$$MOS = \begin{cases} 1 & R_x < 0 \\ 1 + 0.035R_x + R_x(R_x - 60)(100 - R_x) \times 7 \times 10^{-6} & 0 < R_x < 100 \\ 4.5, & R_x > 100 \end{cases},$$

(4)

where $R_x = R$ for narrowband case, and $R_x = R/1.29$ for wideband.

To count $R$ factor from MOS score ITU-T Rec. G.107 also specifies a rule [7]. But it is complex, while in [10] simplified formula is proposed:

$$R_x = 3.026MOS^3 - 25.314MOS^2 + 87.060MOS - 57.336 .$$

(5)

Often it is possible to consider simplified E-model in idealized systems, when $I_s$ and $I_d$ factors are supposed to take default values. In this case R-factor is counted as

$$R = R_0 - I_{e-eff} ,$$

(6)

where factor $I_{e-eff}$ is counted according to the following rule [7]:

$$I_{e-eff} = I_e + (95 - I_e)\frac{Ppl}{\frac{Ppl}{BurstR} + Bpl} .$$

(7)

In expression (7) $I_e$ is a codec-specific equipment impairment parameter that is defined in ITU-T Rec. G.113 [11]. $Bpl$ is also codec-specific factor that characterize its packet loss robustness, i.e. efficiency of a packet loss concealment algorithm. $I_e$ and $Bpl$ are derived using subjective MOS test results and values for several codecs are listed in ITU-T Rec. G.113 [11]. $Ppl$ is the packet loss in percent. $BurstR$ is the burst ratio. In case when packet loss is random, then $BurstR = 1$, and $BurstR > 1$ if packet loss is bursty. In our research random errors are supposed.

## 3 Speech Transmission in 3GPP LTE Standard

Consider a task of speech quality assessment in a packet switched cellular network. Information about speech quality is necessary in such systems for cell work optimization. As an example of such network 3GPP LTE system will be discussed.

For speech coding in LTE AMR-NB [12] and AMR-WB [13], narrowband and wideband codecs respectively, are used. In AMR speech signal is divided into short frames, each frame is analyzed and compressed. After that compressed frames are transmitted as independent packets. If we suppose that delay impairment factor is zero and since in LTE transmitted packets are discarded if they were not decoded correctly, it can be concluded that speech quality is degraded only due to packet losses and codec usage. Thus simplified E-model can be used for speech quality assessment.

AMR specifies not only speech coding algorithm, but also voice activity detection (VAD) and error concealment procedures [12, 13]. Error concealment algorithm restores lost packets using previously received packets. VAD mechanism detects AMR frames that do not contain speech information and compresses them much more effectively. 3GPP LTE transmission mode that uses information about VAD packets is called DTX and is mandatory.

AMR-NB and AMR-WB supports several coding rates and in this work anchor modes 12.2 kbps and 10.2 kbps for AMR-NB, and 23.85 kbps and 12.65 kbps modes for AMR-WB are considered. For different codecs value of $I_{e-eff}$ can be estimated using values provided in [11], but $I_e$ and $Bpl$ for AMR-NB are specified only for 12.2 mode. In [14] authors fill this omission and provide estimated $I_e$ values for other AMR-NB modes. Also authors suppose that codec robustness factor in different modes is the same and that $Bpl = 10$ for all modes, since AMR-NB for all modes has similar structure. For AMR-WB $I_e$ is also specified in ITU-T recommendation [11] and it is specified for all modes, but the values are provisional and $Bpl$ is not given for any mode. In [15] another values of $I_e$ with corresponding $Bpl$'s are provided and they are used in [16] for speech quality assessment. In [15, 16] $R_0 = 93.2$ for wideband codecs is used and in expression (7) constant 129 is used instead of 95. These parameters are used for E-model in our simulations.

# 4 Improved Non-intrusive Speech Quality Assessment Algorithm

Classical E-model does not take into account structure of cellular systems and this fact can be used for it's performance improvement. The network model and speech transmission scenario shown in Fig. 2 is considered. It is supposed that two users UE1 and UE2 that are involved into a speech, and speech quality should be estimated for UE2. BS1 and BS2 are base stations that receive and transmit frames from and to the users considered. As it can be seen UE1 transmits frames to BS1, and $N_{err,1}$ frame errors occur (type 1 errors). After that the frames are passes to BS2. It is supposed that this path is error free. From BS2 speech frames are transmitted to UE2 and additional $N_{err,2}$ frame errors occur (type 2 errors). Sum of $N_{err,1}$ and $N_{err,2}$ form resulting number of error frames $N_{err}$ in the received stream. It is supposed that positions of these two types of errors do not intersect.

Define as $S_{or}$ original speech signal and as $S_{syn}$ speech signal that is obtained by decoding of the AMR stream without errors, i.e. $S_{syn}$ contains quality degradation only due to the vocoder usage. Denote as $S_{err,1}$ decoded AMR stream, where only type 1 errors are taken into account. Also denote as $S_{err,2}$ decoded AMR stream, where only type 2 errors are taken into account. As it will be seen later signal $S_{err,2}$ can be generated only during simulations but not in the real system. Resulting degraded speech, that has both types of errors and that is received by UE2, will be denoted as $S_{dg}$. The goal is to estimate speech quality for this signal.

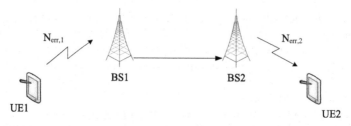

**Fig. 2** Wireless system transmission scheme

Generally E-model can be used for $S_{dg}$ quality estimation on UE2, but we propose another solution. Since in LTE HARQ technique is used and transmitted packets are ACKed and NACKed both in downlink and uplink by UEs and base stations, the following statements are correct:

1. UE1 knows $S_{or}$, the number of errors $N_{err,1}$ (from HARQ), their location and can obtain $S_{err,1}$ signal.
2. BS1 knows the number of errors $N_{err,1}$, their location and can obtain $S_{err,1}$ signal.
3. From protocol headers and HARQ BS2 knows the number of errors $N_{err,1}$, $N_{err,2}$ and their location. Also BS2 can decode signals $S_{err,1}$ and $S_{dg}$.
4. UE2 knows the number of errors $N_{err}$ and their location. $S_{dg}$ is decoded on its side.

First of all our observations show that quality of $S_{dg}$ can be estimated not only on the UE2 side, but also on BS2. Since in LTE a base station takes decision on the resources scheduled to the user and what modulation and coding scheme should be used, estimations made on the base station side can be more preferable and valuable. And since BS2 knows not only $S_{dg}$, but also $S_{err,1}$, it is obvious that more accurate estimations can be performed on its side. Thus it is proposed to make all estimations on the BS2 side.

Consider BS2. In E-model all lost packets has the same influence on the speech quality degradation. But in fact it was noticed that one lost packet can lead to significant quality degradation, while another lost packet will have nearly no influence on the quality. Thus if we can count the "amount" of degradation caused by some errors it is possible to make more accurate speech quality prediction. Suppose that estimation of the "amount" of the degradation due to type 2 errors can be done by $S_{err,1}$ and $S_{dg}$ signals comparison, because the difference of the two signals is connected with the type 2 errors. To make such estimation PESQ algorithm can be used that analyses signal $S_{err,1}$ as an original signal and $S_{dg}$ as a degraded one. It is obvious that resulting MOS score will not contain loss due to the codec usage, since two synthesized signals are analyzed. Also obtained MOS will not contain degradation due to type 1 errors.

To prove our hypothesis the following analysis was made. First of all 28 male and female American English test speech samples 8 seconds each were taken from ITU-T web site [1]. The samples were divided into two groups: 16 for investigations and 12 for accuracy estimations. For 16 mentioned samples random packet errors were modeled, where *Ppl* took values from 1 to 20 % with 1 % step. For each *Ppl* value 30 patterns of errors were generated, where half of the error positions were supposed to be type 1 errors, while the rest were supposed to be type 2 errors. As a result number of $S_{err,1}$, $S_{err,2}$ and $S_{dg}$ signals were obtained and for each pair $(S_{err,1}, S_{dg})$ and $(S_{or}, S_{err,2})$ MOS-LQO values were estimated. Correlation of the two sets of estimates was 97–98 % for AMR-NB and 87–93 % for AMR-WB, depending on the coding rate. Obtained results are shown in Figs. 3 and 4. So it was concluded that our assumption is correct and comparison of $(S_{err,1}, S_{dg})$ reflects degradation due to type 2 errors.

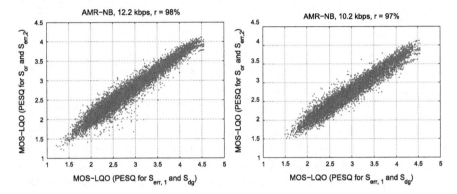

**Fig. 3** Dependency of MOS-LQO values for PESQ($S_{err,1}, S_{dg}$) and PESQ($S_{or}, S_{err,2}$) for AMR-NB modes

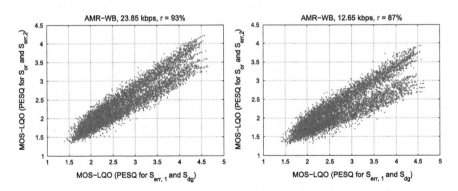

**Fig. 4** Dependency of MOS-LQO values for PESQ($S_{err,1}, S_{dg}$) and PESQ($S_{or}, S_{err,2}$) for AMR-WB modes

Thus resulting algorithm uses both E-model and PESQ algorithms for more accurate non-intrusive quality assessment. BS2 can estimate factual degradation due to type 2 errors using PESQ, while degradation due to type 1 errors will be estimated using E-model. Different heuristic approaches can be considered for combining of E-model and PESQ estimations. In this work we propose following solution. First denote as $Ppl_1 = \frac{N_{err,1}}{N} \cdot 100\%$ and $Ppl_2 = \frac{N_{err,2}}{N} \cdot 100\%$ corresponding to $N_{err,1}$ and $N_{err,2}$ error rates in percent, where $N$ is the total number of transmitted AMR frames. Then the new algorithms looks as follows. First estimate MOS that will be marked as $MOS\_LQO_{syn}$ for $S_{err,1}$ and $S_{dg}$ signals using PESQ. Since this value does not contain degradation due to codec usage it should be mapped (converted) to MOS-LQO values that will be marked as $MOS\_LQO_{or}$ and that contain average degradation value due to codec usage. Mapping rule will be shown below. After that $MOS\_LQO_{or}$ should be converted to R-factor value using expression (5). Using expressions (6) and (7) $Ppl_2$ value can be recounted depending on the influence of type 2 errors. Final expression will look like follows

$$Ppl_{new,2} = \frac{R_0 - R - I_e}{95 - R_0 + R} Bpl \ . \tag{8}$$

Thus for R-factor computation new $Ppl_{new,2}$ value should be used and original $Ppl = Ppl_1 + Ppl_2$ will be changed to $Ppl_{new} = Ppl_1 + Ppl_{new,2}$. $Ppl_1$ is not changed since there is no additional information about type 1 errors on the BS2 side. After that R-factor is estimated using E-model for $Ppl_{new}$ and corresponding MOS is supposed to be the required speech quality. The algorithm scheme is shown in Fig. 5.

To find how $MOS\_LQO_{syn}$ should be mapped to $MOS\_LQO_{or}$ following analysis was performed. Using PESQ algorithm MOS-LQO values for $(S_{or}, S_{dg})$ and $(S_{syn}, S_{dg})$ signals, and for different error rates were estimated. Difference between the sets of received MOS-LQO values is connected with the codec usage. Correlation for the sets of obtained MOS-LQO values was measured and it was 98% for AMR-NB, while for AMR-WB it was about 94%. It was concluded that $MOS\_LQO_{syn}$ and $MOS\_LQO_{or}$ has linear dependency and mapping can be performed using following formula:

$$MOS\_LQO_{or} = A * MOS\_LQO_{syn} + B \ , \tag{9}$$

where $A$ and $B$ are coefficients that differ for different AMR modes. Using obtained MOS-LQO vectors best $A$ and $B$ coefficients in the least-squares sense were found for the considered AMR modes. Obtained values are gathered in Tables 1 and 2 respectively.

**Table 1** $A$ and $B$ values for AMR-NB modes

| AMR-NB mode | $A$ | $B$ |
|---|---|---|
| 12.2 kbps | 0.8770 | 0.1738 |
| 10.2 kbps | 0.8444 | 0.2073 |

**Table 2** $A$ and $B$ values for AMR-WB modes

| AMR-WB mode | $A$ | $B$ |
|---|---|---|
| 23.85 kbps | 0.7721 | 0.2390 |
| 12.65 kbps | 0.6425 | 0.4070 |

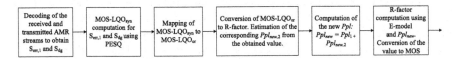

**Fig. 5** Proposed algorithm scheme

## 5 Simulation Results for 3GPP LTE System

Since performance of the proposed algorithm depends on the ratio of $N_{err,1}$ and $N_{err,2}$, several scenarios were taken into account. Three different situations were considered: $\frac{N_{err,1}}{N_{err}} = 0.75$, $\frac{N_{err,1}}{N_{err}} = 0.5$, $\frac{N_{err,1}}{N_{err}} = 0.25$. $Ppl$ values from 0 to 20 % with step 1 % were considered. For every $Ppl$ 30 error patterns were generated and simulations were performed for 12 mentioned above speech samples.

Correlation and RMSE results received for the proposed and PESQ algorithms for both codecs are gathered in Tables 3 and 4. As it can be seen proposed algorithm shows correlation and RMSE results improvement even when $\frac{N_{err,1}}{N_{err}} = 0.75$. This means that information provided by PESQ is quite useful. When $\frac{N_{err,1}}{N_{err}} = 0.25$, i.e. $N_{err,2}$ is higher than $N_{err,1}$, correlation values for AMR-NB modes increase to 97 % and RMSE becomes quite small. For AMR-WB for the case when $N_{err,2} > N_{err,1}$ correlation reaches 94–95 %. It can be seen that gain achieved for AMR-NB is higher that for AMR-WB. This can be explained by the fact that for wideband case PESQ results are much more sensitive to errors.

**Table 3** PESQ and the new algorithm MOS correlation results for AMR-NB

| $\frac{N_{err,1}}{N_{err}}$ | Classical E-model | | 75 % | | 50 % | | 25 % | |
|---|---|---|---|---|---|---|---|---|
| AMR-NB mode | $r$ | RMSE | $r$ | RMSE | $r$ | RMSE | $r$ | RMSE |
| 12.2 kbps | 0.92 | 0.35 | 0.94 | 0.28 | 0.96 | 0.23 | 0.97 | 0.18 |
| 10.2 kbps | 0.92 | 0.31 | 0.94 | 0.29 | 0.96 | 0.24 | 0.97 | 0.18 |

## 6 Conclusion

In this work a task of speech quality assessment for packet switched cellular networks was considered. It was proposed to perform speech quality assessment on the base station side, that gives a number of advantages. It was notices that during speech data transmission a base station has additional information, that can be used for improvement of the accuracy of speech quality assessment. A new algorithm that uses this additional information for E-model performance improvement was proposed. For simulations a model of 3GPP LTE system was considered and obtained results showed that for AMR-NB correlation and RMSE improvement was up to 5 % and 0.17, respectively. For AMR-WB correlation improvement was up to 4 % and RMSE improvement was up to 0.21.

**Table 4** PESQ and the new algorithm MOS correlation results for AMR-WB

| $\frac{N_{err,1}}{N_{err}}$ | Classical E-model | | 75 % | | 50 % | | 25 % | |
|---|---|---|---|---|---|---|---|---|
| AMR-WB mode | $r$ | RMSE | $r$ | RMSE | $r$ | RMSE | $r$ | RMSE |
| 23.85 kbps | 0.92 | 0.46 | 0.93 | 0.39 | 0.94 | 0.31 | 0.95 | 0.25 |
| 12.65 kbps | 0.90 | 0.41 | 0.92 | 0.39 | 0.93 | 0.34 | 0.94 | 0.27 |

Though in simulations simplified E-model was considered, nothing prevents to take into consideration other impairment factors in the proposed algorithm and obtain more common results.

## References

1. ITU-T, Recommendation P.834: Methodology for the derivation of equipment impairment factors from instrumental models
2. ITU-T Recommendation P.862.2: Methods for objective and subjective assessment of quality. Wideband extension to recommendation P.862 for the assessment of wideband telephone networks and speech codecs, Nov 2007
3. ITU-T, Recommendation P.862.1: Mapping function for transforming P.862 raw result scores to MOS-LQO. In: International Telecommunication Union - Telecommunication Standardisation Sector (ITU-T), March 2005
4. ITU-T Recommendation P.862: Perceptual evaluation of speech quality(PESQ): an objective method for end-to-end speech quality assessment of narrow-band telephone networks and speech codecs, Feb 2001
5. Rix, A.W.: Comparison between subjective listening quality and P.862 PESQ score. White Paper, Sept 2003
6. ITU-T, Recommendation P.563: Single-ended method for objective speech quality assessment in narrow-band telephony applications, May 2004

7. ITU-T, Recommendation G. 107: The E-model: a computational model for use in transmission planning, Dec 2011
8. ITU-T, Recommendation G. 107.1: Wideband E-model, Dec 2011
9. Allnatt, J.: Subjective rating and apparent magnitude. Int. J. Man Mach. Stud. **7**, 801–816 (1975)
10. Sun, L.: Speech quality prediction for voice over internet protocol networks. PhD thesis, University of Plymouth, School of Computing, Communications and Electronics Faculty of Technology (2004)
11. ITU-T, Recommendation G. 113: Transmission impairments due to speech processing, Dec 2011
12. 3GPP TS 26.071: Mandatory speech CODEC speech processing functions; AMR speech CODEC; general description
13. 3GPP TS 26.171: Speech codec speech processing functions; adaptive multi-rate - wideband (AMR-WB) speech codec; general description
14. Mertz, F., Vary, P.: Efficient voice communication in wireless packet networks. In: ITG Conference on Voice Communication (SprachKommunikation), pp. 1–4, Oct 2008
15. Moller, S., Raake, A., Kitawaki, N., Takahashi, A., Waltermann, M.: Impairment factor framework for wide-band speech codecs. In: IEEE Transactions on, Audio Speech, and Language Processing. **14**(6):1969–1976, Nov 2006
16. Lee, H.J., Cha, J.R., Kim, J.H.: Rate optimization of G.722.2 AMR-WB voice codec. In: Proceedings of the ICEIC 2013, vol. 2 (2013)

# Block-Permutation LDPC Codes for Distributed Storage Systems

**Evgenii Krouk and Andrei Ovchinnikov**

**Abstract** In the paper the usage of low-density parity-check (LDPC) codes to protect storage systems from failures is considered. These codes are the instance of locally recoverable (LRC) codes which obtain much attention during last years regarding storage systems. The system model of distributed storage system is described, with specific types of failures. The coding schemes based on Reed-Solomon (RS) and LDPC codes are formulated for this model taking into account the specific failures types. These coding schemes are compared using several examples of model parameters. The redundancy and locality provided by different coding schemes are estimated.

**Keywords** LDPC codes · Locally recoverable codes · Bursts-correcting codes · Distributed storage systems

## 1 Introduction

The development of data storage systems (DSS) leads to the creation of systems containing the array of disks rather than simple storage unit. The well-known RAID (Redundant Array of Independent Disks) is the example, but in general such storage system may be considered as distributed system, where disks may be placed far away one from others and connected with communication lines (for example, "cloud" storage systems). Such systems formulate new tasks in performance and fault tolerance.

The traditional way to improve the reliability of transmitted, processed or stored data is usage of error-correcting codes. The task of coding in data storage systems (especially in distributed systems) has specific properties comparing to classical channel coding. Among these properties are [1–4]:

E. Krouk(✉) · A. Ovchinnikov
Saint-Petersburg State University of Aerospace Instrumentation, B.Morskaya 67,
190000 Saint-petersburg, Russia
e-mail: ekrouk@vu.spb.ru

A. Ovchinnikov
e-mail: mldoc@ieee.org

© Springer International Publishing Switzerland 2015
E. Damiani et al. (eds.), *Intelligent Interactive Multimedia Systems and Services*,
Smart Innovation, Systems and Technologies 40,
DOI 10.1007/978-3-319-19830-9_21

- *Erasures/errors*. The failure of storage unit leads to erasure of correspondent data with known fact of existence and positions. However, in some cases the traditional errors also may be considered.
- *Bursts*. The data usually distributed among storage units by blocks, so the failure leads not to single erasure, but rather to block of erasures called burst.
- *Locality*. The read/write operations with storage units are usually expensive, especially in distributed systems, so the decoder should use minimum "alive" storage units to recover information stored by broken one.

These properties lead to the task of constructing new codes especially for storage systems. The ability to correct erasures and bursts is known task in coding theory, while the locality is new specific parameter. During the last years the classes of codes called Locally Recoverable (LRC) Codes were presented, targeting the minimization of locality [1, 2, 5].

We call the linear $(n, k)$-code (where $n$ is codelength and $k$ is the number of information symbols) as $(n, k, \varepsilon)$ LRC-code if any codeword symbol may be recovered using no more than $\varepsilon$ other symbols. In general, the locality satisfies $1 \leq \varepsilon \leq k$. For example, the simple repetition code has the worst redundancy, but the best locality $\varepsilon = 1$. In contrast, the Reed-Solomon (RS) code, providing the best redundancy, has the worst locality $\varepsilon = k$. So there is a trade-off between the redundancy and locality.

Since locality may be estimated for any code (like minimal distance), in wide sense the term LRC does not mean any particular codes, but rather emphasizes the task of minimization the new parameter. So, the low-density parity-check (LDPC) codes, which are considered in this paper regarding storage systems, are also subclass of LRC codes.

The paper is organized as follows. In Sect. 2 some particular model of DSS is formulated, with special failure types. In Sects. 3 and 4 the possibility and features of RS and LDPC codes application for such model are discussed. Section 5 presents numerical examples of constructing coding schemes for some model parameters. Section 6 concludes the paper.

## 2 The Model of Distributed Storage System

Consider the following model of distributed storage system (DSS), see Fig. 1. The system contains $N$ disks $\{d_1, \ldots, d_N\}$, grouped into $L$ nodes $\{D_1, \ldots, D_L\}$. So each node contains $M_d = N/L$ disks. There are three inconsistent failure events possible in the system:

1. failure of $v_d$ or less disks;
2. failure of one node;
3. silent data corruption (SDC) of $t_d$ or less disks.

In terms of coding theory the event "failure" corresponds to the erasure of codeword symbols, the existence and positions of erasures are known to decoder beforehand. "Silent data corruption" means the existence of errors, that is, wrong values of codeword symbols with unknown existence and positions, which may be revealed only through decoder's computations of parity-check relationships. We will assume that $v_d$ disks failures lead to $v$ erasures of codeword symbols (these values may not be equal, since several symbols or only part of the symbol may be written on one disk). Similar, we assume that SDC of $t_d$ disks corresponds to $t$ erroneous codeword symbols.

The error-correcting code can correct any combination of $t$ errors, if its minimal distance is $d_0 \geq 2t + 1$ [6, 7], or any combination of $v$ erasures, if $d_0 \geq v + 1$. The parameter $d_0$ shows only the theoretical possibility of correction, and the efficient practical decoding procedure is needed to implement this possibility.

The "node failure" of $M_d$ disks within one node leads to the erasure burst of length $M$, that is, the first and last erased positions are not far than $M$ positions from each other. Moreover, this burst is fixed, that means that it may occur only on the set of possible positions and not on the arbitrary position. Such fixed burst may be considered as $M$ separate erasures, and the code should possess $d_0 \geq M + 1$ to correct them.

Then the task of coding for such model may be formulated as constructing the code with minimal distance

$$d_0 \geq \max(2t + 1, v + 1, M + 1). \tag{1}$$

However, in practice the value of $v$ is small (in many systems only $t_d = 0$ and $v_d = 1$ are considered), and if disks and nodes failures probabilities are comparable, the maximum in (1) will be always defined by $M + 1$. This leads to unreasonable high requirements on $d_0$ and therefore unreasonable high redundancy.

An alternative approach is to take into account the specificity of fixed burst rather than considering it as separate erasures. Then one may try to select the minimal distance as $d_0 \geq \max(2t + 1, v + 1)$ and to use some additional specific properties of the code to correct fixed burst of length $M$. Thus the attempt to decrease the redundancy can be made.

**Fig. 1** Distributed DSS model

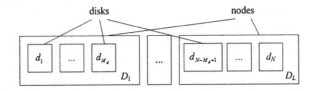

## 3 Reed-Solomon Coding Scheme

In storage systems the usage of Reed-Solomon (RS) codes over finite field $GF(q)$ is traditional [6–8]. If the codelength of RS code is $n$, the number of information symbols is $k$, then the minimal distance $d_0 = n - k + 1$. This is the maximal possible

$d_0$ for given $n$ and $k$ (Singleton bound), and from this point of view RS codes are optimal. Besides, these codes can be viewed as burst-correcting codes, since one symbol of $GF(q)$ may be considered as vector of some elements (usually binary), and by correcting one such element we correct the corresponding vector.

However, coding and decoding procedures should be performed in $GF(q)$, which makes them not very fast (comparing to binary arithmetic). Besides, the locality of RS codes are $k$, which is the worst possible value for $(n, k)$-code.

Consider the application of RS codes to the system model described in Sect. 2. To fulfil the fault tolerance requirements, we should select the RS code with minimal distance satisfying (1).

Straightforward usage of RS code is to write one symbol of the codeword per disk. Then the codelength $n$ is equal to the number of disks $N$, and $v = v_d$, $t = t_d$, $M = M_d$, allowing simple selection of code parameters. However, this scheme may be generalized, if we assume writing $\eta$ codeword symbols per disk (note that the value of $\eta$ may not be integer, if one symbol takes several disks).

Selection of code parameters in this case depends on dominant value in (1). Since the value of $v_d$ is small, we may assume $M_d > v_d$, and in most cases this gives $M > v$, however, some cases exist when the value of $v$ become dominant (but estimation $M_d > v_d$ still holds). Such special cases are considered in Sect. 5.

RS codes are powerful universal coding tool, but their locality is poor and they need calculations in finite fields. These disadvantages may be tried to improve by using another classes of codes by cost of redundancy.

# 4 LDPC Coding Scheme

## 4.1 Block-Permutation Low-Density Parity-Check Codes

Low-density parity-check (LDPC) codes were suggested by Gallager [9] and investigated later in many works [10–13]. During the last decade LDPC codes were intensively analysed and are widely used [14–17].

Traditionally LDPC code is defined by its parity-check matrix $H$, which is sparse, i.e. contains less non-zero elements comparing to matrix size. One of the most widely used constructions are block-permutation LDPC codes, their parity-check matrix consists of sub-blocks which are cyclic permutation matrices [17, 18]. Such structure allows obtaining of analytical estimations of different code parameters (rate, minimal distance, girth, trapping sets etc.), which are usually estimated only by computer simulation.

The simplest case of block-permutation LDPC codes are Gilbert codes suggested in [19] as burst-correcting codes. Burst-correcting capabilities of these codes are later analysed in [15, 20–24]. However, these codes may also be used to correct independent errors.

Gilbert codes are defined by their parity-check matrix $H$,

$$H = \begin{bmatrix} I_m & I_m & I_m & \cdots & I_m \\ I_m & C & C^2 & \cdots & C^{\rho-1} \end{bmatrix}, \tag{2}$$

where $I_m$ is $(m \times m)$ unity matrix, $C$ is $(m \times m)$-matrix of cyclic permutation:

$$C = \begin{bmatrix} 0 & 0 & 0 & \cdots & 0 & 1 \\ 1 & 0 & 0 & \cdots & 0 & 0 \\ 0 & 1 & 0 & \cdots & 0 & 0 \\ \cdots & \cdots & \cdots & \cdots & \cdots & \cdots \\ 0 & 0 & 0 & \cdots & 1 & 0 \end{bmatrix}, \tag{3}$$

and $\rho \leq m$. Clearly, Gilbert codes are LDPC codes with column weight $\gamma = 2$ and row weight $\rho$. The codelength is $n = m\rho$, the number of parity-check symbols (i.e. the rank of $H$) is $r = 2m - 1$ [23].

Simple generalization of these codes may be obtained by adding third "stripe" of blocks to parity-check matrix (2):

$$H = \begin{bmatrix} I_m & I_m & I_m & \cdots & I_m \\ C^0 & C^1 & C^2 & \cdots & C^{\rho-1} \\ C^{i_0} & C^{i_1} & C^{i_2} & \cdots & C^{i_{\rho-1}} \end{bmatrix}, \tag{4}$$

where the degree of matrix $C$ (3) in $j$-th block of third stripe is defined by the integer $i_j \in \{0, \ldots, m-1\}$ for $j = \overline{0, \rho - 1}$. The number of redundant symbols is $r = 3m - 2$ [15]. Extra stripes may be added to the parity-check matrix, but in this paper we consider only $\gamma \leq 3$.

Note that block-permutation LDPC codes with row weight $\rho$ are LRC codes with locality $\varepsilon = \rho - 1$.

There are many decoding algorithms of LDPC codes [15, 17, 25], which are very computationally efficient. Besides, in storage systems the hard-decision decoders should be used, which leads to even more efficiency through usage of only binary operations.

## 4.2 Block-Permutation LDPC Codes for DSS

Consider application of LDPC codes from Sect. 4.1 for the model described in Sect. 2.

We start from $\gamma = 2$, i.e. LDPC codes with matrix $H$ from (2). Then the code is defined by parameters $m$ and $\rho$, where $\rho \leq m$. Unlike RS codes for LDPC codes in general there are no either good estimations of error-correcting parameters (for example, minimal distance), or constructions certainly providing required parameters. Evaluations of LDPC codes are usually made either by computer simulation or using analysis of specific structure for given constructions.

In general we note that the code is able to correct some combination of $v$ erasures, if the matrix formed by columns of $H$ correspondent to erased positions has rank $v$. As it was noted in Sect. 2, the event "node failure" leads to fixed burst of erasures. Usually better LDPC codes have larger lengths, and for considered model the code-length $n$ should be no less than the number of disks $N$. If $n \geq N$, then $b \geq 1$ bits are written to one disk, and the failure of this disk also leads to fixed erasure burst, while the SDC leads to fixed error burst.

For convenience we will assume that one node contains symbols correspondent to one or several block-columns of $H$ (that is $m$ symbols, $2m$ etc.). Consider the case when one block of $m$ bits is written to one node. Then the failure of the node leads to erasure burst covering all columns of one block. Each row of $H$ contains only one non-zero element within the burst, and therefore, such erasure burst can be easily corrected.

But if there are two blocks written in one node, the failure of the node cannot be corrected, because two blocks contain $2m$ columns, and rank of $H$ is only $2m - 1$. To recover from such failures consider the matrix $H$ with three stripes (4). Its rank is $3m - 2$, so the correction of $2m$ erasures is possible at least for some combinations of erasures. Consider the following procedure. We assume that there are $b$ bits written on each disk. Then the $n$-bit codeword may be represented as sequence $(x_1, \ldots, x_{n/b})$ of $b$-bit symbols. By cyclically shifting this sequence we obtain $(x_2, \ldots, x_{n/b}, x_1)$, and write the symbols on the disks in such order. Then the first node will contain $2m$ symbols $x_2, \ldots, x_{2m/b+1}$. This procedure is shown in Fig. 2.

It is easy to verify that the $2m$ columns written in the node by such procedure will be linearly independent, so the node failure may be recovered.

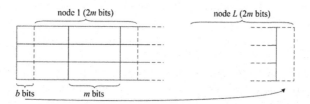

**Fig. 2** Distribution of bits by nodes in LDPC code scheme

The estimation of correction capability against disks erasures and SDC should be done separately. However, we may formulate special case (usually considered in practice).

**Theorem 1** *Code defined by the parity-check matrix (2) is able to correct two fixed erasure bursts of length $1 \leq b < m$, where $b|m$, and start positions of the bursts are divisible by $b$. Alternatively, this code is able to correct one fixed error burst with the same length and starting position.*

*Proof* As we note above, the code is able to correct two erasure bursts of length $b$, if the columns of parity-check matrix correspondent to erased positions are linearly independent.

First consider $b = 1$. In this case the statement of the theorem is clear, since there are no equal columns in $H$.

Lets now $b > 1$. Consider the first stripe of $H$, which contains unity blocks, and two erasure burst satisfying theorem's conditions. Lets the first burst start at position $i_1$, and the second at position $i_2$. Consider two cases. First, let $i_1 \bmod m \neq i_2 \bmod m$. Then each row of $H$ from the first stripe contains no more than one non-zero elements on erased positions, Fig. 3. That means that all erased positions can be recovered using these parity-check rows.

Now lets $i_1 \bmod m = i_2 \bmod m$. Then each row of the first stripe containing non-zero element within one burst also contains non-zero element within other. But since in this case burst belong to different blocks of $H$ and $b < m$, this means that there exists at least one (in fact at least two) parity-check row in the second stripe of $H$ containing only one non-zero element on erased positions. Then the erased bit correspondent to this non-zero element may be recovered. After one bit being recovered we will obtain the row in the first stripe with only one non-zero element on erased positions, this allows recovering on one more symbol, then we move again to second stripe and so on. Regarding the structure of $H$ such procedure allows reconstruction of all erased symbols. An example is shown in Fig. 4.

**Fig. 3** Correction of two fixed erasure packets by one stripe

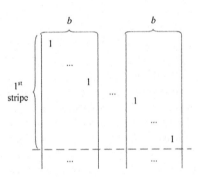

To conclude the proof we note that the conditions on correction ability for one fixed error burst are the same as for two fixed erasure bursts: absence of two bursts of length $b$ correspondent to $2b$ or less linear dependent columns. Since this was already proven, the code can correct one fixed error burst in case of erasure bursts absence. □

Clearly, the theorem's condition is applicable to block-permutation matrices with larger number of stripes.

**Fig. 4** Correction of two fixed erasure packets by two stripes

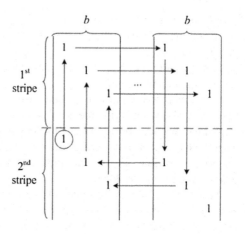

## 5 Numerical Examples

### 5.1 Estimation of RS Codes Parameters

Consider the selection of RS codes parameters for the model described in Sect. 2 regarding notes we have made in Sect. 3.

Usually the case with one disk failure is considered in storage systems. We consider more complex model, allowing one node failure, or $v_d = 2$ disks failures, or $t_d = 1$ SDC errors. We define two example sets of model parameters: $L = 8$ nodes with $M_d = 8$ disks ($N = 64$ disks in total) and $L = 12$ nodes with $M_d = 12$ disks ($N = 144$ disks in total).

If possible, we will assume that RS code is defined over $GF(2^8)$. This field operates with bytes, may be effectively implemented and is most often used in practice. However, if the codelength exceeds 255, we need to consider larger fields, which may lead to some lost in computational efficiency.

*Case 1.* Let $L = 8$, $M_d = 8$. First, we assume that one disk stores one symbol of RS codeword, i.e. $\eta = 1$. Then the code should be able to correct $v = v_d = 2$ erasures, or $t = t_d = 1$ errors, or $M_d = M = 8$ erasures. From (1) such code should have the distance $d_0 = 9$, that gives (64, 56)-code with rate $R = 0.875$. To determine the locality $\varepsilon$ we assume that we estimate it only for 1 disk failure case (which is the common understanding of this term), and we will mean not the number of symbols required for restoration of one symbol, but the number of disks required to read from. For considered case we need reading from $k = 56$ disks.

For $\eta = 2$ there are two symbols per disk, then the codelength $n = 128$, and $v = 4$, $t = 2$, $M = 16$. The distance should be $d_0 = 17$, this gives (128, 112)-code with the same rate $R = 0.875$. To correct an erasure we need 112 symbols, but we may obtain them by reading from any 56 disks, which again gives the locality $\varepsilon = 56$.

The cases $\eta = 1/2$ and $\eta = 1/4$ may be proceeded similarly. Consider the the case $\eta = 1/8$. Here 1 symbol of RS code takes 8 disks, i.e. the whole node. We may still use $GF(2^8)$, but now $v = 2$ and $M = 1$. Thus the dominant value in determining $d_0$ become the number of disks failures rather than node failure. Then $d_0 = 3$ and we have $(8, 6)$-code with rate $R = 0.75$. To recover from failure we need to read from 7 symbols (nodes), or 48 disks. So we obtain the locality decreasing by cost of increased redundancy.

Parameters of considered schemes are given in Table 1.

*Case 2.* Let now $L = 12$, $M_d = 12$. Note that now for $\eta = 2$ the codelength is 288, and we cannot use the field $GF(2^8)$. Also, when using one symbol per node, the symbol should contain no less than 12 bits (since there are 12 disks in the node), this gives the field $GF(2^{12})$. Parameters for other schemes are presented in Table 1.

**Table 1** Comparison of schemes based on RS codes

| $N = 64, L = 8, M_d = 8$ | | | | | |
|---|---|---|---|---|---|
| # | $\eta$ | $GF(q)$ | $(n,k)$ | $R$ | $\varepsilon$ |
| 1 | 1 | $2^8$ | (64,56) | 0.875 | 56 |
| 2 | 2 | $2^8$ | (128,112) | 0.875 | 56 |
| 3 | 1/2 | $2^8$ | (32,28) | 0.875 | 56 |
| 4 | 1/4 | $2^8$ | (16,14) | 0.875 | 56 |
| 5 | 1/8 | $2^8$ | (8,6) | 0.75 | 48 |
| $N = 144, L = 12, M_d = 12$ | | | | | |
| # | $\eta$ | $GF(q)$ | $(n,k)$ | $R$ | $\varepsilon$ |
| 1 | 1 | $2^8$ | (144,132) | 0.916 | 132 |
| 2 | 2 | $2^9$ | (288,264) | 0.916 | 132 |
| 3 | 1/2 | $2^8$ | (72,66) | 0.916 | 132 |
| 4 | 1/4 | $2^8$ | (36,33) | 0.916 | 132 |
| 5 | 1/12 | $2^{12}$ | (12,10) | 0.83 | 120 |

## 5.2 Estimation of LDPC Codes Parameters

When constructing coding schemes based on LDPC codes for the formulated model parameters first note that taking into account Theorem 1 the codes defined by matrices (2) and (4) can correct one SDC error and two disks failures, if we write $1 \le b < m$ codeword bits per disk. So we need to consider only the restoration from node failure.

*Case 1.* First consider the matrix (2) with $\gamma = 2$. As follows from Sect. 4.2, the node failure may be corrected if the node contains $m$ symbols correspondent to block-column in matrix $H$.

Let $m = 8$, $\rho = 8$. We have the code $(64, 49)$ with $R = 0.7656$. Locality is 7, which is much better than using RS code, but the rate of LDPC code is also significantly lower. Increasing of $m$ will further decrease the rate, and increasing of $\rho$ will lead to necessity of using more than one block in the node, and node failure would be impossible to correct.

Now let $\gamma = 3$. To increase the rate we increase the values of $m$ and $\rho$ and write several bits per disk. If $m = 16$, $\rho = 16$ we have $(256, 210)$-code with rate $R = 0.82$, and each disk contains $b = 4$ bits. To estimate the locality note that since we consider the locality regarding one disk failure it is enough to use only first stripe of $H$ to correct such erasure burst. But this stripe contains only unity matrices as blocks, so all symbols needed to recover, for example, second erased symbol in the burst, will be placed at the same disks as symbols which were used to recover first erased symbol. Thus the recovering procedure may be organized in such a way that locality for disks will be the same as locality for bits (surely, if only one failure is assumed). In our case this is 15.

Finally consider $m = 32$, $\rho = 16$. The rate of this code is slightly less than the previous one ($R = 0.816 \approx 0.82$), but for this case $b = 8$, which may be convenient for practical implementation.

Parameters of considered LDPC-based schemes are given in Table 2.

*Case 2.* This case may be processed similarly. Parameters of some LDPC-based schemes for this case are also given in Table 2.

**Table 2**  Comparison of schemes based on LDPC codes

| | | | $N = 64, L = 8, M_d = 8$ | | | | |
|---|---|---|---|---|---|---|---|
| # | $b$ | $\gamma$ | $m$ | $\rho$ | $(n, k)$ | $R$ | $\varepsilon$ |
| 1 | 1 | 2 | 8 | 8 | $(64,49)$ | 0.77 | 7 |
| 2 | 2 | 2 | 16 | 8 | $(128,97)$ | 0.76 | 7 |
| 3 | 4 | 3 | 16 | 16 | $(256,210)$ | 0.82 | 15 |
| 4 | 8 | 3 | 32 | 16 | $(512,418)$ | 0.82 | 15 |
| | | | $N = 144, L = 12, M_d = 12$ | | | | |
| # | $b$ | $\gamma$ | $m$ | $\rho$ | $(n, k)$ | $R$ | $\varepsilon$ |
| 1 | 1 | 2 | 12 | 12 | $(144,121)$ | 0.84 | 11 |
| 2 | 1 | 3 | 12 | 12 | $(144,110)$ | 0.76 | 11 |
| 3 | 4 | 3 | 24 | 24 | $(576,506)$ | 0.88 | 23 |
| 4 | 8 | 3 | 48 | 24 | $(1152,1010)$ | 0.88 | 23 |

Comparing the obtained results, we may note the following. By using the block-permutation LDPC codes with 2 and 3 stripes, it turns to be possible to obtain the coding schemes for considered model, which may be comparative with RS codes or some worse in terms of redundancy (which is shown by the code rate), but they are better in terms of locality and promising in implementation complexity.

# 6 Conclusion

In this paper we considered the model of distributed data storage system with different types of failures. For some example sets of model parameters the coding schemes based on RS and LDPC codes are considered, the numerical parameters of these schemes are obtained and analysed. The possibility of correcting multiple fixed erasure bursts is considered for LDPC codes. However, further research is needed to provide constructions suitable for larger number of failures.

# References

1. Datta, A., Oggier, F.: Storage codes: managing big data with small overheads. In: 2013 International Symposium on, Network Coding (NetCod). pp. 1–6, June (2013)
2. Datta, A., Oggier, F.E.: An overview of codes tailor-made for networked distributed data storage. CoRR, abs/1109.2317 (2011)
3. Oggier, F., Datta, A.: Coding Techniques for Repairability in Networked Distributed Storage Systems. Now Publishers Inc., Hanover (2013)
4. Pamies-Juarez, L., Oggier, F.E., Datta, A.: An empirical study of the repair performance of novel coding schemes for networked distributed storage systems. CoRR, abs/1206.2187 (2012)
5. Tamo, I., Barg, A.: A family of optimal locally recoverable codes. CoRR, abs/1311.3284 (2013)
6. Blahut, R.: Theory and Practice of Error Control Codes. Addison-Wesley, Boston (1984)
7. MacWilliams, F., Sloane, N.: The Theory of Error-Correcting Codes. North-Holland Publishing Company, Amsterdam (1977)
8. Reed, I., Solomon, G.: Polynomial codes over certain finite fields. J. Soc. Ind. Appl. Math. **8**(2), 300–304 (1960)
9. Gallager, R.G.: Low Density Parity Check Codes. MIT Press, Cambridge (1963)
10. MacKay, D.: Good error correcting codes based on very sparse matrices. In: IEEE Transactions on Information Theory **45**, Mar 1999
11. Richardson, T.J., Urbanke, R.L.: The capacity of low-density parity-check codes under message-passing decoding. In: IEEE Transactions on Information Theory, **47**(2), Feb 2001
12. Richardson, T.J., Urbanke, R.L., Shokrollahi, M.A.: Design of capacity-approaching irregular low-density parity-check codes. In: IEEE Transactions on Information Theory, **47**(2), Feb 2001
13. Zyablov, V., Pinsker, M.: Estimation of the error-correction complexity for gallager low-density codes. Probl. Inf. Transm. X **I**(1), 18–28 (1975)
14. Djordjevic, I., Ryan, W., Vasic, B.: Coding for Optical Channels. Springer, New York (2010)
15. Kabatiansky, G., Krouk, E., Semenov, S.: Error Correcting Coding and Security for Data Networks: Analysis of the Superchannel Concept. Wiley, UK (2005)
16. Krouk, E., Semenov, S.: Modulation and Coding Techniques in Wireless Communications. Wiley, New York (2011)
17. Lin, S., Ryan, W.: Channel Codes: Classical and Modern. Cambridge University Press, Cambridge (2009)
18. Diao, Q., Huang, Q., Lin, S., Abdel-Ghaffar, K.: A matrix-theoretic approach for analyzing quasi-cyclic low-density parity-check codes. IEEE Trans. Inf. Theory **58**(6), 4030–4048 (2012)
19. Gilbert, E.N.: A problem in binary encoding. Proc. Symp. Appl. Math. **10**, 291–297 (1960)
20. Arazi, B.: The optimal burst error-correcting capability of the codes generated by $f(x) = (x^p + 1)(x^q + 1)/(x + 1)$. Inform. Contr. **39**(3), 303–314 (1978)
21. Kozlov, A., Krouk, E., Ovchinnikov, A.: An approach to development of block-commutative codes with low density of parity check (in Russian). Priborostroenie **8**, 9–14 (2013)

22. Krouk, E., Ovchinnikov, A.: 3-Stripes Gilbert low density parity-check codes. US Patent 7, 882, 415
23. Krouk, E.A. Semenov, S.V.: Low-density parity-check burst error-correcting codes. In: 2nd International Workshop "Algebraic and combinatorial coding theory", Leningrad, pp. 121–124 (1990)
24. Zhang, W., Wolf, J.: A class of binary burst error-correcting quasi-cyclic codes. In: IEEE Transactions on Information Theory, IT-34:463–479. May 1988
25. Belogolovy, A., Krouk, E., Ovchinnikov, A.: Iterative decoding of concatenated low-density parity-check codes. US Patent 8, 230, 296

# Estimation of the Mean Message Delay for Transport Coding

Dmitrii Malichenko and Evgenii Krouk

**Abstract** This article considers transport coding which is a method for data transmission in a packet switching network. It uses error correcting codes at the transport layer of data network and can help to decrease the mean message delay. Evaluation of the mean message delay is important for estimation of transport coding efficiency. The existing analysis of the mean message delay uses assumption about exponential distribution of packet delay. The network model proposed by L. Kleinrock is considered in this paper. The distribution of the packet delay in Kleinrock network is not proven to be exponential. This work offers calculation of the mean message delay for Kleinrock networks without usage of the assumption about exponential distribution of the packet delay. The accuracy is checked using simulation of the Kleinrock network model.

**Keywords** Transport coding · Message delay · Error correcting code · Kleinrock model · Network simulation · Packet switching

## 1 Introduction

The usage of error correcting codes is a well known method of protection against errors during data transmission over a channel. Usually error correcting coding is applied on the physical layer of the network. In this paper we consider using of error correcting coding at the higher layers together with multipath routing. Applications of redundant data transmission over different routes were proposed in a number of papers [1–4]. The reliable data transmission is considered in [3, 4]. The transmission of priority messages is considered in [5]. The Voice over IP (VoIP) quality

D. Malichenko (✉) · E. Krouk
State University of Aerospace Instrumentation,
Ul. Bolshaya Morskaya, 67, 190000 Saint-petersburg, Russia
e-mail: dml@vu.spb.ru

E. Krouk
e-mail: ekrouk@vu.spb.ru

© Springer International Publishing Switzerland 2015                                        239
E. Damiani et al. (eds.), *Intelligent Interactive Multimedia Systems and Services*,
Smart Innovation, Systems and Technologies 40,
DOI 10.1007/978-3-319-19830-9_22

improvement using transmission over different routes is considered in [6, 7]. The decreasing of the mean message delay is considered in [1, 2, 8, 9] where the authors provide analytical equations for calculation of the delay. They use assumption about exponential distribution of packet delay. The purpose of this work is to propose analytical equations for calculation of the mean message delay when the mentioned assumption couldn't be applied.

The rest of this article is structured as follows. Section 2 describes transport coding. Section 3 describes the network model. Then Sect. 4 gives analysis of the mean message delay. Section 5 describes simulation of Kleinrock network model [10] and contains simulation results for the mean message delay and transport coding gain. The results of the novel and previous calculations of the message delay are compared and discussed. Finally the conclusions are made in Sect. 6.

# 2 Transport Coding

Assume we need to transmit a message which consists of $k$ packets over the data network. The packet length is $m$ bits. We consider each packet as a symbol over Galois Field $GF(2^m)$ [11]. Then we apply a maximum distance separable (MDS) code $(n, k)$ [12] to the packets of the original message in order to obtain an encoded message consisting of $n$ $(n > k)$ packets. After that we transmit $n$ packets of the encoded message instead of $k$ packets of the original message (Fig. 1). The traffic increases by a factor of $1/R$ ($R = k/n$ is a code rate). This should lead to increasing of the mean message delay. However due to properties of MDS codes the receiving node needs to wait for any $k$ packets in order to assemble the message. The works [2, 8] show that this method can lead to decreasing of the message delay. This idea is illustrated in Fig. 2. The figure shows the transmission of encoded message ($n = 5, k = 3$). The 1st and 4th packets are delayed and other packets have arrived. The receiving node

**Fig. 1** Encoded packets are transmitted over the network. Dashed rectangles represent redundant packets obtained by the transport coding

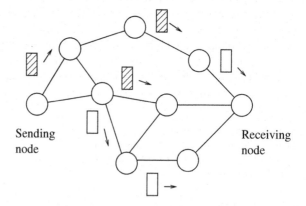

Sending node

Receiving node

**Fig. 2** Transmission of the message ($k = 3$, $n = 5$) using transport coding. Three packets that came earlier than others are used to assemble the message

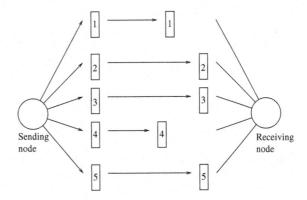

doesn't need to wait for the 1st and 4th packets to assemble the message. In this work we consider that all channels are reliable, there aren't any dropped packets. However the transport coding could also be applied in case of unreliable channels.

## 3 Network Model and Assumptions

### 3.1 Model

In this paper we consider packet switching network with datagram routing. Packet switching is a communication method that supposes grouping of all information before transmission into blocks called packets or datagrams. Packets are considered as separate entities. Each packet follows its own way over the network to the destination node. When a packet arrives at a switching node it waits in a queue for its turn to be transmitted to the next link in its path. The switching node decides which link to choose. The packets which belong to one message may arrive at the destination node out of order. It means that the receiving node has to wait for all the packets that form a message and in case of the wrong order has to reassemble them. Kleinrock network model [10] will be used further on as a model of packet switching network.

Consider a data network which consists of M channels. The capacity of each channel is $C_i$. Let us assume that information messages consist of $k$ packets. The packet transmission time over a channel is an exponentially distributed with expectation $1/\mu$. The traffic arriving at the network forms Poisson flow with an intensity $\gamma$ packets per second. The mean number of packets passing through the $i$th channel is $\lambda_i$ packets per second. The total network traffic is

$$\lambda = \sum_{i=1}^{M} \lambda_i .$$ 

(1)

According to Kleinrock's 'assumption of independence' [10] the packet delays can be regarded as independent random variables. It means that $i$th channel can be represented as a queuing system with a Poisson flow of intensity $\lambda_i$ at the input and exponential servicing time with the mean $\frac{1}{\mu C_i}$. The mean packet delay is

$$\bar{t}(\lambda, \mu) = \sum_{i=1}^{M} \frac{\lambda_i}{\gamma} \frac{1}{\mu C_i - \lambda_i}. \tag{2}$$

Consider that all channel capacities are equal so we denote them as $c$, intensities of the packet flows for all channels are also equal. The Eq. (2) could be rewritten as follows:

$$\bar{t}(\lambda, \mu) = \frac{\bar{l}}{\mu c} \frac{1}{1 - \rho}, \tag{3}$$

where $\bar{l} = \frac{M\lambda_i}{\gamma}$ is the mean route length, and $\rho = \frac{\lambda}{\mu C}$ is the network load.

### 3.2 Assumptions

In order to calculate the mean message delay in [1, 2, 8, 9] some assumptions were made:

1. Packet delay is exponentially distributed.
2. Network load is uniformly distributed over the whole network.
3. Packet delays are independent of each other.
4. All channel capacities are equal.

These assumptions restrict application of the network model. In order to extend the area of the model usage we eliminate the first assumption. This assumption makes it easy to calculate the message delay and to estimate analytically the possible gain of transport coding, but the distribution of the packet delay in Kleinrock network is not proven to be exponential. So the assumption 1 introduces some inaccuracy into calculation of the message delay in Kleinrock networks. Our goal is to introduce the calculation of the mean message delay without usage of the assumption 1.

We consider Erlang-distributed packet delays. Let us explain why this distribution law has been chosen. Let's examine one route in Kleinrock network from a sending node to a receiving node. Let us assume for example that route length is three (see Fig. 3). Channels are represented as queuing systems with exponential servicing time with mean $\frac{1}{\mu C}$. The packet goes through three servicing units of queuing systems in order to reach the receiving node. If the queues are empty the message delay is a sum of three independent random variables distributed as exponentials with parameter $\mu C$. If $Y_1, \ldots, Y_\alpha$ are independently distributed as exponentials with parameter $1/\beta$, then $X = \sum Y_i$ has Erlang distribution with parameters $\alpha$ and $\beta$ [13]. It means that in case of empty queues packet delay will have Erlang distribution. In case of higher network load the queues will have non-zero lengths and delay of different packets will be a sum of different number of exponential random variables. It means that in case of higher network loads packet delay will not have exactly Erlang distribution. Therefore at least for the case of low network loads Erlang distribution looks promising. In Sect. 5 we compare estimations of the mean message delays obtained in different ways. We still retain other assumptions so the network load is still uniformly distributed and all the data flows between the nodes are equal.

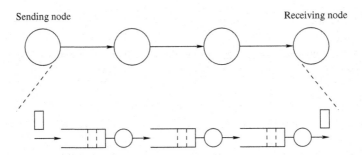

Sending node                                                                    Receiving node

**Fig. 3** Example of one route of the Kleinrock network. All the channels are represented as queuing systems

## 4 Message Delay Analysis

The structure of the message delay analysis performed in this section is the same as in [2, 8].

Let us begin with the case of transmission without transport coding. The assembling node should wait for all $k$ packets to arrive. In this case the message delay is the maximum packet delay among all packets of the given message

$$T = \max \left\{ t_1, \dots, t_k \right\} ,$$

where $t_i$ is the delay of the $i$th packet of the message. Denoting packet delays in increasing order $t_{1:k} \leq \dots \leq t_{k:k}$ we get

$$T = t_{k:k} .$$

The delay of the encoded message is

$$T = t_{k:n} .$$

We use order statistics [14] in order to calculate mathematical expectation of the delay of $i$th packet (for total number of packets $n$)

$$E\left[t_{i:n}\right] = n\binom{n-1}{i-1} \int_{-\infty}^{\infty} t \, [P(t)]^{i-1} \, [1 - P(t)]^{n-i} \, dP(t) , \tag{4}$$

or

$$E\left[t_{i:n}\right] = n\binom{n-1}{i-1} \int_{0}^{1} P^{-1}(u) \, u^{i-1} \, [1 - u]^{n-i} \, du , \tag{5}$$

where $P(t)$ is the distribution function of packet delay and $P^{-1}(u)$ is the inverse function of $P(t)$.

As it was mentioned above exponential distribution for $P(t)$ was used in previous works dedicated to message delay analysis [2, 8]. In this case the Eq. (5) can be rewritten as follows [14]:

$$E[t_{i:n}] = \bar{t} \sum_{j=n-i+1}^{n} j^{-1} , \tag{6}$$

where $\bar{t}$ is the mean packet delay, which is equal to $\bar{t}(\lambda, \mu)$ in our case. The mean delay of the uncoded message is

$$\bar{T} = E[t_{k:k}] = \bar{t}(\lambda, \mu) \sum_{j=1}^{k} j^{-1} . \tag{7}$$

The mean delay of the encoded message is

$$\bar{T}_{cod} = E[t_{k:n}] = \bar{t}(\lambda, \mu) \sum_{j=n-k+1}^{n} j^{-1} . \tag{8}$$

In this paper we consider packet delays distributed as Erlang. If $P(t)$ is an Erlang distribution with parameters $\alpha = \bar{l}$ which is an average route length and $\beta = 1/\bar{t}(\lambda, \mu)$ the Eq. (4) can be rewritten as follows:

$$E\left[t_{i:n}\right] = n\binom{n-1}{i-1} \int_{-\infty}^{\infty} -x\left(1 - e^{-\beta x} \sum_{i=0}^{\bar{l}-1} \frac{(\beta x)^i}{i!}\right)^{i-1} \times$$

$$\times \left(e^{-\beta x} \sum_{i=0}^{\bar{l}-1} \frac{(\beta x)^i}{i!}\right)^{n-i} e^{-\beta x} \sum_{i=0}^{\bar{l}-1} (\beta x)^i \left[\frac{i}{xi!} - \frac{\beta}{i!}\right] dx . \tag{9}$$

The mean delay of the uncoded message is

$$
\bar{T} = E\left[t_{k:k}\right] = n \int_{-\infty}^{\infty} -x \left(1 - e^{-\frac{x}{\bar{t}(\lambda,\mu)}} \sum_{i=0}^{\bar{l}-1} \frac{\left(\frac{x}{\bar{t}(\lambda,\mu)}\right)^k}{k!}\right)^{k-1} \times
$$

$$
\times\ e^{-\frac{x}{\bar{t}(\lambda,\mu)}} \sum_{i=0}^{\bar{l}-1} \left(\frac{x}{\bar{t}(\lambda,\mu)}\right)^k \left[\frac{k}{xk!} - \frac{1}{\bar{t}(\lambda,\mu)k!}\right] dx\ . \tag{10}
$$

The mean delay of the encoded message is

$$
\bar{T}_{cod} = E\left[t_{k:n}\right] = n\binom{n-1}{k-1} \int_{-\infty}^{\infty} -x \left(1 - e^{-\frac{x}{\bar{t}(\lambda,\mu)}} \sum_{i=0}^{\bar{l}-1} \frac{\left(\frac{x}{\bar{t}(\lambda,\mu)}\right)^k}{k!}\right)^{k-1} \times
$$

$$
\times \left(e^{-\frac{x}{\bar{t}(\lambda,\mu)}} \sum_{k=0}^{\bar{l}-1} \frac{\left(\frac{x}{\bar{t}(\lambda,\mu)}\right)^k}{k!}\right)^{n-k} e^{-\frac{x}{\bar{t}(\lambda,\mu)}} \sum_{i=0}^{\bar{l}-1} \left(\frac{x}{\bar{t}(\lambda,\mu)}\right)^k \times
$$

$$
\times \left[\frac{k}{xk!} - \frac{1}{\bar{t}(\lambda,\mu)k!}\right] dx\ . \tag{11}
$$

## 5 Simulation of Kleinrock Network

The goal of this section is to investigate the accuracy of the calculation of the mean message delay proposed in this work (Eq. (11)) and to compare the message delay estimations obtained by simulation and Eqs. (8) and (11).

The network channels are simulated as queuing systems, the packets pass through queues and serving units in its way to the receiving node. The traffic flows between all the pairs of nodes are considered to be equal because of the assumption that the network load is distributed uniformly among the data network. It means that we need to simulate only one pair of nodes communicating with each other.

At first let us analyse the estimations of the mean message delays calculated using Eqs. (8), (11). The relation between the message delay and code ratio is presented in Fig. 4. Three curves are obtained using Eq. (8) and marked as "Exponential model". The other three curves are obtained using Eq. (11) and marked as "Erlang model". Other curves correspond to simulation. As one can see the proposed equation for the mean message delay gives more accurate results than the previous calculations. Even for high network load (Fig. 5) the proposed equation is more accurate. The plot in the Fig. 5 is presented only for code rates $R \geq 0.7$ because the network is overflowed by redundant packets in case of smaller code rates.

**Fig. 4** Mean message delay
of the encoded message
$\rho = 0.1, k = 8$

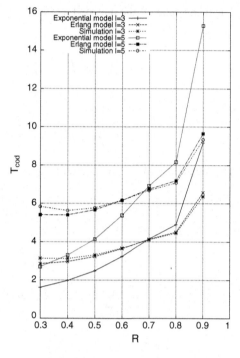

**Fig. 5** Mean message delay
of the encoded message
$\rho = 0.6, k = 8$

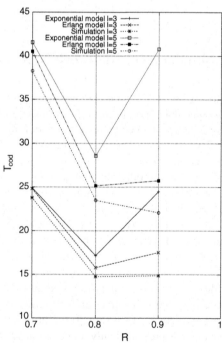

**Fig. 6** Coding gain $T/T_{cod}$ in case of route length $\bar{l} = 3$ and $\bar{l} = 5$, $\rho = 0.1$, $k = 8$

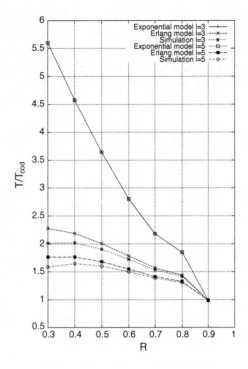

**Fig. 7** Coding gain $T/T_{cod}$ in case of route length $\bar{l} = 3$ and $\bar{l} = 5$, $\rho = 0.2$, $k = 8$

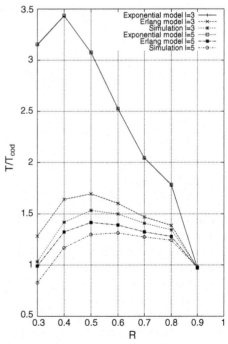

Now let us investigate the estimation of the transport coding gain which is $\bar{T}/\bar{T}_{cod}$. The plots represented in Figs. 6, 7 and 8 show transport coding gain calculated in three different ways. As one can see estimation of the transport coding gain calculated using (11) gives very close value to the transport coding gain obtained by simulation model. The proposed estimation which is marked as 'Erlang model' remains accurate enough in cases of different network loads and different route lengths.

**Fig. 8** Coding gain $T/T_{cod}$ in case of route length $\bar{l} = 3$ and $\bar{l} = 5$, $\rho = 0.4$, $k = 8$

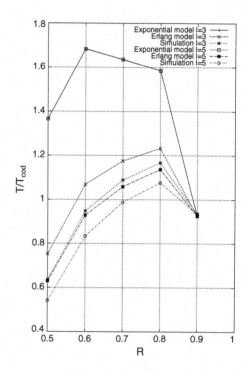

## 6 Conclusions

The transport coding was considered in this paper. The existing analysis of the message delay and transport coding efficiency uses the assumption about exponential distribution of the packet delay. In the paper this assumption was eliminated. Erlang distribution was considered instead of exponential distribution. This led to more complex equations but the accuracy of estimation for the mean message delay in Kleinrock networks became much higher. However there are still other assumptions that restrict practical usage of introduced analysis. For example, the assumption of an uniformly distributed network load. This could be a matter of future works.

# References

1. Kabatianskii, G.A., Krouk, E.A.: Coding decreases delay of messages in networks. In: Proceedings of the IEEE International Symposium on, Information Theory, pp. 255–255, 17–22, Jan 1993
2. Maxemchuk, N.F.: Dispersity Routing. Proceedings of ICC 75, pp. 41–10, 41–13. San Francisco CA, June (1975)
3. Rabin, M.O.: Efficient dispersal of information for security, load balancing, and fault tolerance. J. ACM **36**(2), 335–348 (1989)
4. Mitzenmacher, M.: Digital Fountains: A Survey and Look Forward. Information Theory Workshop (2004)
5. Krouk, E., Semenov, S.: Transmission of priority messages with the help of transport coding. In: IEEE 10th International Conference on, Telecommunications, 2003. ICT 2003, vol. 2, pp. 1273–1277. Feb (2003)
6. Liang, Y.J., Steinbach, E.-G., Girod, B.: Multi-stream voice over IP using packet path diversity. In: Proceedings of the IEEE 4th Workshop on Multimedia Signal Processing, pp. 555–560 (2001)
7. Bettermann, S., Rong, Y.: Effects of fully redundant dispersity routing on VoIP quality. In: Proceedings of the 2011 IEEE International Workshop Technical Committee on Communications Quality and Reliability (CQR). Naples, USA (2011)
8. Krouk, E., Semenov, S.: Application of coding at the network transport level to decrease the message delay. In: Proceedings of Third International Symposium on Communication Systems Networks and Digital Signal Processing, **15–17**, pp. 109–112. Staffordshire University, UK (2002)
9. Kabatiansky, G., Evgenii, K., Sergei, S.: Error Correcting Coding and Security for Data Networks: Analysis of the Superchannel Concept. Wiley, Chichester (2005)
10. Kleinrock, L.: Communication Nets; Stochastic Message Flow and Delay. Dover Publications, Incorporated (1972)
11. Lidl, R., Niederreiter, H.: Introduction to Finite Fields and Their Applications, rev edn. Cambridge University Press, Cambridge (1994)
12. Singleton, R.C.: Maximum distance Q-nary codes. IEEE Trans. Inform. Theory **IT-10**(2), 116–118 (1964)
13. Gentle, J.E.: Random Number Generation and Monte Carlo Methods. Statistics and Computing, 2nd edn. Springer (2003)
14. David, H.A., Nagaraja, H.N.: Order Statistics. Wiley-Interscience, 3rd edn., 488 p. March 2004

# Sustainable Flat Ride Suspension Design

Hormoz Marzbani, Milan Simic, M. Fard and Reza N. Jazar

**Abstract** It was suggested [1] that having natural frequency of the front approximately 80 % of that of the rear suspension in a vehicle will result in a flat ride for the passengers. Flat Ride in this case means that the pitch motion of the vehicle, generated by riding over a bump for instance will fade in to the bounce motion of the vehicle much faster. Bounce motion of the vehicle in mush easier to tolerate and feels more comfortable for the passengers. In a previous study the authors, analytically proved that this situation is not practical. In other words, for any vehicle there will only be one certain velocity, depending on the geometry and suspension system specifications which the flat ride will happen at. The search continued to find a practical method for enjoying the flat ride in vehicles. Solving the equation of motion of the vehicle for different spring rates and road configuration the authors came up with design chart for smart suspension systems. Using the advantages of the analytical approach to the flat ride problem, the chart was established to be used for vehicles with smart active suspension systems. In this paper the mathematical methods used and the resulted criteria for designing a flat ride suspension system which will perform in different speeds is presented.

**Keywords** Flat ride · Maurice olley · Optimal suspension · Vehicle vibrations · Vehicle dynamics · Suspension design · Suspension optimization

## 1 Introduction

The excitation inputs from the road to a straight moving car will affect the front wheels first and then, with a time lag, the rear wheels. The general recommendation was that the natural frequency of the front suspension should be lower than that of the rear. So, the rear part oscillates faster to catch up with the front to eliminate

H. Marzbani (✉) · M. Simic · M.F. Reza · N. Jazar
School of Aerospace, Mechanical and Manufacturing Engineering,
RMIT University, Melbourne, Australia
e-mail: hormoz.marzbani@rmit.edu.au

© Springer International Publishing Switzerland 2015
E. Damiani et al. (eds.), *Intelligent Interactive Multimedia Systems and Services*,
Smart Innovation, Systems and Technologies 40,
DOI 10.1007/978-3-319-19830-9_23

pitch and put the car in bounce before the vibrations die out by damping. This is what Olley called the *Flat Ride Tuning* [1]. Maurice Olley (1889–1983) established guidelines, back in the 1930, for designing vehicles with better ride. These were derived from experiments with a modified car to allow variation of the pitch mass moment. Although the measures of ride were strictly subjective, those guidelines were considered as valid rules of thumb even for modern cars. What is known as Olley's Flat Ride not considering the other prerequisites can be put forward as:

*The front suspension should have around 30 % lower rate than the rear.*

An important prerequisite for flat ride was the uncoupling condition, which was introduced by Rowell and Guest for the first time in 1923 [1]. Rowell and Guest as shown in Fig. 1 used the geometry of a bicycle car model to find the condition which sets the bounce and pitch centers of the model located on the springs. Having the condition, the front and rear spring systems of the vehicle can be regarded as two separate one degree of freedom systems. Passenger comfort, for any seat placement in the vehicle, should be at an acceptable level, which is a personal experience. All of this is important for modern sophisticated cars as well, and for the future autonomous vehicles.

As a result of applying the Olley's conditions the pitch motion of the vehicle will turn in to a bounce motion. Pitch motion provides a much more uncomfortable experience for the passengers compared to bounce motion.

As mentioned earlier ride comfort is a personal experience which is different from one individual to other, but the effects of motion can be similar. Motion in the car can cause fatigue for the passengers. Driver fatigue is a significant cause of accidents on motorways. The fatigue caused by driving extended periods actually impairs driver alertness and performance and therefore can compromise transportation safety [2].

Whole body vibration has been found to correlate with a range of psychological reactions of the human body such as lower back pain and heart rate variability [3, 4]. Disturbance of vision and balance have also been reported to occur [5].

So a better suspension design, will affect the ride comfort, which will result in a safer transportation as well as a more comfortable traveling experience.

Using analytical methods, we investigate the flat ride conditions which have been respected and followed by the car manufacturers' designers since they were introduced for the first time. This article provides a more reliable scientific and mathematical approach into the flat ride design criteria in the vehicle dynamic studies.

## 2 Previous Works

Maurice Olley introduced and studied the concept of flat ride in vehicle dynamics. In his paper [6] he published the results taken from the experiments conducted using the test rig for the first time.Olley explained the relation between vertical acceleration and comfort over a range of frequencies. He generated a curve for passenger comfort, which is very similar to the current *ISO*2631 standard.

Olley as well as other investigators in well-established car companies realized that the pitch and roll modes of the car body are much more uncomfortable than the bounce mode. The investigators' effort focused on the suspension stiffness and damping rates to be experimentally adjusted to provide acceptable vertical vibrations. However, the strategy about roll and pitch modes were to transform them to bounce. Due to usual geometric symmetry of cars, as well as the symmetric excitation from the road, the roll mode is excited much less than the pitch mode. Therefore, lots of investigations were focused on adjustments of the front and rear suspensions such that pitch mode of vibration transforms to the bounce.

**Fig. 1** Bicycle car model used for vibration analysis

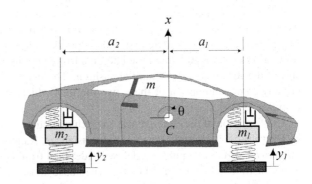

Besides all important facts that Olley discovered during his experiments, the principle known as the Flat Ride Tuning or Olley's Flat Ride proved to be more industry approved and accepted. After his publications [6, 8, 9] in which he advocated this design practice, they became rules of practice.

We can summarize all as the following:

1. The front spring should be softer than the rear for Flat Ride Tuning. This will promote bouncing of the body rather than pitching motions at least for a greater majority of speeds and bump road situations. The front suspension should have a 30 % lower ride rate than the rear suspension, or the spring center should be at least 6.5 % of the wheelbase behind the center of gravity.

2. The ratio $\dfrac{r^2}{a_1.a_2}$ normally approaches unity. This reduces vibration interactions between front and rear because the two suspensions can now be considered as two separate systems.

3. The pitch and bounce frequencies should be close together: the bounce frequency should be less than 1.2 times the pitch frequency.

4. Neither frequency should be greater than 1.3 Hz, which means that the effective static deflection of the vehicle should exceed roughly 6 inches.

5. The roll frequency should be approximately equal to the pitch and bounce frequencies.

Rowell and Guest [10] in 1923 identified the value of $\dfrac{r^2}{a_1.a_2}$ being associated with vehicles in which the front and rear responses were uncoupled. Olley was able to investigate the issue experimentally and these experiments led him to the belief that pitching motion was extremely important in the subjective assessment of vehicle ride comfort.

Deflections are some 30 % greater than the rear then the revolutionary flat ride occurs. Olley's explanation was that because the two ends of the car did not cross a given disturbance at the same instant it was important that the front wheels initiated the slower mode and that the rear wheels initiated the faster mode. This allowed the body movement at the rear to catch up the front and so produce the flat ride.

The condition of Flat Ride is expressed in various detailed forms; however, the main idea states that *the front suspension should have a 30 % lower ride rate then the rear.* The physical explanation for why this is beneficial in reducing pitch motion is usually argued based on the time history of events following a vehicle hitting a bump. First, the front of the vehicle responds approximately- in the well-known damped oscillation manner. At some time later, controlled by the wheelbase and the vehicle speed, the rear responds in similar fashion. The net motion of the vehicle is then crudely some summation of these two motions which minimizes the vehicle pitch response [11].

Confirmation of the effectiveness in pitch reduction of the Olley design was given by Best [7] over a limited range of circumstances. Random road excitation was applied to a half-car computer model, with identical front and rear excitations, considering the time delay generated by the wheelbase and vehicle's speed [12].

Sharp and Pilbeam [13] attempted a more fundamental investigation of the phenomenon, primarily by calculating frequency response for the half-car over a wide range of speed and design conditions. At higher speeds, remarkable reductions in pitch response with only small costs in terms of bounce response were shown. At low speeds, the situation is reversed.

Later on Sharp [12] discussed the rear to front stiffness tuning of the suspension system of a car, through reference to a half-car pitch plane mathematical model. Sharp concluded almost the same facts mentioned by Best and other researchers before him, saying that at higher vehicle speeds, Olley tuning is shown to bring advantage in pitch suppression with a very little disadvantage in terms of body acceleration. At lower speeds, he continues, not only does the pitch tuning bring large vertical acceleration penalties but also suspension stiffness implied are impractical from an attitude control standpoint.

The flat ride problem was revisited by Crolla and King [11]. They generated vehicle vibration response spectra under random road excitations. It was confidently concluded that the rear/front stiffness ratio has virtually no effect on overall levels of ride comfort.

In 2004, Odhams and Cebon investigated the tuning of a pitch-plane model of a passenger car with a coupled suspension system and compared it to that of a conventional suspension system, which followed the Rowell and Guest treatment, [14]. The concluded that the Olley's flat ride tuning provides a near optimum stiffness choice for conventional suspensions for minimizing dynamic tire forces and is very close to optimal for minimizing horizontal acceleration at the chest (caused by pitching) but not the vertical acceleration.

## 3 Uncoupling the Car Bicycle Model

Consider the two degree-of-freedom (*DOF*) system as shown in Fig. 2 . A beam with mass $m$ and mass moment $I$ about the mass center $C$ is sitting on two springs $k_1$ and $k_2$. This is used as a car model for the investigation in bounce and pitch motions. The translational coordinate $x$ of $C$ and the rotational coordinate $\theta$ are the usual generalized coordinates that we use to measure the kinematics of the beam. The equations of motion and the mode shapes are functions of the chosen coordinates.

**Fig. 2** The bicycle model is a beam of mass $m$ and mass moment $I$, sitting on two springs $k_1$ and $k_2$

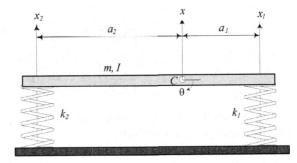

The free vibration equations of motion of the system are:

$$\begin{bmatrix} m & 0 \\ 0 & I \end{bmatrix} \begin{bmatrix} \ddot{x} \\ \ddot{\theta} \end{bmatrix} + \begin{bmatrix} k_1 + k_2 & a_2 k_2 - a_1 k_1 \\ a_2 k_2 - a_1 k_1 & a_2^2 k_2 + a_1^2 k_1 \end{bmatrix} \begin{bmatrix} x \\ \theta \end{bmatrix} = 0 \tag{1}$$

To compare the mode shapes of the system practically, we employ the coordinates $x_1$ and $x_2$ instead of $x$ and $\theta$. The equations of motion of the system would then be:

$$\begin{bmatrix} \dfrac{ma_2^2 + I}{a_1 + a_2{}^2} & \dfrac{ma_1 a_2 - I}{a_1 + a_2{}^2} \\ \dfrac{ma_1 a_2 - I}{a_1 + a_2{}^2} & \dfrac{ma_1^2 + I}{a_1 + a_2{}^2} \end{bmatrix} \begin{bmatrix} \ddot{x}_1 \\ \ddot{x}_2 \end{bmatrix} + \begin{bmatrix} k_1 & 0 \\ 0 & k_2 \end{bmatrix} \begin{bmatrix} x_1 \\ x_2 \end{bmatrix} = 0 \tag{2}$$

Let us define the following parameters:

$$I = mr^2 \quad \Omega_1^2 = \frac{k_1}{m}\beta \quad \Omega_2^2 = \frac{k_2}{m}\beta \quad \beta = \frac{l^2}{a_1 a_2} \quad \alpha = \frac{r^2}{a_1 a_2} \quad \gamma = \frac{a_2}{a_1} \quad l = a_1 + a_2 \tag{3}$$

and rewrite the equations as

$$\begin{bmatrix} \alpha + \gamma & 1 - \alpha \\ 1 - \alpha & \alpha + \dfrac{1}{\gamma} \end{bmatrix} \begin{bmatrix} \ddot{x}_1 \\ \ddot{x}_2 \end{bmatrix} + \begin{bmatrix} \Omega_1^2 & 0 \\ 0 & \Omega_2^2 \end{bmatrix} \begin{bmatrix} x_1 \\ x_2 \end{bmatrix} = 0 \tag{4}$$

Setting

$$\alpha = 1 \tag{5}$$

makes the equations decoupled

$$\begin{bmatrix} \alpha + \gamma & 0 \\ 0 & \alpha + \dfrac{1}{\gamma} \end{bmatrix} \begin{bmatrix} \ddot{x}_1 \\ \ddot{x}_2 \end{bmatrix} + \begin{bmatrix} \Omega_1^2 & 0 \\ 0 & \Omega_2^2 \end{bmatrix} \begin{bmatrix} x_1 \\ x_2 \end{bmatrix} = 0 \tag{6}$$

The natural frequencies $\omega_i$ and mode shapes $u_i$ of the system are

$$\omega_1^2 = \frac{1}{\gamma + 1} \Omega_1^2 = \frac{l}{a_2} \frac{k_1}{m} \quad u_1 = \begin{bmatrix} 1 \\ 0 \end{bmatrix} \tag{7}$$

$$\omega_2^2 = \frac{\gamma}{\gamma + 1} \Omega_2^2 = \frac{l}{a_1} \frac{k_2}{m} \quad u_2 = \begin{bmatrix} 0 \\ 1 \end{bmatrix} \tag{8}$$

They show that the nodes of oscillation in the first and second modes are at the rear and front suspensions respectively.

The decoupling condition $\alpha = 1$ yields

$$r^2 = a_1 a_2 \tag{9}$$

which indicates that the pitch radius of gyration, $r$, must be equal to the multiplication of the distance of the mass canter $C$ from the front and rear axles. Therefore, by setting $\alpha = 1$, the nodes of the two modes of vibrations appear to be at the front and rear axles. As a result, the front wheel excitation will not alter the body at the rear axle and vice versa. For such a car, the front and rear parts of the car act independently. Therefore, the decoupling condition $\alpha = 1$ allows us to break the initial two $DOF$ system into two independent one $DOF$ systems, where:

$$m_r = m\frac{a_1}{l} = m\varepsilon \tag{10}$$

$$m_f = m\frac{a_2}{l} = m(1 - \varepsilon) \tag{11}$$

$$\varepsilon = \frac{a_1}{l} \tag{12}$$

The equations of motion of the independent systems will be:

$$m(1 - \varepsilon)\ddot{x}_1 + c_1\dot{x}_1 + k_1x_1 = k_1y_1 + c_1\dot{y}_1 \tag{13}$$

$$m\varepsilon\ddot{x}_2 + c_2\dot{x}_2 + k_2x_2 = k_2y_2 + c_2\dot{y}_2 \tag{14}$$

The decoupling condition of undamped free system will not necessarily decouple the general damped system. However, if there is no anti-pitch spring or anti-pitch damping between the front and rear suspensions then equations of motion

$$\begin{bmatrix} \alpha + \gamma & 1 - \alpha \\ 1 - \alpha & \alpha + \frac{1}{\gamma} \end{bmatrix} \begin{bmatrix} \ddot{x}_1 \\ \ddot{x}_2 \end{bmatrix} + \begin{bmatrix} 2\xi_1\Omega_1 & 0 \\ 0 & 2\xi_2\Omega_2 \end{bmatrix} \begin{bmatrix} \dot{x}_1 \\ \dot{x}_2 \end{bmatrix} + \begin{bmatrix} \Omega_1^2 & 0 \\ 0 & \Omega_2^2 \end{bmatrix} \begin{bmatrix} x_1 \\ x_2 \end{bmatrix}$$
$$= \begin{bmatrix} 2\xi_1\Omega_1 & 0 \\ 0 & 2\xi_2\Omega_2 \end{bmatrix} \begin{bmatrix} \dot{y}_1 \\ \dot{y}_2 \end{bmatrix} + \begin{bmatrix} \Omega_1^2 & 0 \\ 0 & \Omega_2^2 \end{bmatrix} \begin{bmatrix} y_1 \\ y_2 \end{bmatrix} \tag{15}$$

$$2\xi_1\Omega_1 = \frac{c_1}{m}\beta \tag{16}$$

$$2\xi_2\Omega_2 = \frac{c_2}{m}\beta \tag{17}$$

will be decoupled by $\alpha = 1$.

$$\begin{bmatrix} \alpha + \gamma & 0 \\ 0 & \alpha + \frac{1}{\gamma} \end{bmatrix} \begin{bmatrix} \ddot{x}_1 \\ \ddot{x}_2 \end{bmatrix} + \begin{bmatrix} c_1 & 0 \\ 0 & c_2 \end{bmatrix} \begin{bmatrix} \dot{x}_1 \\ \dot{x}_2 \end{bmatrix} + \begin{bmatrix} \Omega_1^2 & 0 \\ 0 & \Omega_2^2 \end{bmatrix} \begin{bmatrix} x_1 \\ x_2 \end{bmatrix}$$
$$= \begin{bmatrix} 2\xi_1\Omega_1 & 0 \\ 0 & 2\xi_2\Omega_2 \end{bmatrix} \begin{bmatrix} \dot{y}_1 \\ \dot{y}_2 \end{bmatrix} + \begin{bmatrix} \Omega_1^2 & 0 \\ 0 & \Omega_2^2 \end{bmatrix} \begin{bmatrix} y_1 \\ y_2 \end{bmatrix} \tag{18}$$

The equations of motion of the independent system may also be written as

$$m(1 - \varepsilon)\ddot{x}_1 + c_1\dot{x}_1 + k_1x_1 = c_1\dot{y}_1 + k_1y_1 \tag{19}$$
$$m\varepsilon\ddot{x}_2 + c_2\dot{x}_2 + k_2x_2 = c_2\dot{y}_2 + k_2y_2 \tag{20}$$

which are consistent with the decoupled Eq. 18 because of

$$\varepsilon = \frac{1 + \gamma}{\gamma\Omega_2^2} \tag{21}$$

## 4 No Flat Ride Solution for Linear Suspension

The time lag between the front and rear suspension oscillations is a function of the wheelbase, $l$, and speed of the vehicle, $v$. Soon after the rear wheels have passed over a step, the vehicle is at the worst condition of pitching. Olley experimentally determined a recommendation for the optimum frequency ratio of the front and rear ends of cars. His suggestion for American cars and roads of 50s was to have the natural frequency of the front approximately 80 % of that of the rear suspension.

To examine Olley's experimental recommendation and possibly make an analytical base for flat ride, let us rewrite the equation of motion (19) and (20) as:

$$\ddot{x}_1 + 2\xi_1\dot{x}_1 + \frac{k_1}{m(1-\varepsilon)}x_1 = 2\xi_1\dot{y}_1 + \frac{k_1}{m(1-\varepsilon)}y_1 \tag{22}$$

$$\ddot{x}_2 + 2\xi\xi_1\dot{x}_2 + \frac{kk_1}{m\varepsilon}x_2 = 2\xi\xi_1 y_2 + \frac{kk_1}{m\varepsilon}y_2 \tag{23}$$

where,

$$\xi = \frac{\xi_2}{\xi_1} = \frac{c_1}{c_2}\frac{\varepsilon}{1-\varepsilon} \quad k = \frac{k_2}{k_1} = \frac{k_1}{k_2}\frac{\varepsilon}{1-\varepsilon} \quad \xi_1 = \frac{c_1}{m(1-\varepsilon)} \quad \xi_2 = \frac{c_2}{m\varepsilon} \tag{24}$$

Parameters $k$ and $\xi$ are the ratio of the rear/front spring rates and damping ratios respectively.

The necessity to achieve a flat ride provides that the rear system must oscillate faster to catch up with the front system at a reasonable time. At the time both systems must be at the same amplitude and oscillate together afterwards. Therefore, an ideal flat ride happens if the frequency of the rear system be higher than the front to catch up with the oscillation of the front at a certain time and amplitude. Then, the frequency of the rear must reduce to the value of the front frequency to oscillate in phase with the front. Furthermore, the damping ratio of the rear must also change to keep the same amplitude. Such a dual behavior is not achievable with any linear suspension. Therefore, theoretically, it is impossible to design linear suspensions to provide a flat ride, as the linearity of the front and rear suspensions keep their frequency of oscillation constant.

## 5 Nonlinear Damper

The force-velocity characteristics of an actual shock absorber can be quite complex. Although we may express the complex behavior using an approximate function, analytic calculation can be quite complicated with little design information. Furthermore, the representations of the exact shock absorber do not greatly affect the behavior of the system. The simplest linear viscous damper model is usually used for linear analytical calculation

$$F_D = cv_D \tag{25}$$

where $c$ is the damping coefficient of the damper.

The bound and rebound forces of the damper are different, in other words the force-velocity characteristics diagram is not symmetric. Practically, a shock absorber compresses much easier than decompression. A reason is that during rebound in which the damper extends back, it uses up the stored energy in the spring. A high compression damping, prevents to have enough spring compression to collect enough potential energy. That is why in order to get a more reliable and close to reality response for analysis on dampers, using bilinear dampers is suggested. It is similar to a linear damper but with different coefficients for the two directions (Dixon 2008).

$$F_D = \begin{cases} c_{DE}v_D & \text{Extension} \\ c_{DC}v_D & \text{Compression} \end{cases} \tag{26}$$

where $c_{DE}$ is the damping coefficient when damper is extended and $c_{DC}$ is the damping coefficient when the damper is compressed.

An ideal dual behavior damper is one which does not provide any damping while being compressed and on the other hand damps the motion while extending.

After using the nonlinear model for the damper, the motion had to be investigated in 3 steps for the front and same for the rear. Ideally, the unit step moves the ground up in no time and therefore the motion of the system begins when the input $y = 1$ and the suspension is compressed. The first step is right after the wheel hits the step and the damper starts extending. The second step is when the damper starts the compression phase, which the damping coefficient would be equal to zero. The third step is when the damper starts extending again. Each of the equations of motion should be solved for the 3 steps separately in order to find the time and amplitude of the third peak of the motion which have been chosen to be optimal time for the flat ride to happen at.

**Fig. 3** Response of the both suspensions of a near flat ride car with ideal nonlinear damper to a unit step

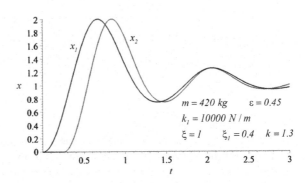

Figure 3 illustrates the behavior of the car equipped with a nonlinear damper when going over a unit step input.

## 6 Near Flat Ride Solution for Ideal, Nonlinear Damper

The conditions that $x_1$ and $x_2$ meet after one and a half oscillations can be shown by Eqs. (18) and (27).

**Fig. 4** Spring ratio $k = k_2/k_1$ versus $\tau = l/v$ for near flat ride with ideal nonlinear damping, for different $\varepsilon = a_1/l$

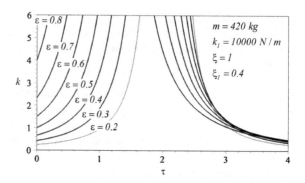

$$x_1 = x_2 \quad t_{p1} = t_{p2} \tag{27}$$

The equation resulted from $x_1 = x_2$, (Eq. 28) has got $\xi$ and $\xi_1$ as its variables and could be plotted as an explicit function of the variables which interestingly shows that the value for $\xi = \xi_2/\xi_1$ must equal to 1 for any value for damping coefficient of the front suspension $\xi_1$.

$$
\begin{aligned}
EQ1 = \frac{-0.8 \times 10^{-18}}{(\zeta_1^2 - 1)(\zeta_1^2 \zeta^2 - 1)} \Big( &- 0.3172834025 \times 10^{18} V \zeta_1^4 \zeta^2 \\
&+ 0.3172834025 \times 10^{18} V \zeta_1^2 + 0.3172834025 \times 10^{18} V \zeta_1^2 \zeta^2 - 0.3172834025 \times 10^{18} V \\
&+ 0.130151797 \times 10^{18} V \sqrt{1 - \zeta_1^2} \zeta_1^3 \zeta^2 - 0.3172834025 \times 10^9 V \sqrt{1 - \zeta_1^2} \zeta_1 \\
&+ 0.3172834025 \times 10^{18} V - 0.3172834025 \times 10^{18} V \zeta_1^2 \zeta^2 - 0.3172834025 \times 10^{18} V \zeta_1^2 + 0.3172834025 \\
&\times 10^{18} V - 0.130151797 \times 10^9 V \sqrt{1 - \zeta_1^2 \zeta^2} \zeta_1^3 \zeta + 0.130151797 \times 10^9 V \sqrt{1 - \zeta_1^2 \zeta^2} \zeta_1 \zeta \Big) \tag{28}
\end{aligned}
$$

where $V$ is equal to $e^{\frac{-\pi \zeta_1}{\sqrt{1-\zeta_1^2}}}$.

Therefore, regardless of the value of $\xi_1$ the rear suspension should have an equal coefficient for the damper. The equation resulted from $t_{P_1} = t_{p_2}$ generates Eq. 29 to determine $k = k_2/k_1$. Figure 4 illustrates the spring ratio $k = k_2/k_1$ versus $\tau = l/v$, to have near flat ride with ideal nonlinear damping, for different $\varepsilon = a_1/l$.

$$
EQ2 = \frac{\pi \sqrt{\frac{-k_1}{m(\varepsilon - 1)}} (2\sqrt{1 - \zeta_1^2} + 1 - \zeta_1^2) m(\varepsilon - 1)}{(\zeta_1^2 - 1)k_1}
$$
$$
- \tau + \frac{\pi(2\sqrt{1 - \zeta^2 \zeta_1^2} + 1 - \zeta^2 \zeta_1^2)}{\sqrt{\frac{kk_1}{m\varepsilon}}(\zeta^2 \zeta_1^2 - 1)} \tag{29}
$$

EQ1 and EQ2 are resulted after following the conditions mentioned in Eq. (27).

**Table 1** Specification of a sample car

| Specification | Nominal value |
|---|---|
| $m$ [kg] | 420 |
| $a_1$ [m] | 1.4 |
| $a_2$ [m] | 1.47 |
| $l$ [m] | 2.87 |
| $k_1$ [N/m] | 10000 |
| $k_2$ [N/m] | 13000 |
| $c_1$ [Ns/m] | 1000 |
| $c_2$ [Ns/m] | 1000 |
| $\beta$ | 4.00238 |
| $\gamma$ | 1.05 |
| $\Omega_1$ | 95.2947 |
| $\Omega_2$ | 123.8832 |
| $\xi_1$ | 0.05 |
| $\xi_2$ | 0.0384 |

The average length of a sedan vehicle has been taken 2.6 m with a normal weight distribution of a front differential vehicle 56/44 heavier at the front. Using the given information some other values can be calculated as: $a_1 = 1144$ mm and $a_2 = 1456$ mm which yields to $\varepsilon = 0.44$. Considering the existing designs for street vehicles, only the small section of $0.1 < \tau < 0.875$ is applied. The mass center of street cars is also limited to $0.4 < \varepsilon < 0.6$.

Figure 5 shows how $k$ varies with $\tau$ for $\xi_1 = 0.5$ and different $\varepsilon$ for a near flat ride with ideal nonlinear damper. For any $\varepsilon$, the required stiffness ratio increases for higher $\tau$. Therefore, the ratio of rear to front stiffness increases with lower car speed. Figure 6 also provide the same design graphs for $\xi_1 = 0.4$. Same graphs were plotted for $\xi_1 = 0.3, 0.2,$ and $0.1$.

There will be a possibility of using the $\tau$ vs. $\varepsilon$ diagrams as a design chart, which has been illustrated using the values in Table 1, by Fig. 7.

**Fig. 5** $\varepsilon$ versus $\tau$, for different $k$ for $\xi_1 = 0.5$ to have near flat ride with ideal nonlinear damping

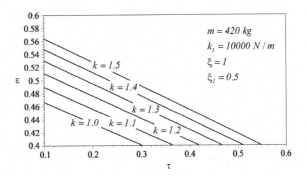

**Fig. 6** $\varepsilon$ versus $\tau$, for different $k$ for $\xi_1 = 0.4$ to have near flat ride with ideal nonlinear damping

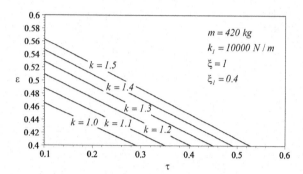

**Fig. 7** Design chart for a smart suspension with a non-linear damper [15]

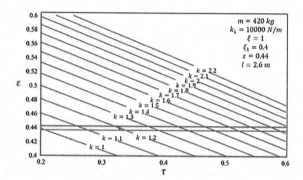

The box in Fig. 7, is indicating the values that the spring rate should be having as the travelling speed of the vehicle changes to provide the passengers with a flat ride, in case of having a smart active suspension. The point on the figure is an example for a passive suspension vehicle. It is showing the required spring rate, for getting a flat ride in a car with a wheelbase of 2.6 m, traveling at 28 km/h.

# 7 Conclusion

Olley's flat ride tuning has been regarded as a rule for designing chassis. The fact that these rules were based on experimental results, motivated many researchers to study and validate them. We have introduced a novel approach and investigated flat ride, analytically for the first time.

As a result of the dual behavior of the suspension which is required to get the optimal flat ride, more accurate results were researched using a nonlinear suspension system for this analysis. The results prove that the forward speed of the vehicle affects the flat ride condition, which agrees with previous researchers' results. In a passive suspension system flat ride can be achieved at a certain speed only, so the suspension system of a car should be designed in a way which provides the flat ride at a certain forward speed [16–19].

A design chart based on the nonlinear analysis, for smart active suspension systems has been provided which enables a car with smart suspension system to provide flat ride at any forward speed of the vehicle. The design chart can be used for designing chassis with passive suspension for a specified speed as well. Examples of the above mentioned conditions have been reviewed and discussed, by using some numerical values from a sample car. The research proves the effectiveness of Olley's flat ride for getting a more comfortable ride in cars, and considering the shortcomings of the principles suggests better ways of implementing them to the design of suspension systems for a better and more effective flat ride tuning. These issues are even more important for the future autonomous vehicles, where car occupants will have even more importance for the future autonomous vehicles, where car occupants will have no influence on driving, and their comfort is one of the extremely important criteria for the autonomous driving algorithm's application.

# References

1. Milliken, W.F., Milliken, D.L., Olley, M.: Chassis Design, Professional Engineering Publ (2002)
2. Azizan, M.A., Fard, M., Azari, M.F.: Characterization of the effects of vibration on seated driver alertness. Nonlinear Engineering 3(3), 163–168 (2014)
3. Helmkamp, J.C., Talbott, E.O., Marsh, G.M.: Whole body vibration-a critical review. Am. Ind. Hyg. Assoc. J. 45(3), 162–167 (1984)
4. Tanaka, M., Mizuno, K., Tajima, S., Sasabe, T., Watanabe, Y.: Central nervous system fatigue alters autonomic nerve activity. Life Sci. 84(7), 235–239 (2009)
5. Griffin, M.J.: Subjective equivalence of sinusoidal and random whole body vibration. J. Acoust. Soc. Am. 60(5), 1140–1145 (1976)
6. Olley, M.: Independent wheel suspension its whys and wherefores. Soc. Automot. Eng. J. 34, 73–81 (1934)
7. Best, A.: Vehicle ride-stages in comprehension. Phys. Technol. 15(4), 205 (2002)
8. Olley, M.: National influences on American passenger car design. Proc. Inst. Automobile Eng. 32(2), 509–572 (1938)
9. Olley, M.: Road manners of the modern car. Proc. Inst. Automobile Eng. 41(1), 147–182 (1946)
10. Rowell, H.S., Guest, J.J.: Proc. Inst. Automobile Eng. 18, 455 (1923)
11. Crolla, D., King, R.: Olley's Flat Ride revisited (2000)
12. Sharp, R.: Wheelbase filtering and automobile suspension tuning for minimizing motions in pitch. Proc. Inst. Mech. Eng. Part D: J. Automobile Eng. 216(12), 933–946 (2002)
13. Sharp, R., Pilbeam, C.: Achievability and value of passive suspension design for minimum pitch response. Veh. Ride Handling 39, 243–259 (1993)
14. Odhams, A., Cebon, D.: An analysis of ride coupling in automobile suspensions. Proc. Inst. Mech. Eng. Part D: J. Automobile Eng. 220(8), 1041–1061 (2006)
15. Dai, L., Jazar R.N. (eds.): Nonlinear Approaches in Engineering Applications. Springer, New York. Chapter 1: Smart Flat Ride Tuning (2013). http://www.springer.com/materials/mechanics/book/978-1-4614-6876-9
16. Marzbani, H., Jazar R.N., Fard, M.: Hydraulic engine mounts: a survey. J. Vibr. Control (2013)
17. Marzbani, H., Jazar, R.N., Khazaei, A.: Smart passive vibration isolation: requirements and unsolved problems. J. Appl. Nonlinear Dyn. 1(4), 341–386 (2012)

18. Marzbani, H., Jazar, R. N.: Smart flat ride tuning. Nonlinear Approaches in Engineering Applications 2, pp. 3–36. Springer New York (2014)
19. Marzbani, H.: Flat ride; problems and solutions in vehicle. Nonlinear Eng. 1(3–4), 101–108 (2012)

# Better Road Design for Autonomous Vehicles Using Clothoids

Hormoz Marzbani, Milan Simic, M. Fard and Reza N. Jazar

**Abstract** During the project for generating a mathematical algorithm for autonomous vehicles, a sample road which included different turns and scenarios was required. Studying different types of roads and their line equations Euler spirals, also known as Clothoids were found to be the best solution for designing new roads suitable for autonomous vehicles. During 19th century Arthur Talbot derived the equation of Clothoids to be used as an easement curve for the purpose of avoiding shock and disagreeable lurch of trains, due to instant change of direction. The Euler Spiral is a curve whose degree-of-curve increases directly with the distance along the curve from the start point of the spiral. This provides a linear change in the steering angle required by the driver to go through the turn. In other words for a car traveling on a Clothoid transition road curve there is no need for sudden changes in the steering angle of the wheels. The angle required starts from zero and increases to a maximum value and goes back to zero linearly. This provides a very comfortable ride for the passengers of the vehicle. The use of these curves for road design have been investigated, and a design chart have been proposed to be used for finding the best suitable transition curve for different applications.

**Keywords** Road design · Clotoids · Autonomous vehicles · Vehicle dynamics

## 1 Introduction

Clothoid or "Cornu Spiral", also known as "Euler's Spiral" shown in Fig. 1 has been introduced as the answer to the question of finding the best transition curve. Primary function of a transition curve or easement curve is to accomplish gradual transition from a straight line to a circular curve, so that curvature changes from zero to a finite

H. Marzbani (✉) · M. Simic · M. Fard · R.N. Jazar
School of Aerospace, Mechanical and Manufacturing Engineering,
RMIT University, Melbourne, Australia
e-mail: hormoz.marzbani@rmit.edu.au

© Springer International Publishing Switzerland 2015
E. Damiani et al. (eds.), *Intelligent Interactive Multimedia Systems and Services*,
Smart Innovation, Systems and Technologies 40,
DOI 10.1007/978-3-319-19830-9_24

given value. Spiral between a straight and a curve as a valid transition curve, it has to satisfy the following conditions:

- One end of the spiral should be tangential to the straight line
- The other end should be tangential to the curve
- Spiral's curvature at the intersection point with the circular arc should be equal to arc curvature
- The rate of curvature change along the transition curve should be the same as that of the increase of cant
- Its length should be such that full cant is attained at the beginning of circular cant

The research on trajectory design for mobile robots', can be used to help designing better roads. Several trajectory design methods have been presented using simple curve elements, such as straight lines, circles, clothoids, and their combinations. It is important that a trajectory forms a smooth curve because the mass of a robot or a vehicle must move along the designed road curve. If the curve is not smooth enough, undesirable effects occur such as side slip or deviation from the course at points where the curvature along the curve is not continuous. The history of trajectory design can be viewed as attempts to avoid this inconvenience by using smoother curves [1].

Kanayama [2] believes that including Clothoids in the sure designing, makes curvature control feasible. He used a clothoid pair to make curves with zero curvature at their junctures with line segments in order to produce continuous drive-velocity functions for a differential drive cart. Clothoid pairs have the advantage of providing the minimum-length curves for a given limit on jerk; they have the disadvantage, however, that the $(x, y)$ coordinates of the curve have no closed-form expressions, but must be derived by integrating along the path length, $s$, with the curvature $c(s)$ linearly increasing from zero during the first half of the turn, then linearly decreasing to zero during the second half, in such a way the end-point of the curve precisely matches the position and slope constraints of the succeeding line segment.

It is known today, based on the experience of the past 50 years, that the alignment of modern highways has to be consistent and efficient from a driving dynamic and convincing from a driving psychological point of view. In this connection the Clothoid, as a transition curve, offers good solutions. The clothoid satisfies esthetical solutions and by being flexible, enables a good adaptation to the topography and existing local constraints. The clothoid guarantees an economically efficient ride for motor vehicles, and saves, through its appropriate insertion into the local environment, considerable construction costs [3].

## 2 Definition of Clothoids

Cornu Spirals were introduced, long before being used for path generation for different applications. " The underlying mathematical equation is also most commonly known as the Fresnel Integral. The profusion of names reflects the fact that the curve

**Fig. 1** Euler Spiral and its
parametric equation, i.e.
Fresnel Integrals

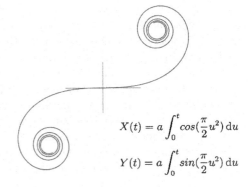

$$X(t) = a \int_0^t cos(\frac{\pi}{2}u^2)\, du$$

$$Y(t) = a \int_0^t sin(\frac{\pi}{2}u^2)\, du$$

has been discovered several different times, each for a completely different application: first, as a particular problem in the theory of elastic springs; second, as a
graphical computation technique for light diffraction patterns; and third, as a railway
transition spiral which is the closest application to the interest of the topic of this
study.

The Euler spiral is defined by Eq. 1 as the curve in which the curvature increases
linearly with arc length, parameter $s$. Equations 1 and 2 show the general form of the
parametric equation of this curve, which are known as the Fresnel integrals. Considering curvature as a signed quantity, it forms a double spiral with odd symmetry, a
single inflection point at the center, as shown in Fig. 1. The first appearance of the
Euler spiral is as a problem of elasticity, posed by James Bernolli in the same 1694
publication as his solution to a related problem, that of elastics.

The elastica is the shape defined by an initially straight band of thin elastic material (such as spring metal) when placed under load at its endpoints. The Euler
spiral can be defined as something of the inverse problem; the shape of a pre-curved
spring, so that when placed under load at one endpoint, it assumes a straight line,
as shown in Fig. 2. In 1744, Euler rephrases the problem as: What shape must the
lamina $amB$ -see Fig. 2- take so that it is flattened in to an exactly straight line when
the free end is pulled down by weight $P$? The answer derived by Euler appeals to

**Fig. 2** Reconstruction of
Euler's figure, with complete
spiral superimposed

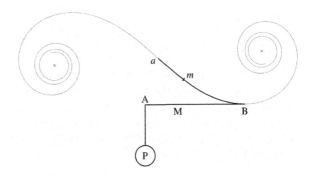

the simple theory of moments: The moment at any point $M$ along the lamina is the force $P$ times the distance $s$ from $A$ to $M$. It took Euler about thirty-eight years to solve the problem of the integral's limits. Around 1818, Augustine Fresnel considered a problem of light diffracting through a slit, and independently derived integrals equivalent to those defining the Euler spiral. The third completely independent discovery of the Euler spiral is in the context of designing railway tracks to provide a smooth riding experience. Over the course of the 19th century, the need for a track shape with gradually varying curvature became clear. Arthur Talbot was among the first to approach the problem mathematically, and derived exactly the same integrals as Bernoulli and Fresnel before him. His introduction to "The Railway Transition Spiral" [4] describes the problem and his solution articulately.

## 3 Clothoids in Road Design

The main area of interest in the existing study is the use of these type of planar curves in designing roads. Roads are made by continuously connecting straight and circular paths by proper transition turning sections. The transition curve is represented by the Euler spiral curve applied to effect the transition between two circular curves or between a circular curve and a tangent.

– Provide a linear gradual increase or decrease of the centrifugal acceleration for the transition from one design element to the other when passing through curves
– Serve as a transition section for a convenient desirable arrangement for the super elevation runoff
– Make possible through the gradual change of the curvature a consistent alignment and through this a consistent operating speed
– Create a satisfactory optical appearance of the alignment

As mentioned earlier the *clothoid spiral* is the best smooth transition connecting curve in road design which is expressed by parametric equations called *Fresnel Integral*:

$$X(t) = a \int_0^t cos(\frac{\pi}{2}u^2)\,du \tag{1}$$

$$Y(t) = a \int_0^t sin(\frac{\pi}{2}u^2)\,du \tag{2}$$

The curvature of the clothoid curve varies linearly with arc length and this linearity makes clothoid the smoothest driving transition curve. The scaling parameter $a$ is only a magnification factor that shrinks or magnifies the curve. Parameter $\mu$ is representing the range of $t$ determines the variation of curvature within the clothoid, as well as the initial and final tangent angles of the clothoid curve. The *arc length, s*, of a clothoid for a given value of $t$ is

$$s = at \tag{3}$$

If the variable $t$ indicates time then, $a$ would be the speed of motion along the path. The *curvature k* and *radius of curvature R* of the clothoid at a given $t$ is

$$k = \frac{\pi t}{a} \tag{4}$$

$$R = \frac{1}{k} = \frac{a}{\pi t} \tag{5}$$

The tangent angle $\theta$ of a clothoid with a given value $t$ is

$$\theta = \frac{\pi}{2} t^2 \tag{6}$$

Having a road with linearly increasing curvature is equivalent to entering the path with a steering wheel at the neutral position and turning the steering wheel with a constant angular velocity. This is a desirable and natural driving action that ensures passengers' comfort, saves on tyres wear and tear expenses and saves energy. One of the most common scenarios in road design is connecting a straight road to a circle as presented in the nest section.

## 3.1 Connecting a Straight Road to a Circle Using a Clothoid

Assume that we need to define a clothoid road section to connect a straight line to a circle. We need to find a clothoid which has a zero curvature to start with, and meet up with a circular curve at the end which means it need to have the same curvature as the curve at the end point. For simplifying we can assume the straight part of the road is the X-axis, and then if needed transfer or rotate it to the required place and orientation. The circle used in this example have a radius of $R = 100$ m at the center point $C(62.811, 106.658)$.

$$(X - 62.811)^2 + (Y - 106.658)^2 = 100^2 \tag{7}$$

As mentioned earlier the transition road must begin with $k = 0$ on the X-axis and touch the circle at a point when its curvature is $k = 1/R = 1/100$. Using

$$k = \frac{\pi s}{a^2} = \frac{\pi t}{a} \tag{8}$$

Combining Eqs. 2 and 3 with 9 we can write parametric equation of the clothoid to be used as the following:

$$X(t) = a \int_0^{\frac{ka}{\pi}} cos(\frac{\pi}{2} u^2) \, du \tag{9}$$

$$Y(t) = a \int_0^{\frac{ka}{\pi}} sin(\frac{\pi}{2}u^2)\,du \tag{10}$$

Knowing the clothoid curvature at the end point, $k = 0.01$, we can find the coordinates of the end point as a function of $a$.

$$X(t) = a \int_0^{\frac{0.01a}{\pi}} cos(\frac{\pi}{2}u^2)\,du, \quad Y(t) = a \int_0^{\frac{0.01a}{\pi}} sin(\frac{\pi}{2}u^2)\,du \tag{11}$$

In the next step, we need to find the magnifying factor, $a$, such that the clothoid meets the circle with the same slope. The slope of the circle at $(X, Y)$ is

$$\acute{Y} = tan\theta = -\frac{X - 62.811}{Y - 106.658} \tag{12}$$

and the slope angle of the clothoid at the end point, where it meets the circle is

$$\theta = \frac{\pi}{2}t^2 = \frac{1}{2\pi}a^2k^2 = 1.5915 \times 10^{-5}a^2 \tag{13}$$

By equating the slope angles of the circle with the end point of the clothoid we'll end up with an equation with respect to $a$.

$$arctan\frac{-(X - 62.811)}{Y - 106.658} = 1.5915 \times 10^{-5}a^2 \tag{14}$$

Using Eqs. 7, 11, and 14, we get and equation which $a$ can be driven from.

$$arctan\frac{-(X - 62.811)}{Y - 106.658} - 1.5915 \times 10^{-5}a^2$$
$$= arctan\frac{-(X - 62.811)}{\sqrt{100^2 - (X - 62.811)^2}} - 1.5915 \times 10^{-5}a^2$$
$$= arctan\frac{-(a\int_0^{\frac{0.01a}{\pi}} cos(\frac{\pi}{2}u^2)\,du - 62.811)}{\sqrt{100^2 - (a\int_0^{\frac{0.01a}{\pi}} cos(\frac{\pi}{2}u^2\,du - 62.811)^2)}} - 1.5915 \times 10^{-5}a^2 \tag{15}$$

solving Eq. 21, $a$ will be found, which in this case:

$$a = 200 \tag{16}$$

Having $a$, the equations of the clothoid can be defined. As mentioned before having the curvature of the end point of the clothoid one can find the point which the clothoid and the circle intersect.

$$X_0 = 122.2596310 \qquad Y_0 = 26.24682756 \tag{17}$$

The slope of the road at the point, the tangent line to the road and the normal line to the road are

$$\theta = \frac{1}{2\pi}a^2k^2 = 0.64\,\text{rad} = 369.475\,\text{deg} \tag{18}$$

$$Y = -64.14007833 + 0.7393029502X \tag{19}$$

$$Y = 191.6183183 - 1.352625469X \tag{20}$$

Figure 3 illustrates the clothoid, tangent line, normal line and the tangent circle to the clothoid at point $(X_0, Y_0)$. Here is another example for finding the right transition curve, to connect a straight road to a circle. This time we will try to find the appropriate transition curve to connect a straight road as *X-axis*, with zero curvature to a circle of radius $R = 80$ m at center $c(100,100)$.

**Fig. 3** The tangent line, normal line, and the tangent *circle* to the clothoid at the point where $k = 0.01$ for a given $a = 200$. The clothoid, black line, is plotted *up* to $k = 0.025$

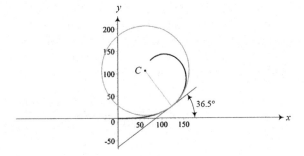

$$(X - 100)^2 + (Y - 100)^2 = 80^2 \tag{21}$$

Using

$$X(t) = a \int_0^{\frac{ka}{\pi}} cos(\frac{\pi}{2}u^2)\,du \tag{22}$$

$$Y(t) = a \int_0^{\frac{ka}{\pi}} sin(\frac{\pi}{2}u^2)\,du \tag{23}$$

and knowing $k = 1/R = 0.0125$ at the destination point, determines the coordinates of the end of the clothoid as function of $a$.

$$X(t) = a \int_0^{\frac{0.0125a}{\pi}} cos(\frac{\pi}{2}u^2)\,du \tag{24}$$

$$Y(t) = a \int_0^{\frac{0.0125a}{\pi}} \sin(\frac{\pi}{2}u^2)\,du \tag{25}$$

The slope of the tangents to the circle 21, at a point $(X,Y)$ is

$$\acute{Y} = tan\theta = -\frac{X-100}{Y-100} \tag{26}$$

and the slope angle of the clothoid at its end point as a function of $a$ is,

$$\theta = \frac{\pi}{2}t^2 = \frac{1}{2\pi}a^2k^2 = 2.4868 \times 10^{-5}a^2 \tag{27}$$

It is important that the clothoid and circle have the same slope angle at the end point of the clothoid, which leads us to the following equation by equating the two slope angles.

$$arctan\frac{-(X-100)}{Y-100} = 2.4868 \times 10^{-5}a^2 \tag{28}$$

Equations 21 and 28 along with the equation of the clothoid, gives us as equation with respect to $a$. However, substituting $Y = Y(X)$ and replacing $tan$ and $arctan$ generates four equations to be solved for possible $a$. To visualize the possible solutions, let us define four error equations.

$$e = arctan\frac{-(X-100)}{\pm\sqrt{80^2 - (X-100)^2} - 2.4868 \times 10^{-5}a^2} \tag{29}$$

$$e = \frac{-(X-100)}{\pm\sqrt{80^2 - (X-100)^2} - tan(2.4868 \times 10^{-5}a^2}) \tag{30}$$

Figure 4 depicts Eq. 29 and Fig. 5 shows Eq. 30. From the first one we get the solutions,

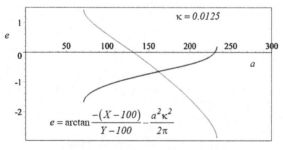

**Fig. 4** Plot of $e = arctan\dfrac{-(X-100)}{\pm\sqrt{80^2-(X-100)^2}-2.4868\times10^{-5}a^2}$ versus $a$

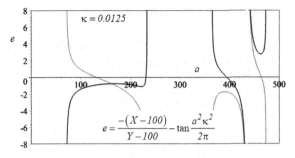

**Fig. 5** Plot of $e = \dfrac{-(X-100)}{\pm\sqrt{80^2-(X-100)^2}-tan(2.4868\times10^{-5}a^2}$ ) versus $a$

$$a = 230.7098693 \qquad a = 130.8889343 \tag{31}$$

and the second equation gives the following as the solutions,

$$a = 230.7098693 \quad a = 130.8889343 \quad a = 394.0940573 \quad a = 463.5589702 \tag{32}$$

**Fig. 6** The transition road starting on X-axis and goes to a *circle* of radius R = 80 m at center c(100,100)

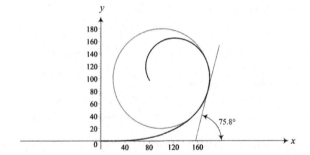

The correct answer is $a = 230.7098693$ and Fig. 6 depicts the circle and the proper clothoid, Having $a$ defines the clothoid equation which at $k = 0.0125$ reaches to

$$X_0 = 177.5691613 \qquad Y_0 = 82.38074640 \tag{33}$$

at angle, $\theta = \dfrac{1}{2\pi}a^2k^2 = 1.3236\,\text{rad} \approx 75.84\,\text{deg}$. There are exceptions using the method which has been introduced here, as well. It is sometimes needed to shift the clothoid in order to meet the given circle. It is not generally possible to design a clothoid starting at the origin and meet a given circle at an arbitrary center and radius. However, it is possible to start the clothoid from other points on the X-axis rather than the origin to meet the given circle. Following is the equation of the given circle for which we need to find a transition clothoid curve,

$$(x - x_c)^2 + (y - y_c)^2 = R^2 \tag{34}$$

where $(x_c, y_c)$ is the circle center and the following equations indicate the clothoid that should have the same radius of curvature as the circle.

$$X(t) = a \int_0^{\frac{a}{R\pi}} cos(\frac{\pi}{2}u^2)\, du \tag{35}$$

$$Y(t) = a \int_0^{\frac{a}{R\pi}} sin(\frac{\pi}{2}u^2)\, du \tag{36}$$

where $t$ has been expressed in terms of $a$ and $R$.

$$t = \frac{a}{R\pi} = \frac{ak}{\pi} \tag{37}$$

Equating the slope angle of the clothoid

$$tan\theta = tan(\frac{\pi t^2}{2}) = tan(\frac{a^2}{2\pi R^2}) \tag{38}$$

and the slope angle of the circle

$$tan\theta = -\frac{x - x_c}{y - y_c} \tag{39}$$

at a point $x$, $y$, we get

$$tan(\frac{a^2}{2\pi R^2}) = -\frac{x - x_c}{y - y_c} \tag{40}$$

Searching for a match point in the upper half of the circle

$$y - y_c = \sqrt{R^2 - (x - x_c)^2} \tag{41}$$

makes the slope equation to be a function of $a$

$$tan(\frac{a^2}{2\pi R^2})\sqrt{R^2 - (a \int_0^{\frac{a}{R\pi}} cos(\frac{\pi}{2}u^2)\, du - x_c)^2} + a \int_0^{\frac{a}{R\pi}} cos(\frac{\pi}{2}u^2)\, du - x_C = 0 \tag{42}$$

Solution of this equation provides us with an $a$ for which the clothoid ends at a point with the same curvature as the circle. At the same $y$ of the end point, the slope of the clothoid is also equal to the slope of the circle. A proper shift to the clothoid of the X-axis will match the clothoid and the circle. The following example will clarify the problem. Assume the given circle is

$$(x - 60)^2 + (y - 60)^2 = 50^2 \tag{43}$$

and therefore, the slope equation will be

$$\tan(\frac{a^2}{2\pi2500})\sqrt{2500 - (a\int_0^{\frac{a}{50\pi}} \cos(\frac{\pi}{2}u^2)\,du - 60)^2} + a\int_0^{\frac{a}{50\pi}} \cos(\frac{\pi}{2}u^2)\,du - 60 = 0 \tag{44}$$

The Numerical solution of the equation is

$$a = 132.6477323 \tag{45}$$

The plot of the clothoid and the circle with this scenario are shown in Fig. 7. For the calculated $a$, the value of $t$ at the end point of the clothoid is

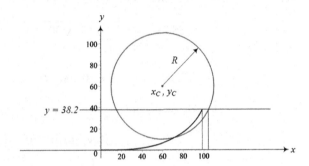

**Fig. 7** clothoid starting at origin ends at a point with the same slope and curvatureand $y$ as a given circle

$$t = \frac{a}{R\pi} = \frac{2.652954646}{\pi} = 0.84446 \tag{46}$$

and therefore, the coordinates of the end point are

$$x(k) = 132.6\int_0^{0.84} \cos(\frac{\pi}{2}u^2)\,du = 98.75389126 \tag{47}$$

$$y(k) = 132.6\int_0^{0.84} \sin(\frac{\pi}{2}u^2)\,du = 38.22304651 \tag{48}$$

At the point, the radius of curvature of the clothoid is

$$R = \frac{1}{k} = \frac{a}{t\pi} = \frac{132.6477323}{0.84446\pi} = 50 \tag{49}$$

**Fig. 8** Shifted clothoid
from a point on *x*-axis ends
on a given *circle* with same
slope and curvature

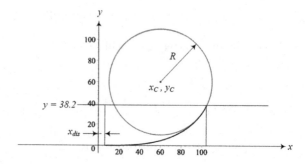

and the slope is

$$\theta = \frac{\pi}{2}t^2 = \frac{\pi}{2}0.84446^2 = 1.1202\text{rad} \tag{50}$$

The *x*-coordinate of the circle at the same $y = 38.22304651$,

$$38.22304651 - 60 = \sqrt{50^2 - (x - 60)^2} \tag{51}$$

is

$$x_{circle} = 105.0084914 \tag{52}$$

If we shift the clothoid by the difference between $x_{circle}$ and $x_{clothoid}$

$$x_{dis} = x_{circle} - x_{clothoid} = 105.0084919 - 98.75389126 = 6.2546 \tag{53}$$

Then the clothoid and circle meet at a point on the circle with all requirements to have a smooth transition. Figure 8 illustrates the result.

## 3.2 Design Chart

Figure 9 illustrates a design graph of the relationship between the clothoid and parameters of magnification factor *a*, curvature *k*, and slope *θ*. The higher the magnification factor the larger the clothoid will be. The clothoid curves of different *a* are intersected by the constant slope lines of *θ*. The curves of constant curvature *k* intersect both, the constant *a* and constant *θ* curves. Assume we are trying to find a transition road with the clothoid shape to connect a straight road to a circle of radius $R = 58.824$ m. Having *R* is equivalent to have the destination curvature $k = 1/R = 0.017$. Desired circle must be tangent to a clothoid with a given *a* at the point that the clothoid is intersecting the curve of $k = 0.017$. The clothoid for $a = 250$ m hits the curve of $k = 0.017$ at a point for which we have

**Fig. 9** Design graph of
relating clothoid and
parameters: magnification
factor $a$, curvature $k$, slope $\theta$

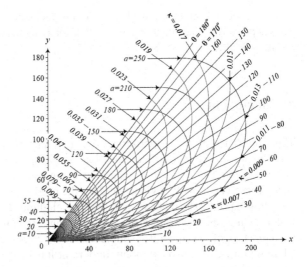

$$X = 147.39 \, \text{m} \qquad Y = 176.44 \, \text{m} \tag{54}$$

$$\theta = 164.71 \, \text{deg} \qquad s = 338.2 \, \text{m} \tag{55}$$

The clothoid for $a = 210$ m hits the curve of $k = 0.017$ at a point for which we have

$$X = 157.47 \, \text{m} \qquad Y = 119.71 \, \text{m} \tag{56}$$

$$\theta = 116.22 \, \text{deg} \qquad s = 238.64 \, \text{m} \tag{57}$$

**Fig. 10** Few clothoid
transition road sections
connecting a straight road to
a *circle* of radius
$R = 58.824$ m

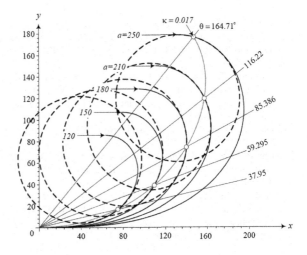

and so on. Figure 10 illustrates these solutions and some more. The number of solutions is practically infinite and the best solution depends on safety cost, and physical constraints of the field [5–8].

# 4 Conclusion

The importance of using appropriate transition curves, for sustainable road design was explained. An appropriate transition curve should help with reducing the distance between the two sections which are being connected in the shortest time possible. These factors will obviously reduce the need for higher speeds which reduces fuel consumption. The other advantage of the clothoids besides the above mentioned factors is the linear change in the steering angle required by the driver. This will make the task of steering much simpler and more consistent while travelling through Clothoid turns. It is also extremely important for the passenger comfort. This prevents sudden and excessive changes in acceleration/deceleration and motion direction chenages. Following that we have less tyre wear and tear costs and save on energy. Examples were used to show the best and easiest ways for finding the appropriate equation of a Clothoid which connects different types of road sections to each other. As the final stage, a design chart generated as a result of calculation for many different scenarios. This design chart can be used to find appropriate numerical values which are required to be substituted into the equation of a Clothoid with respect to the sections on its both sides.

# References

1. Komoriya, K., Tanie, K.: Trajectory design and control of a wheel type mobile robot using B-spline curve. Intelligent Robots and Systems '89. The Autonomous Mobile Robots and Its Applications. IROS '89. In: Proceedings of the IEEE/RSJ International Workshop on, pp. 398–405, 4–6 Sep 1989
2. Kanayama, Y., Miyake, N.: Trajectory generation of mobile robots. Robotics Research: The Third International Symposium (1985). ISBN: 0262061015
3. Lamm, R., Psarianos, B., Mailaender, T.: Highway Design and Traffic Safety Engineering Handbook, McGraw-Hill (1999)
4. Talbot, A.N.: The Railway Transition Spiral. U. Illinois, 1899. Reprinted from the Technograph No. 13
5. Jazar, R.N.: Vehicle Dynamics: Theory and Application. Second Edition Springer (2014)
6. Marzbani, H., Jazar, R.N., Fard, M.: Hydraulic engine mounts: a survey. J. Vibration Control (2013)
7. Marzbani, H., Reza, N., Jazar, R.N., Khazaei, A.: Smart passive vibration isolation: requirements and unsolved problems. J. Appl. Nonlinear Dyn. 1(4), 341–386 (2012)
8. Marzbani, H., Jazar, R.N.: Smart flat ride tuning. Nonlinear Approaches in Engineering Applications 2, pp. 3–36. Springer New York (2014)

# Towards Formal Modelling of Autonomous Systems

**Maria Spichkova and Milan Simic**

**Abstract** Autonomous systems perform decision making without human intervention. They collect the data from the environment, process it to build the awareness and perform the actions. Consequently, the adaptivity functions can be seen as the core parts of these systems. In this paper we introduce a formal framework for modelling and analysis of autonomous systems and their compositions, especially focusing on the adaptivity modelling aspects and reasoning about adaptive behaviour.

**Keywords** Autonomous and adaptive systems · UGV · UAV · Modelling · Formalisation · Verification

## 1 Introduction

As the term *autonomous systems* (AS) already suggests, autonomy is one of the essential characteristics of such systems. Although that AS can refer to various systems and environments, like Internet, or biosystems, we will put emphasis on robotics. Autonomous computing, applied in robotics, aims at unburdening human administrators from complex and hazardous tasks. This allows to use the system in areas, or situations, which could be dangerous to humans or which are remote and hardly accessible. Another crucial advantage of this systems is the reliability of the systems in the cases when it could be complicated for the human to solve some tasks quickly, precisely and safely, e.g. in the case of Parking Assists Systems [8] or more demanding tasks like autonomous Unmanned Aerial Vehicle (UAV) landing [17].

M. Spichkova (✉)
School of Computer Science and Information Technology, RMIT University, Melbourne, Australia
e-mail: maria.spichkova@rmit.edu.au

M. Simic
School of Aerospace, Mechanical and Manufacturing Engineering, RMIT University, Melbourne, Australia
e-mail: milan@rmit.edu.au

© Springer International Publishing Switzerland 2015
E. Damiani et al. (eds.), *Intelligent Interactive Multimedia Systems and Services*,
Smart Innovation, Systems and Technologies 40,
DOI 10.1007/978-3-319-19830-9_25

279

Human factors, targeted by the Engineering Error Paradigm [20], typically include the design of the user interfaces, human machine interface (HMI), as well as, any corresponding automation. In this paradigm people are seen as being equivalent, i.e. seen as software and hardware components, in the sense of operations with data and other components. At the same time humans are seen as being the most unreliable constituent of the whole system. They are often inconsistent and unreliable. This implies that designing humans role out of the main system actions, through automatisation of some key system functions, is considered a good proposal for reducing risk.

Thus, *adaptivity* functions are the core part of these systems. The simplest form of autonomy is a rule engine following a predefined set of instructions represented by conditional statements (e.g. *if_then_else*). In many cases, such a simple rule-based mechanism, however, may not suffice. The adaptation and context-awareness can have many forms: navigation applications to guide users to a given destination, auto-connection of the mobile-phone functions with the headset functions, robot motion [10], keyless entry systems [13], etc. A general roadmap of the research fields for adaptive systems is given in [6].

As already mentioned, autonomous systems perform complex tasks without human intervention. In comparison to the decision support systems (DSSs), autonomous systems not only provide an assistance for the users of the system, but need to determine their behaviour based only on the predefined behaviours patterns and the adaptivity algorithms. DDSs, however, also can be viewed as a special way of partially designing humans out of the main system actions, particularly if we assume that the human will follow the decision recommended by the support system, cf., e.g., [3, 14, 18]. Driver assistance applications, as given in [21] are a step forward in that direction. The first steps are just informing messages, or warning messages in critical situations, like blind spot warning, line change detection and similar. The next step, which is, also, already there, is giving full control to the vehicle, like in autonomous parking application, available today with many modern vehicles. The process of transition from the human control of the vehicle, to the fully autonomous driving, goes through the following stages:

1. Human Driving (HD) without automated assistance. Level of autonomy is 0 %;
2. HD with warning messages: Blind spot warning, line change warning, parking sensors, vehicular ad hoc network (VANET) applications. In a VANET network [22], vehicles are communicating with each other and exchange warning messages. Sensory data from one vehicle are available to other as well. Vehicle has perception about environment well ahead and behind its instant location.
3. Partial Autonomy: Adaptive cruse control, automatic parking.
4. Full Autonomy: Unmanned Ground Vehicle (UGV), Unmanned Aerial Vehicle (UAV) and other. Level of autonomy is 100 %.

*Contribution:* We propose the development of a formal framework that supports the modelling and analysis of autonomous systems and their compositions. In our approach, we especially focus on the adaptivity-modelling aspects and dependencies between the autonomous systems, as well as on reasoning about adaptive behaviour.

*Outline:* This paper is organised as follows. After this Introduction, in Sect. 2, we present the main contribution of our approach. It is in setting the foundations for the formal definition of autonomous system behaviour and performances. In the Sect. 3 we discuss the application of our approach using selected examples of the autonomous systems. In the Sect. 4 we present conclusions and suggestion for the future work.

## 2 System Behaviour

Broy et al. [4] investigated the notion of adaptive system behaviour also presenting a mathematical foundations to their intuition. They classified adaptive behaviour to three groups: *non-transparent* (the user can not observe the implicit inputs from the environment that influence the system behaviour), *transparent* (the user observes implicit inputs from the environment, but is not able to influence or control the resulting system reactions), and *diverted* (the user is also able to influence or control the system reactions, resulting from the implicit inputs).

Autonomous systems continuously exchange information with their environment, to adapt to the changing conditions and to choose the most appropriate behaviour strategy.[1] In contrast to a non-adaptive behaviour, where the system behaviour is determined only by *explicit* user inputs, the adaptivity characteristics means that behaviour of the considered system is determined by its environment (which is especially important for autonomous systems) or by *implicit* user inputs (as such inputs we can see, e.g., patterns in the user interaction with the system). At a certain abstraction level, it is formally insignificant for the autonomous system, whether the input comes directly from a user or a data acquisition sensor: the system just reacts on the received inputs, according to the predefined behavioural patterns. However, it could make a difference for the users, whether they can observe the inputs (and know/derive an abstract behaviour pattern) which come from the environment.

A concise survey of concepts, architectural frameworks, and design methodologies that are relevant in the context of self-adapting and self-optimizing systems is presented in [1]. Following the ideas presented in [1] for adaptive systems in general, we specify the behavioural architecture of an autonomous system as presented on Fig. 1. The set of communication channels $I_U$ is optimal in the sense that it is restricted (e.g., in some cases it contains only channels for system initialization/start). Thus, direct human interaction is very restricted, in contrast to the adaptive systems in general.

In this section, we present a formal approach to specify the behaviour of autonomous systems. The modelling language that we use in our approach is FOCUS$^{ST}$ [29]. It allows us to create concise but easily understandable specifications and is appropriate for application of the specification and proof methodology presented in [24, 31]. This methodology allows writing specifications in a way that

---

[1] We distinguish here the terms *user* (human being) and *environment* (environment exempt humans) to have a clear separation between autonomous and non-autonomous systems.

carrying out proofs is quite simple and scalable to practical problems. Compared with testing, *verification* means a correctness proof for system *properties* (which may hold for an infinite number of inputs), but requires significant effort, especially if we refer to verification of the code. Thus, verification should be performed only for safety-critical systems, e.g., the most critical parts of automotive systems [15, 16, 23].

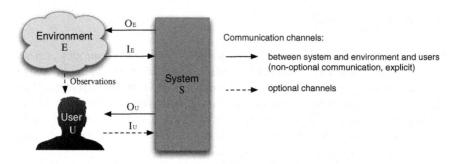

**Fig. 1** Autonomous system behaviour: Architecture of the system

## 2.1 Focus<sup>ST</sup>

The FOCUS<sup>ST</sup> language was inspired by FOCUS [5], a framework for formal specification and development of interactive systems. In both languages, specifications are based on the notion of *streams* and channels (a channel is in effect a name for a stream). However, in the original FOCUS input and output streams of a component are mappings of natural numbers $\mathbb{N}$ to single messages, whereas a FOCUS<sup>ST</sup> stream is a mapping from $\mathbb{N}$ to lists of messages within the corresponding time intervals. Moreover, the syntax of FOCUS<sup>ST</sup> is particularly devoted to specify spatial (S) and timing (T) aspects in a comprehensible fashion, which is the reason to extend the name of the language by <sup>ST</sup>. The FOCUS<sup>ST</sup> specification layout also differs from the original one: it is based on human factor analysis within formal methods [26, 27].

In particular, a FOCUS<sup>ST</sup> specification can be translated to a Higher-Order Logic and verified by the interactive semi-automatic theorem prover Isabelle [19] also applying its component Sledgehammer [2]. Sledgehammer employs resolution based first-order automatic theorem provers (ATPs) and satisfiability modulo theories (SMT) solvers to discharge goals arising in interactive proofs. Another advantage is a well-developed theory of composition as well as the representation of processes within a system [25]. The collection of FOCUS<sup>ST</sup> operators over timing aspects and their properties specified and verified using the theorem prover Isabelle is presented in the Archive of Formal Proofs [28]. In this work we focus on modelling of spatial aspects.

We describe the behaviour of a component (or a system) by relations between its inputs and outputs. For a system $S$, we denote its syntactic interface by $(I_S \triangleright O_S)$, where $I_S$ and $O_S$ are sets of timed input and output streams respectively. We specify every component using assumption-guarantee-structured templates. This helps avoiding the omission of unnecessary assumptions about the system's environment since a specified component is required to fulfil the guarantee only if its environment behaves in accordance with the assumption. In a component model, one often has transitions with local variables that are not changed. Also, outputs are often not produced, e.g., when a component gets no input or some preconditions necessary to produce a nonempty output are violated. In many formal languages this kind of invariability has to be defined explicitly in order to avoid underspecified component specifications. To make our formal language better understandable for programmers, we use in FOCUS$^{ST}$ so-called *implicit else-case* constructs. That means, if a variable is not listed in the guarantee part of a transition, it implicitly keeps its current value. An output stream not mentioned in a transition will be empty. Further, we do not require to introduce auxiliary variables explicitly: The data type of a not introduced variable is universally quantified in the specification such that it can be used with any data value.

## 2.2 Autonomous Systems: Formal Definitions

As shown on Fig. 1, an autonomous system $S$ has the syntactic interface $(I_E \cup I_U \triangleright O_E \cup O_U)$, whereby the set of input channels $I_U$ might be empty. The syntactic interfaces of user and environment are denoted by $(O_U \cup Observations \triangleright I_U)$ and $(O_E \triangleright I_E)$, respectively. The set *Observations* might be empty too, which leads to a non-transparent behaviour of the system (from the user point of view), i.e., the interaction takes place only between the system and its environment and between the subject and the considered system.

The system $S$ can be seen as a composite component. It is important to trace back which parts of the system provide information to a certain (sub)component, and on which input charnels from the set $I_E \cup I_U$ depends each of channels from $O_U$. To answer these questions, we extend the approach on formal analysis of dependencies between services [30] to be applicable for components (the basic difference between services and components here is in their partiality/totality: a service is defined as a partial function from input streams to output streams, where a component is represented, in general, by a total function). While modelling communication between components on a certain abstraction level $L$ (i.e. level of refinement/decomposition), we specify the dependencies by the function

$$Sources^L : CSet^L \rightarrow (CSet^L \; set)$$

$CSet^L$ denotes here the set of components at the level $L$. $Sources^L$ returns for any service $A$ the corresponding (possibly empty) set of components that are the sources for the input streams of $A$. More precisely, all the dependencies can be divided into the direct and indirect ones.

Values of an output channel $y \in \mathbb{O}(C)$ of a component $C$ do not necessarily depend on the values of all its input streams. This means that an optimisation of system architecture may be needed in order to localise these dependencies as well as to analyse the relations between influence grade of user's and environment's input on the system behaviour. To express any restrictions we use the following notation: $\mathbb{I}^D(C, y)$ denotes the subset of $\mathbb{I}(C)$ that $y$ depends upon. There are three possible cases to consider:

1. $y$ depends on all input streams of $C$: $\mathbb{I}^D(C, y) = \mathbb{I}(C)$;
2. $y$ depends on some input streams of $C$: $\mathbb{I}^D(C, y) \subset \mathbb{I}(C)$;
3. $y$ is independent of any input stream of $C$, i.e., $\mathbb{I}^D(C, y) = \emptyset \neq \mathbb{I}(C)$.

If the input part of the service's interface is specified correctly in the sense that the service does not have any "unused" input channels, the following relation will hold:

$$\forall x \in \mathbb{I}(C).\ \mathbb{O}^D(C, x) \neq \emptyset.$$

On each abstraction level $L$ of logical architecture, we can define a function $Acc^L$ : $CSet^L \rightarrow (CSet^L\ set)$. For any service (name) $A$ the function $Acc^L$ returns the corresponding (possibly empty) set of services (names) $B_1, \ldots, B_{AN}$ that are the acceptors for the output streams of $A$. This function dual to the function $Sources^L$: $x \in Acc^L(y) \iff y \in Sources^L(x)$.

This model allows us to analyse the influence of a service's failure on the functionality of the overall system.

## 3 Examples of Autonomous Systems

In this section, we present a number of examples of autonomous systems developed at the RMIT University, and discuss the behavioural modelling aspects for them. In all presented cases we deal with the full autonomy (level of autonomy is 100 % and focus on the development of the Unmanned Ground Vehicles (UGVs). Currently, there is a huge research interest to the autonomy introduced in UGVs. There are much more intelligent AS to analyse, if we consider other mobile robots, and/or vehicles as Unmanned Aerial Vehicle (UAV), Autonomous Underwater Vehicle, Space Vehicles, or Unmanned Underground Mining Vehicle.

**(a)**        **(b)**

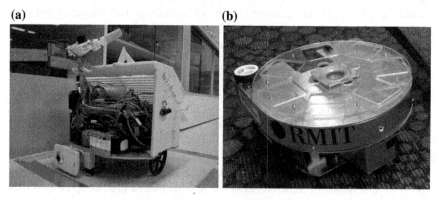

**Fig. 2** Intelligent mobile robots designed at the RMIT University. **a** AS for mining. **b** AS for agriculture

In Fig. 2 we can see intelligent, mobile robots designed to perform various tasks. The first one, as shown in the Fig. 2a, is designed as mini model of an autonomous system that could be used in mining operations, while the second one (Fig. 2b) is a model for agriculture AS applications. More on AS projects in agriculture can be found in [9]. An UGV with control system, designed by RMIT University students, is given in Fig. 3. The vehicle is used for the research in autonomous systems path planning. Figure 4 is another step in the systems investigation, where the logical architecture of this UGV and the dependencies within the system are presented. Global Positioning System (GPS) module and Compass are used for global AS positioning. Laser, i.e. Light Detection and Ranging (LIDAR) module, and stereo vision with two cameras are used for obstacle detection. The main data processing is conducted on laptop using MATLAB and Python programming languages and environment. Speed tracking and motion control, through throttle and steering control, are performed by a dedicated controller. It also gives error messages through diagnostics lights. Finally, user interaction is enabled by the application of mechanical and wireless emergency stops. They are in series connection so that any of those can safely stop the system independently.

**Fig. 3** An electrical UGV with the control designed at the RMIT University

Building the *Sources*$^L$ and *Acc*$^L$ sets for each component, we can analyse dependencies between the components and automatically detect possible sources of errors and trace the influences of the adaptivity functions. As we can see from Fig. 4, the data flow from the *Throttle Controller* to the *Drive Motor* is independent from the data flow from the *Steering Controller* to the *Steering Motor*. Thus if the system has problems with the driving function only, we can automatically exclude from the possible fault causes the following components: *Steering Motor*, *Relay1*, *Relay2*,

*Steering Controller*, *E-Stop*, *Wireless E-Stop*, *Relay Board*, and *Lights*. We can also exclude the *Compass* component, as its outputs do not influence any local variables of the *Laptop* and *C Controller* components, which are used to compute the input for the *Throttle Controller*.

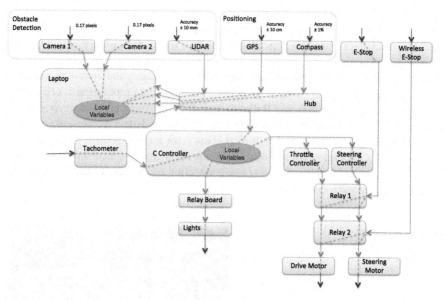

**Fig. 4** Analysis of the dependencies within the UGV, cf. also Fig. 3

Schematic diagram that can represent just motion control, for all of the shown examples, and more in the category of the mobile robots, is shown in Fig. 5. Mobile robot becomes aware of the environment through the process of perception, i.e. data acquisition and processing. Simultaneous Localisation and Mapping (SLAM)

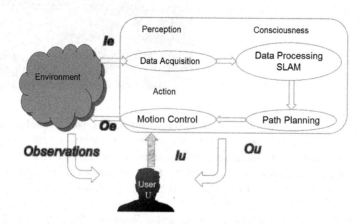

**Fig. 5** Mobile robots motion control of the electrical UGV, cf. also Fig. 3

is performed using data from various sensors. Diagrams for various task executions, like mining samples collections, or seeds planting, are not shown as they are performed by other algorithms, which are task-specic.

Communication channels $I_E$, $O_E$, $I_U$, $O_U$ and *Observations* are labeled the same and have the same meaning as shown in Fig. 1. System $S$ is presented with more details about the functionality and data flow directions internally. After the awareness about the environment and the self placement in it is obtained, autonomous systems can perform action, i.e. path planning and motion. More on path planning and motion control could be found in [7, 11, 12].

# 4 Conclusions and Future Work

In this paper we have introduced a formal framework for modelling and analysis of autonomous systems and their compositions. The focus of our work is on the adaptivity-modelling aspects and dependencies between the autonomous systems, as well as on reasoning about adaptive behaviour. The proposed framework allows automatically detect possible sources of errors and trace the influences of the adaptivity functions, based on the dependencies between the components.

In the future work, we intent to automatise the construction of the functions that specify the dependencies for a concrete case. One of the other possible directions of our future work is analysis the data/control flow dependencies between components with respect to the performance and reliability properties of AS.

# References

1. Bauer, V., Broy, M., Irlbeck, M., Leuxner, C., Spichkova, M., Dahlweid, M., Santen, T.: Survey of modeling and engineering aspects of self-adapting & self-optimizing systems. Technical Report TUM-I1324, TU München (2011)
2. Blanchette, J.C., Böhme, S., Paulson, L.C.: Extending Sledgehammer with SMT solvers. In: Børner, N., Sofronie-Stokkermans, V.: Automated Deduction, vol. 6803, pp. 116–130. LNCS. Springer (2011)
3. Bonczek, R.H., Holsapple, C.W., Whinston, A.B.: Foundations of Decision Support Systems. Academic Press, Oakland (1981)
4. Broy, M., Leuxner, C., Sitou, W., Spanfelner, B., Winter, S.: Formalizing the notion of adaptive system behavior. In: Proceedings of the 2009 ACM Symposium on Applied Computing, pp. 1029–1033. ACM (2009)
5. Broy, M., Stølen, K.: Specification and Development of Interactive Systems: Focus on Streams, Interfaces, and Refinement. Springer, New York (2001)
6. Cheng, B.H.C., de Lemos, R., Giese, H., Inverardi, P., Magee, J.: Software Engineering for Self-Adaptive Systems: A Research Roadmap. LNCS (2009)
7. Elbanhawai, M., Simic, M.: Continuous-curvature bounded trajectory planning Using parametric splines. Frontiers in Artificial Intelligence and Applications (2014)
8. Elbanhawai, M., Simic, M.: Examining the use of B-splines in parking assist systems. Applied Mechanics and Materials **490491**, 1025–1029 (2014)

9. Elbanhawai, M., Simic, M.: Randomised kinodynamic motion planning for an autonomous vehicle in semi-structured agricultural areas. Biosystems Engineering (2014)
10. Elbanhawai, M., Simic, M.: Sampling-based robot motion planning: a review. IEEE Access **30**(99), (2014)
11. Elbanhawai, M., Simic, M., Jazar, R.N.: Continuous path smoothing for car-like robots using B-spline curves. J. Intell. Robotic Syste. **77**, (2015)
12. Elbanhawi, M., Simic, M.: Sampling-based robot motion planning: a review. IEEE Access **2**, 56–77 (2014)
13. Feilkas, M., Hölzl, F., Pfaller, C., Rittmann, S., Schätz, B., Schwitzer, W., Sitou, W., Spichkova, M., Trachtenherz, D.: A refined top-down methodology for the development of automotive software systems - The Keyless Entry System Case Study. Technical Report TUM-I1103, TU München (2011)
14. Fick, G., Sprague, R.H.: Decision Support Systems: Issues and Challenges. Pergamon Press, Oxford (1980)
15. Kühnel, C., Spichkova, M.: Upcoming automotive standards for fault-tolerant communication: FlexRay and OSEKtime FTCom. In: Proceedings of EFTS (2006)
16. Kühnel, C., Spichkova, M.: Fault-tolerant communication for distributed embedded systems. In: Software Engineering and Fault Tolerance, Series on Software Engineering and Knowledge Engineering (2007)
17. Lu, K., Li, Q., Cheng, N.: An autonomous carrier landing system design and simulation for unmanned aerial vehicle. In: IEEE Chinese on Guidance, Navigation and Control Conference (CGNCC), pp. 1352–1356 (2014)
18. Marakas, G.M.: Decision Support Systems, 2nd edn. Prentice-Hall, Upper Saddle River (2003)
19. Nipkow, T., Paulson, L.C., Wenzel, M.: Isabelle/HOL—a proof assistant for higher-order logic, vol. 2283. LNCS, Springer (2002)
20. Redmill, F., Rajan, J.: Human Factors in Safety-Critical Systems. Butterworth-Heinemann, Oxford (1997)
21. Simic, M.: Vehicle and public safety through driver assistance applications. In: Proceedings of the 2nd International Conference Sustainable Automotive Technologies (ICSAT 2010), vol. 490491, pp. 281–288 (2010)
22. Simic, M.N.: Vehicular ad hoc networks. In: 11th International Conference on Telecommunication in Modern Satellite, Cable and Broadcasting Services (TELSIKS), vol. 02, pp. 613–618 (2013)
23. Spichkova, M.: FlexRay: verification of the FOCUS specification in Isabelle/HOL. A Case Study. Technical Report TUM-I0602, TU München (2006)
24. Spichkova, M.: Specification and seamless verification of embedded real-time systems: FOCUS on Isabelle. PhD thesis, TU München (2007)
25. Spichkova, M.: Focus on processes. Technical Report TUM-I1115, TU München (2011)
26. Spichkova, M.: Human factors of formal methods. In: In IADIS Interfaces and Human Computer Interaction 2012, IHCI 2012 (2012)
27. Spichkova, M.: Design of formal languages and interfaces: "Formal" does not mean "unreadable". IGI Global (2013)
28. Spichkova, M.: Stream processing components: Isabelle/HOL formalisation and case studies. Archive of Formal Proofs, 2013. Formal proof development
29. Spichkova, M., Blech, J., Herrmann, P., Schmidt, H.: Modeling spatial aspects of safety-critical systems with focus ST. In: 11th Workshop on Model Driven Engineering, Verification and Validation (MoDeVVa) (2014)
30. Spichkova, M., Schmidt, H.: Towards logical architecture and formal analysis of dependencies between services. In: APSCC'14: Proceedings of the 2014 IEEE Asia-Pacific Services Computing Conference, CPS (2014)
31. Spichkova, M., Zhu, X., Mou, D.: Do we really need to write documentation for a system? In: International Conference on Model-Driven Engineering and Software Development (MODELSWARD'13) (2013)

# Developing a Navigation System
# for Mobile Robots

Jeffery Young, Mohamed Elbanhawi and Milan Simic

**Abstract** Design solution of a novel mobile robot navigation system, presented here, is used to control robot's locomotion across slippery surfaces. Usually, motion control strategies, are based on assumption of sufficient traction between tyres and the road. Motion across slippery surfaces can endanger the robot and its surroundings. Our solution combines Light Detection and Ranging (LIDAR) measurements with odometry data. It performs well on any surface, regardless of sensing, localization and navigation errors, within an indoor environment, in real-time. An accelerated feature detection method is used to improve LIDAR localization update rate and improve localization accuracy. Experiments conducted validate proposed approach.

**Keywords** Real-time · Localization · LIDAR · Hough transform · Gyroscope · Motion

## 1 Introduction

Majority of mobile robots move on various surfaces with limited knowledge of the surface properties, such as friction coefficient. Wheel slipping can cause poor localization results [1] and could lead to task failure and collision. Often, techniques employed in mobile robotics lack to provide reliable data for safe autonomous navigation. Traditionally, robots perform pre-programmed sequences of operations

J. Young (✉) · M. Elbanhawi · M. Simic
School of Aerospace Mechanical and Manufacturing Engineering, RMIT University,
Melbourne, Victoria, Australia
e-mail: S3314581@student.rmit.edu.au

M. Elbanhawi
e-mail: mohamed.elbenhawi@rmit.edu.au

M. Simic
e-mail: milan.simic@rmit.edu.au

© Springer International Publishing Switzerland 2015     289
E. Damiani et al. (eds.), *Intelligent Interactive Multimedia Systems and Services*,
Smart Innovation, Systems and Technologies 40,
DOI 10.1007/978-3-319-19830-9_26

in constrained and predicted environments, like path following, and are not able to operate in new environments, or to face unexpected dynamic situations. There is an emerging need for truly autonomous systems. Applications include intelligent service robots for offices, hospitals and factory floors. In such conditions, an additional capability is needed, to maintain a sufficiently accurate onboard position estimate.

Several technologies for indoor localization [2] were suggested, such as inertial measurement [3], visual odometry, LIDAR [4] and dead-reckoning. However, none of the existing technologies, by itself, could perform reliable solutions. LIDAR is an attractive technology due to its high accuracy in ranging, its wide-area view and low data-processing requirements. Unfortunately, LIDAR can be inaccurate with respect to turns and rotational movements. On the other hand, inertial odometry, based on accelerometer and gyroscope readings, is intended for the short distances, and so it suffers from drift. The distance traveled is obtained by double integration of the accelerometer sensor signal. Bias offset drift exhibited on the acceleration signal is accumulative and the accuracy of the distance measurement can deteriorate with time due to the integration. A novel method is used to solve the problem by correcting raw inertial odometry with LIDAR measurements, which allow robot to navigate accurately on any indoor slippery surface.

## 2 Environment and Task

We assume that the vehicle is mostly travelling in the horizontal plane [5]. The test field was set up, as shown in Fig. 1, to simulate a food manufacturing workshop, separated into two, 16 m$^2$ areas. The left workshop (area **I**) had dynamic obstacles and slippery surface, and the right workshop (area **II**) had icy surface similar to cold working environment. A vertical lift, auto roller shutter Door (G), jointed the two areas together. No guiding lines were available in both workshops. Mobile robot had 3 tasks:

- Run from A (charging point) to loading area B by wireless instruction order.
- Then deliver parts to nine robot arm work cell, $W_i$ $(i = 1–9)$.
- Finally, mobile robot transports the collected products from robot arm work stations back to charging point.

Navigation system aims to enable the robots to move safely and avoid all unknown obstacles in the area I and robot arm work stations in the area II. It has to park at loading area B accurately for automatic loading, and transport parts to all work stations.

In the Fig. 1, letter $M$ represents mobile robot, $A$ stands for the charging, start and finishing point, $G$ is the automatic door gate, red dots represent unknown obstacles in the area I, and W1–W9 are stationary robots' working cells.

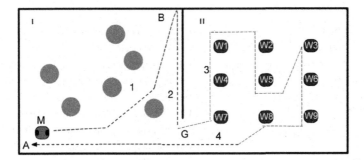

**Fig. 1** Test environment and mission path layout

# 3 System Components

The National Instruments MyRio-1900 [6] was chosen as on board microprocessor for its capabilities, that comes from the concurrent use of the microcontroller, for processing and Field Programmable Gate Area (FPGA) for fast input/output (IO). It has good Electrostatic Discharge (ESD) and over-voltage protection, with many external interfaces: analog/digital IO, Ethernet, and USB. Hokuyo LIDAR URG-04LX-UG01 was selected as it has optimal performances suitable for the localization applications in indoor environments. An appropriate off shelf gyroscope is equipped on board too.

# 4 LIDAR and Gyroscope Based Motion Control Solution

This section presents detailed algorithms for LIDAR based motion control solution in LabView environment. The measurement parameters are listed in the Table 1. Solution enables a precise, fast update for robust localization, with the focus on: two walls LIDAR based detection, Hough Transform (HT) [7] localization, wall alignment, wall following, and obstacle avoidance.

## 4.1 Obstacle Free Environment Localization

The perpendicular lines from LIDAR to walls were used to update robot with precise localization data ($X$, $Y$, and heading direction) in wall enclosure scenario, without obstacles. The coarse heading direction, shown on Fig. 2, as a tick, black arrow, from either preset scenario, or from gyroscope; the perpendicular distances to two walls which were most exposed to LIDAR, would be the $X$ and $Y$ in global coordinate. As each LIDAR's output was a paired data of object's magnitude and

**Table 1** Measurement parameters

| | |
|---|---|
| ——————— LIDAR full range | $\theta$: The corrected orientation |
| ············ LIDAR searching range | $\alpha$: The raw orientation from gyroscope |
| ············ The reference line/direction | $\rho$: Line perpendicular distance from origin |
| ◄ – – – – Raw gyroscope orientation | $\varnothing$ Angle between $\rho$ and X axis |
| ◄——————— Corrected orientation | Am: the angle to first line in H Transform |
| ▭ | Am1: the angle to 2nd line in H Transform |
| | Xm: the perpendicular distance to first line |
| | Xm1: the perpendicular distance to 2nd line |
| The test track enclosed by 4 Walls, 4 m each side. | $\mu1$, $\mu2$: is the turning ratio vobtained experimentally for different surface and chassis. |

angle related to LIDAR. The difference between angle paired with $X$ and LIDAR 0 degree would be the exact orientation of LIDAR also mobile robot in global coordinate.

In Fig. 2(a), based on rough heading, entire LIDAR scan region has been minimized into 2 sub-regions, each was 20 degrees range. The shortest distance magnitude in sub-regions **A** and **B** would be $X$ and $Y$; see Fig. 2(b). In Fig. 2(c), $\varnothing$, the angular difference between $X$'s angle and LIDAR's 0 degree would be the exact mobile robot's orientation in the global coordinate.

In the programming code example given in Fig. 3, the preset angle was set as 45°, and LIDAR range was set from −60° to 60°. In the region between −40° to −60°, the shortest distance i.e. magnitude was selected. Meanwhile, the angle between robot's current heading and the shortest magnitude to left wall could be calculated. Then we were able to find exact heading direction.

## 4.2 Unknown Cluttered Environment Localization

In this scenario, the perpendicular directions were blocked by obstacles. The previous method could not be reliably utilized for solving localisation and heading direction calculation. The Hough transform (TH) [7], also called Standard Hough Transform (SHT), was applied as it is capable of detecting straight line segments in the presence of noise.

Standard LabVIEW library, used for static images, could not be applied into real time target like MyRio. Subsequently, a real time, wall detection method, needed to be developed to work within unknown obstacle scenario.

After the Hough Transform method was implemented as shown in Fig. 4 the precise $X$, $Y$ coordinates and heading could be obtained in obstacle scenario, Figs. 5 and 6. On the other hand, due to noisy measurements and false readings of LIDAR, the Hough Transform becomes more time consuming. Using hardware, as

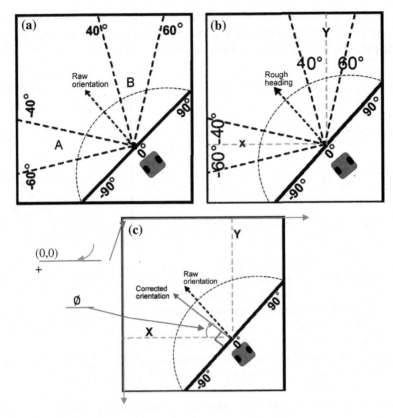

**Fig. 2** LIDAR self-localization in obstacle free and wall enclosure

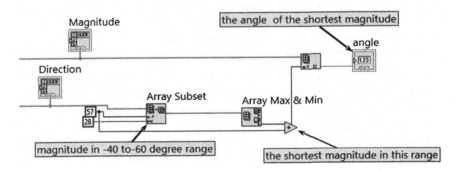

**Fig. 3** LabVIEW code that calculates the perpendicular distance to a wall in obstacle free scenario

already defined for this research, it takes 7 s to calculate location, based on the two walls. For a mobile robot moving at 2 m/s speed, in the real-time scenario, localization time cost should be under 0.5 s for each update.

To accelerate SHT performance, the abundant literature about the HT for straight line have been researched. HT could be classified into two large groups depending on the parameterization used for expressing the lines [8].

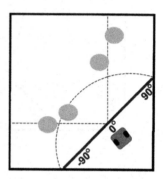

**Fig. 4** Scenario of Hough transform applications with obstacles

On one hand, the most abundant group corresponds to those using the parameters $\rho$ and $\theta$, where $\rho$ is the distance from the line to the origin and $\theta$. the angle from a vector which was normal to the line to the abscissas axis. In this group is included the work about the combinatorial Hough transform (CHT) [9] and the piecewise-linear Hough function (PLHT) [10]. Another group used the parameters $m$ and $c$ to express the lines, where $m$ was the slope and $c$ was the point of intersection with the ordinate axis. In the second group we have randomized Hough transform (RHT) [11], and fast Hough transform (FHT) [12].

In the case like Fig. 6, there were no complex image or noise, therefore all those algorithm would not improve SHT to be fast enough to achieve the mobile robot real-time localization requirements. The solution for the problem of recognition, presented here, is in using a method that minimize detecting range in Hough domain (an accumulator array) by combining estimation orientation from gyroscope and given indoor coordinate to save computation load.

There were 5 methods applied to achieve improvement:

1. Reduce searching range in Hough domain using gyroscope, or pre-calculation. Raw orientation $\pm 10$ degree searching range for 1st wall and $\pm 5$ degree searching range for 2nd wall were applied instead of searching in full Hough domain of 180 degrees twice.
2. Use median filter and screen to reduce the noise and false readings. Median filters were used for removing noise in image processing. They set each pixel value to the median, in a two-dimensional neighborhood mesh. This allowed us to remove spike noise.
3. Increase the bin size. The time to fill and search into the Hough domain is proportional to the size of Hough domain [13]. Considering the mobile robot's size was $300 \times 300$ mm and the test field was $4000 \times 4000$ mm, centimeter was the most suitable minimum length unit to represent robot's localization.

4. Calculate with fix point instead floating point arithmetic.
5. Rewrite program into FPGA. In contrast to traditional processors, programming
   an FPGA rewires the chip itself to implement algorithms in hardware.

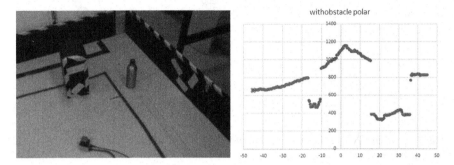

**Fig. 5** HOUGH experiment scenario (*left*) and raw reading (*right*)

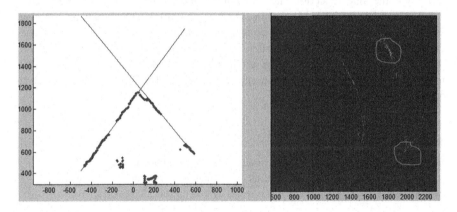

**Fig. 6** HOUGH transform results in Matlab. *Left* Cartesian and *Right* Hough Domain

Results Comparison

The least squares method was used to measure the accuracy of the proposed filter.
The coefficient of determination compares the estimated and actual values, where a
perfect correlation would be 1 and no correlation would be 0. As the wall was a
straight line, the accuracy of the applying different HOUGH domain sizes and filters
were compared using the coefficient of determination, $r^2$. In Fig. 7 (a) $r^2 = 0.47$
which was smaller than Fig. 7 (b)'s $r^2 = 0.56$. This indicated that $180 \times 560$
(degrees × centimeters) HOUGH domain's bin size gave much better representation
of the wall and therefore it improved the localization of the robot.

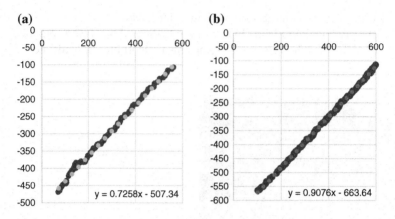

**Fig. 7** Comparing the line detection accuracy. **a** 180 × 5600 without filter. **b** 180 × 560 with filtering

Table 2 explained results obtained by applying method (1–5) one by one. It was clear that the line detection time was improved, when system required fewer computational operations. Meanwhile, the accuracy of line detection is not affected, as the proper filter and HOUGH domain have been applied. There were 15 successful runs to evaluate the accuracy and efficiency of the obtained localization data. An instance after 1.5 s is given in Fig. 8. LIDAR localization with all 5 methods applied, has improved the update rate to 10 Hz. This was fast enough to allow real-time localisation.

**Table 2** Time results

| Method | Time (seconds) to detect walls | | | Improvement | |
|---|---|---|---|---|---|
| | First wall | Second wall | Total | Seconds | Percentage (%) |
| Baseline | 3 | 4 | 7 | – | – |
| 3 | 2 | 3 | 5 | 2 | 28 |
| 2 | 1.5 | 2.5 | 4 | 1 | 20 |
| 1 | 1 | 1 | 2 | 2 | 50 |
| 4 | 0.5 | 0.5 | 1 | 1 | 50 |
| 5 | 0.05 | 0.05 | 0.1 | 0.9 | 90 |

## 5 Supplementary Files

The robot demonstration video could be visited by:
https://www.youtube.com/watch?v=-4t5nPGBhHU
https://www.youtube.com/watch?v=fcTDMds9Mmg

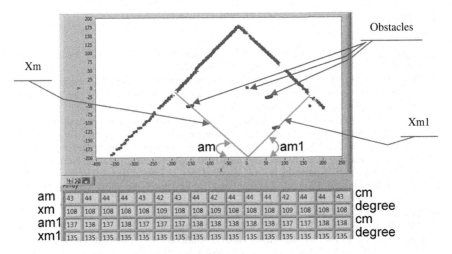

| am | 43 | 44 | 44 | 44 | 43 | 42 | 43 | 44 | 42 | 44 | 44 | 44 | 42 | 44 | 44 | 43 | cm |
| xm | 108 | 108 | 108 | 108 | 108 | 109 | 108 | 108 | 109 | 108 | 108 | 108 | 109 | 108 | 108 | 108 | degree |
| am1 | 137 | 138 | 137 | 138 | 137 | 137 | 137 | 138 | 138 | 138 | 138 | 138 | 137 | 137 | 138 | 138 | cm |
| xm1 | 135 | 135 | 135 | 135 | 135 | 135 | 135 | 135 | 135 | 135 | 135 | 135 | 135 | 135 | 135 | 135 | degree |

**Fig. 8** Results of localization acquired in 1.5 s

## 6 Conclusion

A motion control approach, for mobile robots' locomotion across slippery surfaces, is presented here. Novel algorithm developed and tested, is efficient and tolerant to noise in the scenario without obstacles on the road, as well as, in the case with obstacles. The methods of minimizing the detecting range, by applying raw orientation, extracted from gyroscope, were applied in both environments. After course direction is obtained, the LIDAR, i.e. robot's $(X, Y)$ coordinates in global positioning system were obtained finding the shortest magnitudes through Hough Transform application. Various HT transforms were considered. Finally exact orientations were calculated from the difference between perpendicular angle to the reference wall and $0°$.angle of LIDAR scan range. Those two localization solutions removed much of the bias on the single gyroscope method, including the drifting at low speed conditions. Following that, a novel motion control over slippery surfaces was established and tested comprehensively.

In practice, the algorithm presented here was efficient and tolerant to noise. The Hough transform iterates, over points of interest, in a given data cloud, only in a reduced range of $2 \times 20°$ instead of full range of $2 \times 180°$, which requires less amount of memory and is fast enough to be used in real-time applications with the given time constraints as defined by the project requirements.

Following all of that, we have better motion control over slippery surfaces. Due to the intense computational loads, the proposed method was limited to speeds below 2 m/s. Under the time constraints, inside that given speed limit, test robot, with the hardware available, was easily performing real time operations. Future work includes a more powerful processor and algorithms, to deal with the intense computational loads that would be required for real-time operation if the motion is performed at the higher speeds.

# References

1. Lingemann, K., Nüchter, A., Hertzberg, J., Surmann, H.: High-speed laser localization for mobile robots. Robot. Auton. Syst. **51**, 275–296 (2005)
2. Uijt de Haag, M., Venable, D., Smearcheck, M.: Integration of an inertial measurement unit and 3D imaging sensor for urban and indoor navigation of unmanned vehicles. In: Proceedings of the Institute of Navigation National Technical Meeting, pp. 829–840 (2007)
3. Titterton, D.H., Weston, J.L. (ed.): E. Institution of Electrical, A. American Institute of, and Astronautics, Strapdown inertial navigation technology. In: IEE radar, sonar, navigation, and avionics series 17, 2nd edn. United Kingdom, Institution Of Engineering & Technology (IET)
4. Harrap, R., Lato, M. (ed.): An overview of LIDAR: Collection to application, Norway (2010)
5. Iqbal, U., Okou, A.F., Noureldin, A.: An integrated reduced inertial sensor system—RISS/GPS for land vehicle. In: Position, Location and Navigation Symposium. IEEE/ION 2008, 1014–1021 (2008)
6. National_Instruments, http://www.ni.com/myrio/what-is/, 2014
7. Duda, R.O., Hart, P.E.: Use of the Hough Transformation to Detect Lines and Curves in Pictures. In: C. Sri International Menlo Park Ca Artificial Intelligence, Ed., ed, 1971
8. Guil, N., Villalba, J., Zapata, E.L.: A fast Hough transform for segment detection. Image Process. IEEE Trans. **4**, 1541–1548 (1995)
9. Ben-Tzvi, D., Sandler, M.B.: A combinatorial Hough transform. Pattern Recogn. Lett. **11**, 167–174 (1990)
10. Koshimizu, H., Numada, M.: On a fast Hough transform method PLHT based on piecewise-linear Hough function. Syst. Comput. Japan **21**, 62–73 (1990)
11. Xu, L., Oja, E., Kultanen, P.: A new curve detection method: randomized Hough transform (RHT). Pattern Recogn. Lett. **11**, 331–338 (1990)
12. Li, H., Lavin, M.A., Le Master, R.J.: Fast Hough transform: a hierarchical approach. Comput. Vision Graph. Image Proc. **36**, 139–161 (1986)
13. Haidekker, M.A.: The Hough transform. Hoboken, NJ, USA, Wiley (2010)

# Experimental Flight Test for Autonomous Station-Keeping of a Lighter-Than-Air Vehicle

**Cees Bil, Jeroen Zegers, Leroy Hazeleger, Liuping Wang and Milan Simic**

**Abstract** The benefits of stationary aerial platforms for continuous observation of a fixed area are evident in cases such as natural disasters, environmental monitoring, intelligence gathering, etc. Lighter-Than-Air (LTA) technology has advanced significantly, for example high-pressure balloons that would make such a platform a potential solution for long, uninterrupted and persistent observation. An adaptive nonlinear control system was designed for LTA station-keeping, with an adaption mechanism to accommodate uncertainties in system parameters. The controller was subsequently implemented on a micro controller and tested in a flight experiment using a helium balloon. The tests were conducted in a 4.5 m by 4.5 m enclosure using infrared sensors to determine position by measuring distance from the walls.

**Keywords** Adaptive nonlinear control · Lighter-than-air · Station-keeping · Flight test experiments

C. Bil (✉) · M. Simic
School of Aerospace, Mechanical and Manufacturing Engineering, RMIT University,
Melbourne, Victoria, Australia
e-mail: cees.bil@rmit.edu.au

M. Simic
e-mail: milan.simic@rmit.edu.au

J. Zegers · L. Hazeleger
Mechanical Engineering Dynamics, Systems and Control, Eindhoven University
of Technology, Eindhoven, Netherlands
e-mail: j.c.zegers@student.tue.nl

L. Hazeleger
e-mail: l.hazeleger@student.tue.nl

L. Wang
School of Electrical and Computer Engineering, RMIT University,
Melbourne, Victoria, Australia
e-mail: liuping.wang@rmit.edu.au

© Springer International Publishing Switzerland 2015                                   299
E. Damiani et al. (eds.), *Intelligent Interactive Multimedia Systems and Services*,
Smart Innovation, Systems and Technologies 40,
DOI 10.1007/978-3-319-19830-9_27

# 1  Introduction

Lighter-Than-Air (LTA) systems have been studied for various applications [1, 2]. An LTA design is proposed that is able to observe a fixed point continuously and for an extended period of time [3, 4]. The balloon is a super-pressure balloon with a closed envelope and therefore its volume is relatively constant with changing temperature, which allows the balloon to keep a relatively stable altitude without the need to release gas or remove ballast. Therefore it is assumed that a two-dimensional control design in the horizontal $(x,y)$-plane suffices. The LTA vehicle has 4 fixed, thruster-type propellers $P_i$ mounted around the perimeter of the balloon as shown in Fig. 1. The target position of the LTA observation platform is depicted by the coordinates $(x_{ref}, y_{ref})$ relative to a global coordinate frame. The actual position of the observation platform is given by the coordinates $(x_p, y_p)$ relative to a global coordinate frame. A body-fixed frame is defined which is rotated around the global $z$-coordinate.

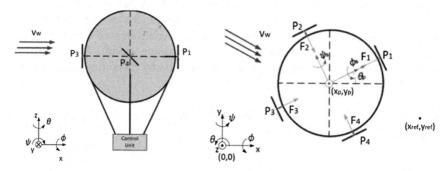

**Fig. 1** Lighter-Than-Air vehicle with 4 propellers in a global coordinate and body-fixed frame

The ratio between the input voltage to the propellers and the force delivered by the propeller is assumed to be linear such that:

$$F_i = K_i u_i \quad \text{for } i = 1, \ldots, N_p \tag{1}$$

with the ratio $K_i$ assumed to be constant. The following nonlinear control laws were derived for $F_x$ and $F_y$ forces in the x- and y-direction respectively [5–7]:

$$F_x = -mC_x x_1 + \frac{1}{2}\rho_{air}C_D S x_2 |x_2| - mD_x x_2 \tag{2}$$

$$F_y = -mC_y x_3 + \frac{1}{2}\rho_{air}C_D S x_4 |x_4| - mD_y x_4 \tag{3}$$

where $C_x$, $C_y$, $D_x$ and $D_y$ are positive constants and the drag coefficient $C_D$ is based on the frontal area S. The variables $C_x$, $C_y$, $D_x$ and $D_y$ can be chosen freely, where $C_x$

and $C_y$ represent a proportional gain and $D_x$ and $D_y$ represent a derivative gain. The adaptation mechanism was added using again Lyapunov's Stability Theorem [5].

## 2  Experimental Flight Tests

The purpose of the experimental flight tests was to investigate the effectiveness of the adaptive nonlinear controller on real hardware. A helium-filled balloon with a 0.6 m radius and 0.9 kg mass was used as shown in Fig. 2. A control unit was designed and built to measure and processes signals and to manipulate the horizontal motion of the vehicle using the propeller motors. To control the position of the LTA observation platform the actual position and orientation of the platform $(x_p, y_p, \theta_p)$ must be determined. To obtain the orientation of the substructure $\theta_p$ a combination of multiple sensors were integrated in the IMU-device. This measurement device contains three gyroscopes, three accelerometers and three magnetometers. A complementary filter is used to determine an estimate for the orientation $\theta_p$, that combines magneto- and accelerometer measurement data for low frequency orientation estimation $\hat{\theta}_{m/a}$ with gyroscope measurement data for high frequency orientation estimation $\hat{\theta}_g$. The tilt angles $\phi_s$ and $\psi_s$ can be estimated in a similar fashion, if required. It is assumed that the design of this prototype ensures that tilting of the substructure is zero.

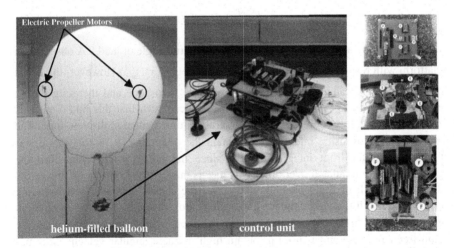

**Fig. 2**  Experimental prototype of a LTA vehicle

A sensor-fixed Cartesian coordinate frame is defined in Fig. 3.

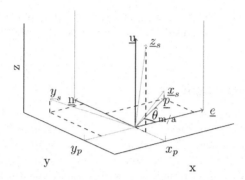

**Fig. 3** Sensor-fixed reference frame $\left\{\underline{x}_s, \underline{y}_s, \underline{z}_s\right\}$, the global reference frame $\{\underline{e}, \underline{n}, \underline{u}\}$. The vector $\underline{p}$ is a projection of $\underline{x}_s$ to the horizontal plane. the angle $\hat{\theta}_{m/a}$ is the angle between the projected vector $\underline{p}$ and the east-vector $\underline{e}$

The directions of this coordinate frame coincide with the directions of the orthogonal sensor configuration. Using one of the gyroscopes, accurate low noise measurement data of the angular velocity of the sensor around its $z_s$-axis $\dot{\theta}_s$ is obtained. Since the tilt is assumed to be zero and the sensor is mounted approximately horizontal on the substructure, the integration of this measured angular velocity $\dot{\theta}_s$ is used as one of two approximations of the orientation angle $\theta_p$. This estimation is defined as:

$$\hat{\theta}_g = \int_0^T \dot{\theta}_s dt + \theta_{\text{init}} \qquad (4)$$

where $T$ is the measurement time and $\theta_{\text{init}}$ is the initial angle of the LTA observation platform with respect to the global reference frame $\left\{\underline{x}_s, \underline{y}_s, \underline{z}_s\right\}$. However, the initial orientation angle cannot be determined using the gyroscope and therefore during experiments the LTA observation platform is placed such that this initial angle is zero. In practice, this continuous-time integral is implemented in discrete-time. The drawback of only using integration to obtain the orientation $\hat{\theta}_g$ is that this orientation will suffer from integration drift since the measurement data from the gyroscope is accumulated over time. This means that errors (from integration errors due to slight tilt of the sensor, limitations of the sensor, mathematical rounding errors, etc.) will also accumulate and if it is not corrected for, the orientation estimate $\hat{\theta}_g$ is not accurate. Note that the gyroscope data is less contaminated by high-frequent noise and is used to obtain high frequent angular velocity information. To overcome the problem of drift due to the integration of the gyroscope data, three accelerometers and three magnetometers are used to determine a second approximation of the orientation angle $\theta_p$. These two sets of sensors are also oriented in the orthogonal configuration $\left\{\underline{x}_s, \underline{y}_s, \underline{z}_s\right\}$ such that using these six sensors a second global reference frame $\{\underline{e}, \underline{n}, \underline{u}\}$ can be determined using the measured gravitational vector and magnetic north vector. The north-vector $\underline{n}$ is measured using the magnetometers and points towards the Earth's magnetic north. The acceleration readings are used to determine the up-vector $\underline{u}$ since gravity is measured as an upward acceleration. The cross product of the north- and up-vector gives the east-vector $\underline{e}$. Since the north-vector and the up-vector are not guaranteed to be orthogonal, the north-vector is re-determined using the cross product of the obtained east-vector $\underline{e}$ and the up-vector $\underline{u}$. The vectors are normalized such that an orthonormal coordinate frame is obtained. Note that the vectors $\{\underline{e}, \underline{n}, \underline{u}\}$ are expressed in the sensor-fixed $\left\{\underline{x}_s, \underline{y}_s, \underline{z}_s\right\}$ directions since these vectors are measured using the sensors. Also note that in Fig. 3 the directions of the second global coordinate frame $\{\underline{e}, \underline{n}, \underline{u}\}$ coincide with the directions of the global coordinate frame $\left\{\underline{x}_s, \underline{y}_s, \underline{z}_s\right\}$, but this is not always the case. When present, the angular difference is corrected using an angular shift which is the determined initialisation.

As an accelerometer not only measures the acceleration due to gravity but also other applied forces, the measured up-vector $\underline{u}$ is not exactly identical to the inverse of the gravity vector. Every force working on the LTA observation platform, such as propeller and wind force, will induce acceleration of the LTA observation platform which disturbs the estimated up-vector $\underline{u}$. However, this induced acceleration is small compared to the gravity vector and therefore the up-vector $\underline{u}$ can be determined quite accurately. In contrast to the measured angular velocity, the measured accelerations and magnetic field strength is not subjected to drift.

To obtain the second approximation $\hat{\theta}_{m/a}$ of the orientation angle $\theta_p$, first the vector $\underline{x}_s$ is projected on the horizontal plane spanned by the east-vector $\underline{e}$ and the north-vector $\underline{n}$ which results in the vector $\underline{p}$. The projected vector $\underline{p}$ is decomposed as a vector:

$$\underline{p} = \underline{p}_e + \underline{p}_n = \frac{\underline{x}_s.\underline{e}}{\underline{e}.\underline{e}}\underline{e} + \frac{\underline{x}_s.\underline{n}}{\underline{n}.\underline{n}}\underline{n} \tag{5}$$

The angle $\hat{\theta}_{m/a}$ is determined using the lengths of the vectors $\underline{p}_e$ and $\underline{p}_n$ as arguments for the arctangent function $atan2$. The lengths of these vectors $\underline{p}_e$ and $\underline{p}_n$ are determined using the square root of the inner product of the vector with itself.

$$\hat{\theta}_{m/a} = \arctan2\left(\sqrt{\underline{p}_n.\underline{p}_n}, \sqrt{\underline{p}_e.\underline{p}_e}\right) \tag{6}$$

and using the following orthonormality property $\underline{e}.\underline{e} = 1$ and $\underline{n}.\underline{n} = 1$ results in

$$\hat{\theta}_{m/a} = \arctan2(\underline{x}_s.\underline{n}, \underline{x}_s.\underline{e}) \tag{7}$$

This is the definition of the estimation angle $\hat{\theta}_{m/a}$ obtained using the magneto- and accelerometer measurements which is a second estimation of the orientation angle $\theta_p$ and is not subjected to drift. Two methods to obtain an estimate for the orientation angle $\theta_p$ are described above. Using a complementary filter these two estimations are combined to obtain one single estimate for $\theta_p$ which is not subjected to drift. The complementary filter is an estimation technique that is often used in the flight control industry to combine different measurements and is related to the more complex Kalman filter [8, 9]. The block diagram of this complementary filter is shown in Fig. 4.

The expression for this complementary filter in the Laplace domain is:

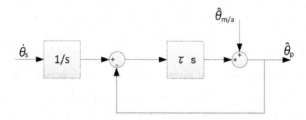

**Fig. 4** Block diagram of complementary filter

$$\hat{\theta}_p = \frac{\hat{\theta}_{m/a} + \tau\dot{\theta}_s}{1 + \tau s} = \frac{1}{1 + \tau s}\hat{\theta}_{m/a} + \frac{\tau s}{1 + \tau s}\frac{1}{s}\dot{\theta}_s \tag{8}$$

where $\tau$ is the filter time constant. Using the backward difference expression:

$$s = \frac{1}{\Delta t}\left(1 - z^{-1}\right) \tag{9}$$

with $\Delta t$ being the sample time, the expression in (8) can be transformed from the Laplace domain to the discrete-time domain:

$$\hat{\theta}_p^k = \alpha\left(\hat{\theta}_p^{k-1} + \dot{\theta}_s^k\Delta t\right) + (1 - \alpha)\hat{\theta}_{m/a}^k \quad \forall k = 1, 2, 3, \ldots \tag{10}$$

with $\hat{\theta}_p^0$ being the initial angle estimation at $t = 0$ and $\alpha$ being the filter constant defined as $\frac{\tau}{\Delta t}/\left(1 + \frac{\tau}{\Delta t}\right)$. The complementary filter has the advantages of the gyroscope, magneto- and accelerometers: low noise measurements and no drift. The filter constant $\alpha$ is a value between 0 and 1 which determines the cut-off frequency of the first-order high-pass and low-pass filter. For practical incorporation of the complementary filter in this application $\alpha = 0.98$ was chosen. Experiments were performed in which the sensor was rotated rapidly and randomly while measuring $\hat{\theta}_g$ and $\hat{\theta}_{m/a}$, and the complementary filter is implemented to determine the orientation estimate $\hat{\theta}_p$.

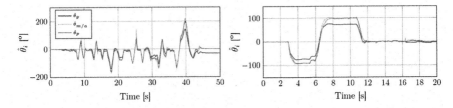

**Fig. 5** Estimations for the orientation angle $\theta_p$ determined using integration of the gyros (*blue*), determined using the magneto- and accelerometer measurement data only (*green*) and determined using the complementary filter (*red*). $\alpha = 0.98$ is used in the complementary filter

Figure 5 shows the influence of the complementary filter. The estimated orientation angles determined using pure integration of the gyroscope data $\hat{\theta}_g$, determined using the gravitational vector and magnetic north vector $\hat{\theta}_{m/a}$ and determined using all measurement devices combined via the complementary filter $\hat{\theta}_p$ are shown. It is experimentally observed that a filter constant of $\alpha = 0.98$ gives the best results for the orientation estimation. The upper figure shows the estimated orientation determined during fifty seconds of random (fast) rotation of the substructure around the z-axis. It can be observed that the orientation angle $\hat{\theta}_g$ drifts over time. Furthermore can be observed that the estimation angles $\hat{\theta}_{m/a}$ and $\hat{\theta}_p$ do not drift and are approximately the same, except for some additional noise in $\hat{\theta}_{m/a}$. This result shows that for this application the use of the gyroscope to estimate the orientation is not strictly necessary since applying a low-pass filter to $\hat{\theta}_{m/a}$ will result in approximately the same estimation. However, in this prototype the complementary filter using the gyroscope measurement data is applied.

Another motion test is performed, of which the results are shown in Fig. 5, to check the accuracy of the estimated orientation angle. During this test the substructure is rotated approximately 90 degrees in negative and positive direction around the z-axis. The results show that the estimation angle determined using the complementary filter becomes approximately $-90$ and $+90$ degrees. In addition to estimating the orientation of the LTA observation platform $\hat{\theta}_p$, an estimate for its position must also be derived from measured data. This estimated position is defined as $(\hat{x}_p, \hat{y}_p)$. This estimation cannot be obtained by double integration of the accelerometer data since the position will drift very fast over time (due to high noise ratio) and there is no possibility to correct for this drift.

Since the orientation of the substructure of the LTA observation platform is estimated, multiple distance sensors placed on the platform could provide enough information to determine its position. Ultra-sonic distance sensors, laser range finders and IR distance sensors are easy to implement, therefore four SHARP GP2Y0A710YK0F Infra-Red distance sensors were fitted.

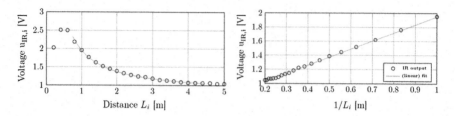

**Fig. 6** Voltage sensor $u_{IR,i}$ as a function of the distance $L_i$ and the inverse of the distance $1/L_i$, for i = 1, 2, 3, 4

The IR distance sensor output was measured and is depicted in Fig. 6. The voltage from the IR distance sensor is linear with respect to the inverse of the distance for the region for 1–5 m. Using this linear relation between $u_{IR,i}$ and $1/L_i$ the distance between the LTA vehicle and the wall is approximated.

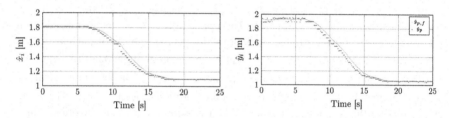

**Fig. 7** Estimated position $(\hat{x}_p, \hat{y}_p)$ as function of time

In Fig. 8 the estimated velocity is depicted which is determined as the time derivative of the filtered estimated position. The time derivative of the filtered estimated position is obtained by using numerical differentiation and is implemented as a finite difference equation. The noise which is still present is amplified by the derivative operation and therefore an additional first order low-pass was applied in a similar fashion having a filter constant of $\alpha_v = 0.975$. This filter constant was determined by analyzing the filtered data for different filter constants. By comparing Figs. 7 and 8, it can be observed that the filtered velocity profile is a reasonable approximation.

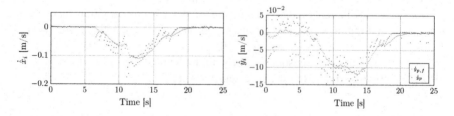

**Fig. 8** Estimated velocity $\dot{x}_p$ and $\dot{y}_p$ in as a function of time, using the filtered position

The microcontroller could only send control signals in the form of PWM signals with a resolution of 8-bit. Two PWM signals are send to each dual motor driver and the output voltages from these motor drivers, which are send to the low-voltage brushed DC motors, are determined by the power supply voltage of a 9 V battery and the controlled PWM signals send from the microcontroller. Therefore, not only the relation between the voltage over the propeller and the thrust force must be found, but also the relation between the PWM signal send by the microcontroller and the voltage over the propeller. Due to the low frequency of the PWM pins of the microcontroller, a phenomena occurred which is called ripple current. Due to ripple current, the maximum allowable current of the motor drivers was exceeded and consequently the motor drivers were shut down. To prevent ripple current inductors of approximately 200 µH were added. Another solution was to increase the frequency of the PWM pins of the microcontroller. The relation between the thrust force of the propellers and the control output $u_i$ was measured as well as the relation between the thrust force of the propellers and the PWM signal send by the microcontroller as depicted in Fig. 9.

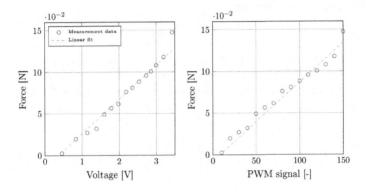

**Fig. 9** Measured thrust force versus voltage (left) and the PWM signal (right)

**Fig. 10** Control output voltage $u_i$ versus PWM signal

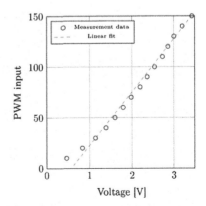

From these measurements the relation between the control output $u_i$ and the PWM signal is obtained and is given in Fig. 10 as a linear fit. Figure 9 shows that the relation is approximately linear, but has a small offset. The same holds for the relation between the control output $u_i$ and the PWM signal. Therefore the relation between the input voltage $u_i$ to the propeller and the delivered force by the propeller $F_i$ being linear is not correct and a small offset was added to the control laws. The adaptation gain matrix is evaluated beforehand during off-line simulations [6]. Since we were interested in the behaviour of the prototype with respect to its position, the parameter adaptation was not logged. From simulation results it was shown that the parameter adaptation is very slow and is not expected to oscillate. Therefore it is assumed that the parameter adaptation has no great contribution to the performance of the controlled LTA observation platform and therefore it is chosen to not save experimental data of the parameter estimation. Within 50 s the LTA observation platform is expected to have reached the desired position, and therefore after 50 s the control unit is turned off.

## 3   Experimental Results

In Fig. 11 the trajectory of the LTA vehicle for three different controller settings is shown. The black dots indicate the initial positions, two initial positions at (1, 1) and one at (4, 4). All trajectories move to the desired target position. The values used for these three measurements show the benefit of the controller values.

**Fig. 11** Trajectory for different controller settings for $C_x$, $C_y$, $D_x$ and $D_y$

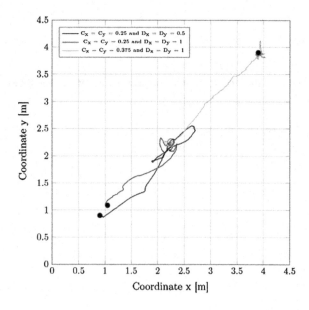

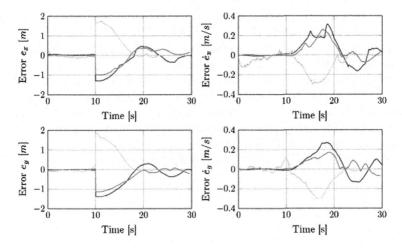

**Fig. 12** Errors $e_x$, $e_y$, $\dot{e}_x$ and $\dot{e}_y$ versus time

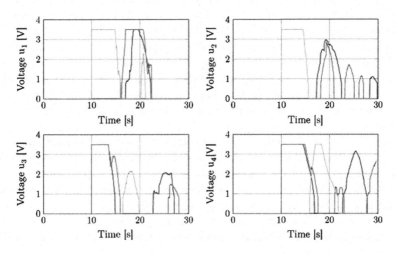

**Fig. 13** Control inputs $u_i$ versus time

The trajectory where $C_x = C_y = 0.25$ and $D_x = D_y = 0.5$ shows oscillation around the target position which is less for the trajectory where $D_x = D_y = 1$ with identical $C_x$ and $C_y$ (Fig. 12). Due to the lack of damping, the LTA vehicle with $D_x = D_y = 0.5$ will oscillate around the target position. The energy efficiency can be explained by observing Fig. 13 where the control effort for the damping terms $D_x$ and $D_y$ equals 0.5 is higher than when the damping terms $D_x$ and $D_y$ equals 1. Increasing the damping terms $D_x$ and $D_y$ reduces the oscillation around the desired position, decreases the control effort and increases the endurance of the LTA vehicle. The trajectory where $C_x = C_y = 0.375$ and $D_x = D_y = 1$ shows almost no oscillation, which indicates that the controller values $D_x = D_y = 1$ are high enough. Compared to the trajectory where $C_x = C_y = 0.25$ and $D_x = D_y = 1$, the trajectory is faster due to the higher gains $C_x$ and $C_y$. The controller values are a trade-off, but are easy to tune. They have a direct influence on the performance of the controlled LTA vehicle and as long as the values are positive, the unperturbed controlled system is asymptotically stable. It was observed that the assumption of –s and s being small was justified. The experimental results indicate that the controller with $C_x = C_y = 0.375$ and $D_x = D_y = 1$ reaches the desired output.

# 4 Conclusions

Using Lyapunov's Stability Theorem and LaSalle's Invariance Principle, a non-linear controller was designed that asymptotically stabilizes an LTA observation platform with no wind disturbances and system parameters exactly known. Adding additional control terms (values) to the Lyapunov function, the performance of the nonlinear controlled LTA observation platform can be tuned and the control effort can be reduced. To allow estimated system parameters, an adaptive mechanism was developed to ensure asymptotic stability. Flight experiments were conducted to demonstrate the effectiveness of the controller. The prototype consists of a helium balloon lifting an instrumentation package, consisting of a microcontroller, gyro-scopes, accelerometers, IR distance sensors, and a data logger, to a stationary height.

The obtained signals were processed to obtain the position and orientation of an experimental LTA vehicle and the designed adaptive nonlinear controller was implemented to control the voltage to the propellers. The experiments were per-formed in a test environment with given dimensions. The experimental results show that the designed adaptive nonlinear controller is able to control the position of the LTA vehicle such that it drifts to a given target position given subject to wind disturbance. The control values can be used to increase the performance or reduce the control effort.

From a practical point of view, the adaptation mechanism may not be necessary as the systems parameters $m$, $S$, $C_d$ can be measured. This will reduce the com-putational effort, the demand for persistently exciting signals $\dot{q}_x$ and $\dot{q}_y$ and intro-ducing a time-varying adaptation gain matrix $\Gamma(t)$ are not needed. The next phase is to test the control system performance outdoors. A different positioning system, e.g. GPS, has to be implemented. More measurements should be performed such that it can be checked whether the experimental data matches the theoretical model.

# References

1. Cathey, H.M. Jr., Pierce, D.L.: Development of the NASA ultra-long duration balloon. In: Paper C3P3—NASA Science Technology Conference (NSTC2007), 19–21 June (2007)
2. Androulakakis, S.P., Judy, R.A.: Status and Plans of High Altitude Airship (HAA$^{TM}$) Program, AIAA Lighter-Than-Air Systems Technology (LTA) Conference, Daytona Beach 25–28 March (2013)
3. Karnadi, J., Haulder, N., Masakazu Kotake, N.: Lighter-Than-Air observation platform. Undergraduate Thesis, RMIT University (2012)
4. Bil, C.: Lighter-Than-Air stationary unmanned observation platform concept. In: R. Neves-Silva et al. (eds.) Smart Digital Futures, pp. 523–532. ios press (2014)
5. Khalil, H.K.: Nonlinear Systems, 3rd edn, Prentice Hall (2002)
6. Slotine, J.E., Li, W.: Applied Nonlinear Control, Prentice Hall (1991)
7. Jager, B. de: Applied Nonlinear Control (4J820), Eindhoven University of Technology
8. Walter T. Higgins, Jr.: A comparison of complementary and Kalman filtering, IEEE Trans. Aerosp. Electron. Syst. **11**(3) (1975)
9. Eusten, M. et al.: A complementary filter for attitude estimation of a fixed-wing UAV. University of Sydney (2008)

# Experimental Evaluation of Multi-key Content-Based Image Retrieval

**Hideki Sato and Shigemi Nagata**

**Abstract** A user of existing Content-Based Image Retrieval (CBIR) systems suffers from single query key selection, in case that he/she might begin retrieval process toward an ambiguous goal. In the other case, he/she cannot judge whether one image is more adequate than the other exactly for retrieval. To overcome the difficulty, this paper proposes Multi-Key CBIR (MK-CBIR) which differs from Single-Key CBIR (SK-CBIR) in retrieving similar images from image databases with a multi-key query, not a single-key one. To implement MK-CBIR, *skyline* query based MK-CBIR ($\Phi_{sky}$) and aggregate $k$-Nearest Neighbor query based MK-CBIR ($\Omega_{sum}$, $\Omega_{max}$, and $\Omega_{min}$) are presented. Experimental results on retrieval performance of MK-CBIR are as follows:

- *Recall* of $\Omega_{max}$ is over 0.3 higher than that of $k$-NN search and the highest among those of MK-CBIR. It is comparatively stable against noisy keys.
- *Recall* of $\Omega_{min}$ is roughly the same as that of $k$-NN search and the lowest among those of MK-CBIR.
- *Recall* of $\Omega_{sum}$ and *recall* of $\Phi_{sky}$ are roughly the same and between those of $\Omega_{min}$ and $\Omega_{max}$.

It is concluded that MK-CBIR is better than SK-CBIR in retrieval performance.

**Keywords** Content-based image retrieval · Multi-key · Single-key · *Skyline operator* · *Summation* · *Maximum* · *Minimum*

## 1 Introduction

As the network and the development of image capturing devices such as digital cameras, image scanners, are becoming more popular, the volume of digital image collec-

H. Sato (✉)
School of Informatics, Daido University, 10-3 Takiharu-cho, Nagoya, Minamiku 457-8530, Japan
e-mail: hsato@daido-it.ac.jp

S. Nagata
College of Engineering, Kanazawa Institute of Technology, Nonoichi, Japan

© Springer International Publishing Switzerland 2015       313
E. Damiani et al. (eds.), *Intelligent Interactive Multimedia Systems and Services*,
Smart Innovation, Systems and Technologies 40,
DOI 10.1007/978-3-319-19830-9_28

tion is increasing rapidly. Efficient image retrieval systems are required by users from various domains. For this purpose, there are two kinds of retrieval methods: text-based and content-based. The text-based method can be tracked back to 1970s. With the text-based method, images are manually annotated by text descriptors, which are then used to perform image retrieval. However, there are two disadvantages with this method. The first is that manual annotation requires a human labor of considerable level. The second is that manual annotation might be inadequate due to the subjectivity of human perception.

To overcome the disadvantages with the text-based method, the content-based method was introduced in the early 1980s. Content-Based Image Retrieval (CBIR) is a technique to retrieve similar images from large image databases based on the query image. In CBIR, visual features, such as color, texture, and shape are extracted from images and the query image is compared with each of images in the databases by using a similarity function on visual features.

To the best of our knowledge, existing CBIR systems rely upon the assumption that a user can explicitly give a single query key, though search engines for Web pages allow a user to input multiple keywords to express his/her retrieval requests. However, it is difficult for a user to choose a single query key among a set of candidates confidently. In a case, a user might begin retrieval process toward an ambiguous goal. In the other case, a user cannot judge whether one image is more adequate than the other exactly for retrieval.

To overcome the above-mentioned difficulty, this paper proposes Multi-Key CBIR (MK-CBIR) which retrieves similar images from image databases with multi-key queries, in contrast with Single-Key CBIR (SK-CBIR) which relies upon single-key queries. Additionally, there might be potential that MK-CBIR brings higher retrieval performance than that of SK-CBIR. To make MK-CBIR concrete, two implementation methods, that is to say, *skyline* query [1] based MK-CBIR and Aggregate $k$-Nearest Neighbor ($k$-ANN) query [5, 6] based MK-CBIR are presented. From the experimental results, it is concluded that MK-CBIR is better than SK-CBIR.

The rest of this paper is organized as follows. Section 2 mentions the related work. Section 3 proposes the implementation methods of MK-CBIR. Section 4 experimentally evaluates MK-CBIR. Section 5 presents the consideration on the experimental results. Finally, Sect. 6 concludes the paper.

## 2 Related Work

Group spatial queries retrieve objects based on an aggregate distance function with respect to a set of query points. The work described in [2–4] is concerned with Aggregate $k$-Nearest Neighbor ($k$-ANN) queries. First, it has been dedicated to the case of *summation* as an aggregate function and *euclidean* distance between an object and a query point [2]. Then, it has been extended to the cases of *summation* and *maximum* as an aggregate function and the network distance between an object and a query point [3, 4]. Another work has proposed the regular polygon based search algorithm to answer group spatial queries over remote spatial databases in the cases of *sum-*

*mation* and *maximum* as an aggregate function and *euclidean* distance between an object and a query point [5, 6]. It has been applied to $k$-ANN queries [5] and aggregate range queries [6]. Although the above-mentioned works use an aggregate distance function, our work is the first which applies an aggregate distance function to CBIR. Also, the visual feature of images being targets in CBIR is any dimensional, while the location of spatial objects being targets in group spatial retrieval is 2-dimensional.

Relevance Feedback (RF) is a powerful technique to improve retrieval performance. RF was initially built up for document retrieval [7]. RF was distorted and introduced into CBIR [8] to reduce the gap between low-level visual features and high-level semantic concepts. Since then, a variety of RF approaches have been widely developed in the field of CBIR [9]. The usual process of RF is as follows:

1. From the retrieved images, the user labels a number of relevant samples as positive and a number of non-relevant samples as negative.
2. Based on these labeled feedback, CBIR system improves its retrieval process.

RF employs multiple samples as feedback to improve retrieval process.

Estimation and visualization of ambiguous retrieval intentions are proposed [10]. For the purpose, the retrieval system presents image samples to the user and he/she selects some images among them. Next, the system estimates the most appropriate visual feature common to the selected images, by comparing means and standard deviations of all the visual features one another. Then, the system presents the estimated feature to the user and he/she continues to issue the following until intended images are obtained. It employs multiple samples selected by a user for estimating the visual feature of which he/she takes notice. The purpose to make use of multiple images is different from that of MK-CBIR.

# 3 Multi-key Content-Based Image Retrieval

In this section, CBIR is described first and then two implementation methods of MK-CBIR are presented.

## 3.1 Content-Based Image Retrieval System

In CBIR systems (See Fig. 1), a visual feature of each image in the databases is extracted and described by a multi-dimensional feature vector. The feature database is composed of feature vectors extracted. To retrieve images, a user provides the SK-CBIR (MK-CBIR) system with a single-key (multi-key) query. The system then alters the key image(s) with the use of feature vector(s). The similarity, between the feature vector(s) of the query image(s) and those of the images in the database is computed and retrieval is performed. In SK-CBIR systems, distance functions $d(,)$ such as *euclidean* are used to measure similarity between two images.

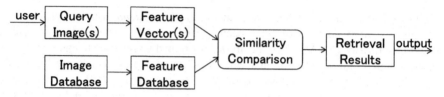

**Fig. 1** CBIR system model

## 3.2 Skyline Query Based MK-CBIR

*Skyline* queries [1] have received considerable attention in the database and data mining fields. Given a set $S$ of *skyline* attributes, a tuple $t$ is said to dominate another tuple $t'$, denoted by $t \succ_S t'$, if Eq. (1) is satisfied. It is assumed that smaller values are preferable over larger ones. Here, $t[A_i]$ is used to represent the value of the attribute $A_i$ of the tuple $t$. Given a set $D$ of tuples, Eq. (2) defines *skyline* operation $\Psi$ on $D$. In other words, a tuple $t$ belongs to *skyline* result set if no other tuples dominate it.

$$(\exists A_i \in S, t[A_i] < t'[A_i]) \wedge (\forall A_j \in S, t[A_j] \leq t'[A_j]) \tag{1}$$

$$\Psi(D, S) = \{t \in D | \nexists t' \in D, t' \succ_S t\} \tag{2}$$

Given a set $P$ of objects and a set $Q(= \{q_1, \ldots, q_m\})$ of query keys, Eq. (3) defines *skyline* query based MK-CBIR, $\Phi_{sky}$ on $P$. Here, $<, \ldots, >$ is a tuple constructor and $d(,)$ is a distance function. Note that $D'$ is $\{< p, d(p, q_1), \ldots, d(p, q_m) > | p \in P\}$ and $S'$ is $\{S_0, S_1, \ldots, S_m\}$, where distance $d(p, q_i)(i = 1, \ldots, m)$ is the value of *skyline* attribute $S_i$. As the cardinality of *skyline* result set is uncertain, $\Phi_{sky}(P, Q)$ also returns uncertain number of objects.

$$\Phi_{sky}(P, Q) = \{t[S_0] | t \in \Psi(D', S')\} \tag{3}$$

## 3.3 Aggregate k-Nearest Neighbor Query Based MK-CBIR

k-ANN queries have been used to retrieve objects located nearest to a set of query keys. Given an objects $p$ and a set $Q$ of query keys, aggregate distance function $d_{agg}(p, Q)$ is defined to be $agg(\{d(p, q) | q \in Q\})$, where $agg()$ is an aggregate function and $d(,)$ is a distance function. Each aggregate distance function of *summation*, *maximum*, and *minimum* is defined in Eqs. (4), (5), and (6) respectively.

$$d_{sum}(p, Q) = \sum_{i=1}^{|Q|} d(p, q) \tag{4}$$

$$d_{max}(p, Q) = maximum(\{d(p,q)|q \in Q\}) \tag{5}$$

$$d_{min}(p, Q) = minimum(\{d(p,q)|q \in Q\}) \tag{6}$$

Given a set $P$ of objects, a set $Q$ of query keys, and aggregate distance function $d_{agg}(p, Q)$, $k$-ANN query based MK-CBIR, $\Omega_{agg}(k, P, Q)$ retrieves a set $I$ of objects satisfying Eq. (7).

$$(I \subset P) \wedge (|I| = k) \wedge (k < |P|) \wedge \forall p \in I \, \forall p' \in (P - I)(d_{agg}(p, Q) \leq d_{agg}(p', Q)) \tag{7}$$

## 4 Experimental Evaluation

In this section, MK-CBIR and SK-CBIR are experimentally compared. First, the experimental overview is introduced. Then, the experimental results are shown.

### 4.1 Overview of Experiments

Figure 2 shows the overview of experiments to make comparison between MK-CBIR and SK-CBIR with regard to retrieval performance. For explanation, let preliminary retrieval in the upper part be (a), reminding retrieval with MK-CBIR in the lower-left part be (b), and reminding retrieval with SK-CBIR in the lower-right part be (c) (See Fig. 2). Let an image in the image database $P$ be a key image $t$ (hereafter called target image). First, (a) is performed by using $k$-NN search with $t$ from $(P-\{t\})$. The result set $I$ consists of $k$ images similar to $t$, which are then used as query keys in (b) and (c). Given the number $r$ of query keys for (b), the set of distinct query key sets over $I$, $\Upsilon(I, r)$ is defined in Eq. (8). $\Upsilon(I, r)$ collects each subset $K$ of $I$, whose cardinality is $r$. Evidently, the cardinality of $\Upsilon(I, r)$ is $_{|I|}C_r$. For $Q(\in \Upsilon(I, r))$, a retrieval trial with query keys $Q$ from $(P - Q)$ is conducted. On the other hand, each retrieval trial with query key $q(\in I)$ from $(P - \{q\})$ is conducted in (c). By comparing the result sets of (b) and those of (c) in *recall*, MK-CBIR is experimentally to be measured.

$$\Upsilon(I, r) = \{K|K \in 2^I \wedge |K| = r\} \tag{8}$$

Retrieval performance of $\Phi_{sky}$, $\Omega_{sum}$, $\Omega_{max}$, $\Omega_{min}$, and $k$-NN search are measured by the criterion *recall*. *Recall* is defined in Eq. (9). It specifies the ratio with which a target image is included in a query result. While the size of query results can be controlled by specifying it with a parameter value in each case of $\Omega_{sum}$, $\Omega_{max}$, $\Omega_{min}$, and $k$-NN search, that in case of $\Phi_{sky}$ cannot be managed. To cope with this problem in fairly comparing them, each *recall* is derived as follows:

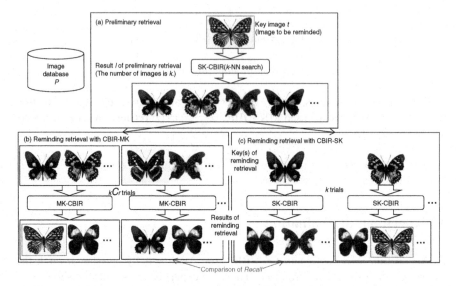

**Fig. 2** Experiments on comparison of MK-CBIR and SK-CBIR

1. Trials regarding $\Phi_{sky}$ are conducted and *recall* is calculated.
2. The average size $s$ of the query results is obtained by calculating equation $(\sum_{Q \in \Upsilon(I,r)} |\Phi_{sky}(P - Q, Q)|)/_{|I|}C_r$ where $I$ is the query results of (a) and $r$ is the number of keys.
3. Trials of $\Omega_{sum}$, $\Omega_{max}$, $\Omega_{min}$, and $k$-NN search are conducted to obtain query results of size $\lfloor s \rfloor$ and those of size $\lfloor s \rfloor + 1$, both of which are used to calculate *recall* by specifying the value[1] of the numerator in the right side of Eq. (9) according to the rules shown in Table 1.

$$Recall = \frac{number\ of\ trials\ returning\ target\ image\ in\ query\ result}{number\ of\ trials} \tag{9}$$

**Table 1** Rules specifying appearance value for a target image in query results

| Value | Condition |
| --- | --- |
| 1 | If the target image is within the $\lfloor s \rfloor$-th |
| s-$\lfloor s \rfloor$ | If the target image is the ($\lfloor s \rfloor + 1$)-th |
| 0 | If the target image is over the ($\lfloor s \rfloor + 1$)-th |

The experiments have been conducted by combining two kinds of datasets with two kinds of visual features. One kind of datasets is composed of 200 butterfly

---

[1] This adjustment is necessary, because $s$ is a real number.

images and another kind is composed of 400 landscape images.[2] For each combination, average *recalls* of $\Phi_{sky}$, $\Omega_{sum}$, $\Omega_{max}$, $\Omega_{min}$, and $k$-NN search are measured by conducting trials, each of which uses a distinct image in the dataset as a key of a preliminary retrieval. One of visual features is color and a color histogram regarding 64 reduced colors is prepared to represent it. Another visual feature is texture and a set of wavelet coefficients is extracted to represent it, as shown in Fig. 3. First, an original color image ($2950 \times 2094$ pixels in size) is converted into a reduced gray scale image ($22 \times 16$ pixels in size). Then, the converted image is partitioned into hierarchically resolved images with wavelet transform, multi-resolution analysis technique. It filters an image horizontally first and vertically next. As a result, the image is resolved into 4 regions, low frequency (LL), horizontal high frequency (LH), vertical high frequency (HL), and diagonal high frequency (HH) component. Given $n$ levels, the process is recursively repeated $n$ times. *Haar* wavelet transform [11] is employed to extract texture information, because it is speedy and easily implementable. In Fig. 3, 2-level *Haar* wavelet transform is conducted first. Then, the shaded regions are extracted for the texture information, because they are especially considered to represent the feature.

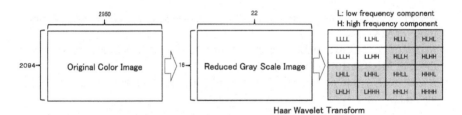

**Fig. 3** Extraction of wavelet coefficients

## 4.2 Experimental Results

Figure 4(a–d) respectively show *recalls* of $\Phi_{sky}$, $\Omega_{sum}$, $\Omega_{max}$, $\Omega_{min}$, and $k$-NN search for every combination of two datasets and two kinds of visual features. The experiments are conducted under the condition that the size of the result set, namely the parameter $k$, is 10 for preliminary retrievals and the number of keys for $\Phi_{sky}$, $\Omega_{sum}$, $\Omega_{max}$, and $\Omega_{min}$ is varied from 2 to 10 (See (a) in Fig. 2). The more the number of query keys becomes, the larger the result set of $\Phi_{sky}$ tends to grow. Since the size of the result sets of $\Omega_{sum}$, $\Omega_{max}$, $\Omega_{min}$, and $k$-NN search is set to that of $\Phi_{sky}$ (See (b) and (c) in Fig. 2), *recalls* and their graphs tend to rise to the right. Although *recalls* for the landscape dataset and the texture feature are relatively lower than those for other combinations, the tendency of each graph is roughly the same as follows:

---

[2]Various image datasets are obtainable from the site whose URL address is http://sozaijiten.net/search_title/. The dataset of No.012 is concerned with butterflies, while the datasets of No.006 and No.122 are concerned with landscapes.

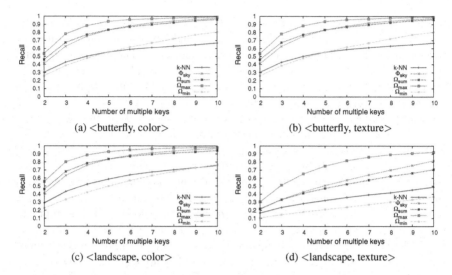

**Fig. 4** Recall of $k$-NN and MK-CBIR for varying number of query keys (<dataset, feature>)

- *Recall* of $\Omega_{min}$ is roughly the same as that of $k$-NN search and the lowest among those of MK-CBIR.
- *Recall* of $\Omega_{max}$ is over 0.3 higher than that of $k$-NN search and the highest among those of MK-CBIR.
- *Recall* of $\Omega_{sum}$ and *recall* of $\Phi_{sky}$ are roughly the same and between those of $\Omega_{min}$ and $\Omega_{max}$.

Figure 5(a–d) respectively show *recalls* of $\Phi_{sky}$, $\Omega_{sum}$, $\Omega_{max}$, and $\Omega_{min}$ with 10 keys for every combination of two datasets and two kinds of visual features. The experiments are conducted under the condition that the size of the result set is 10, namely the parameter $k$, for preliminary retrievals and some of $k$ keys are replaced with noisy keys which are randomly chosen among the nearest neighbors between the $(k+1)$-th and the $(2 \times k)$-th of preliminary retrievals (See (a) in Fig. 2). The number of noisy keys is varied from 0 to $k$. *Recall* data of $k$-NN search without a noisy key is shown for comparison, whose size of the result set is the average of $\Phi_{sky}$ with 10 keys. The tendency of each graph is roughly the same as follows:

- *Recall* of $\Omega_{min}$ rapidly falls to the right.
- *Recall* of $\Omega_{sum}$ and *recall* of $\Phi_{sky}$ gradually falls to the right.
- *Recall* of $\Omega_{max}$ is comparatively stable against noisy keys.

Figure 6(a–d) respectively show *recalls* of $\Phi_{sky}$, $\Omega_{sum}$, $\Omega_{max}$, and $\Omega_{min}$ with the varying number of keys for the butterfly dataset and the texture feature. The experiments are conducted under the condition that the size of the result set is $k$, the parameter of preliminary retrievals and some of $k$ keys are replaced with noisy keys which are randomly chosen among the nearest neighbors between the $(k+1)$-th and the $(2 \times k)$-th of preliminary retrievals (See (a) in Fig. 2). The experimental results are as follows:

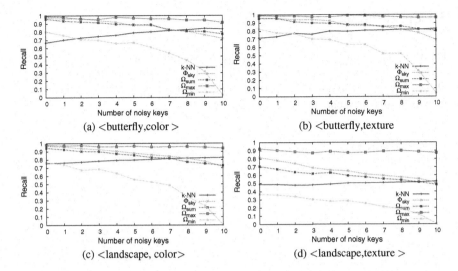

(a) <butterfly,color >

(b) <butterfly,texture

(c) <landscape, color>

(d) <landscape,texture >

**Fig. 5** Recall of $k$-NN and CBIR with 10 keys for varying number of noisy keys (<dataset, feature>)

- *Recall* of $\Omega_{sum}$ and *recall* of $\Phi_{sky}$ are roughly the same and gradually falls to the right (See Fig. 6a, b).
- *Recall* of $\Omega_{max}$ is comparatively stable against noisy keys.
- *Recall* of $\Omega_{min}$ gradually falls to the right and finally becomes nearly zero.

These tendencies are almost common to the other combinations of two datasets and two kinds of visual features.

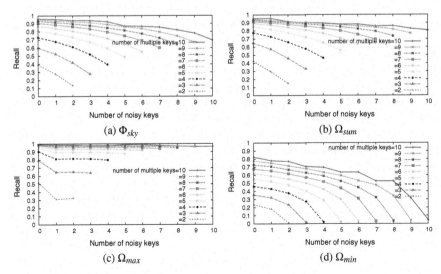

(a) $\Phi_{sky}$

(b) $\Omega_{sum}$

(c) $\Omega_{max}$

(d) $\Omega_{min}$

**Fig. 6** Recall of MK-CBIR with varying number of noisy keys (<butterfly, texture>)

# 5 Consideration

Figure 7 shows a query key $t$ and its result set $I$ of a preliminary retrieval (See Fig. 2). $I$ consists of $k$ images similar to $t$, which are used as query keys of $\Phi_{sky}$, $\Omega_{sum}$, $\Omega_{max}$, $\Omega_{min}$, and $k$-NN search in the experiments of Sect. 4. Considerations on the experimental results are mentioned in the followings.

**Fig. 7** Key image and
results of preliminary
retrieval

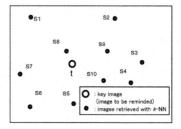

(1) Why is recall of $\Omega_{min}$ roughly the same as that of k-NN search? $\Omega_{min}$ executes multiple retrievals, each of which seeks for the target image around the corresponding key. However, each retrieval is independently done in non-cooperation with others and the size of its result set is on the average $1/(number\ of\ keys)$ of $k$-NN search whose key is single. Supposedly, this leads to the *recall* of $\Omega_{min}$, roughly the same as that of $k$-NN search.

(2) Why is recall of $\Omega_{max}$ the highest among those of MK-CBIR? Fig. 8(a) plots a contour graph of $d_{max}(p, Q)$ over a set of 10 keys whose coordinates are randomly generated. For explanation, each key corresponds to a 2-dimensional visual feature vector. Since the function is convex, there certainly exists a single point at which function value is the lowest. Given a set $Q$ of keys, $d_{max}(p, Q)$ partitions the plane into a set of farthest-point Voronoi regions [12]. Equation (10) defines the farthest-point Voronoi region $FV(q_i)$ with regard to $q_i(\in Q)$ (called *seed* in mathematics). A set of $FV(q_i)(q_i \in Q)$ is called a farthest-point Voronoi diagram. For point $p$, $d_{max}(p, Q)$ computes its distance from key $q_i(\in Q)$, if it belongs to the farthest-point Voronoi region $FV(q_i)$ with regard to $q_i$ (See Fig. 8(b)). Note that $FV(q_i)$ only exists with regard to $q_i$ which is a vertex of the convex hull of $Q$. Minimum Covering Circle (MCC) is the minimum circle which contains every key of $Q$ [12] (See Fig. 8(b)). The center of MCC corresponds to the lowest point of the contour graph of $d_{max}(p, Q)$. $\Omega_{max}$ retrieves the surface of the contour graph upward from this point to seek the target image. Given a set of query keys, some keys surrounding the target image cooperatively tend to make MCC whose center is located near the target image. The dimension is extensible over 2 and MCC is to be generalized to Minimum Covering Sphere (MCS)[12].

$$FV(q_i) = \bigcap_{i \neq j} \{p | d(p, q_i) > d(p, q_j), q_j \in Q\} \tag{10}$$

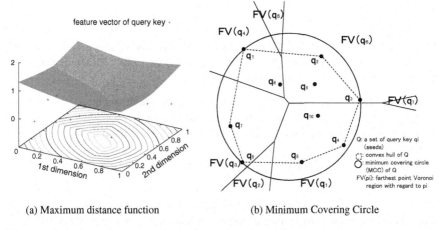

(a) Maximum distance function                    (b) Minimum Covering Circle

**Fig. 8**  Relation between feature vector of key and maximum distance

(3) Why is recall of $\Omega_{sum}$ between those of $Omega_{min}$ and $\Omega_{max}$? For a set of numbers, *average* is between *minimum* and *maximum*. Also, it is synonymous with *summation* for a fixed set. Accordingly, it is supposed that *recall* of $\Omega_{sum}$ is between those of $\Omega_{min}$ and $\Omega_{max}$, because each of $\Omega_{sum}$, $\Omega_{min}$, and $\Omega_{max}$ is based on $d_{sum}(p, Q)$, $d_{min}(p, Q)$, and $d_{max}(p, Q)$.

(4) Why is recall of $\Phi_{sky}$ roughly the same as that of $\Omega_{sum}$? Fig. 9 depicts *skyline* retrieval over image set $\{p_1, \ldots, p_9\}$ with 2 keys. For explanation, each image is plotted as a point whose coordinate is a pair of distance from *key1* and distance from *key2*. *Skyline* set consists of $p_1, p_2, p_3$, and $p_5$. From *skyline* definition of Eq. (2), it is definite that each of $p_4, p_8$, and $p_9$ in the first quadrant of *skyline* $p_1$ is non-*skyline*. In other words, each of $p_4, p_8$, and $p_9$ is larger than $p_1$, with regard to *summation* of distance from *key1* and distance from *key2*.[3] Actually, OSP algorithm computes *skyline* based on this property, by comparing each tuple in the ascending order of *summation* of *skyline* attribute values [13]. The number of keys is extensible over 2.

**Fig. 9**  Skyline retrieval
with 2 keys

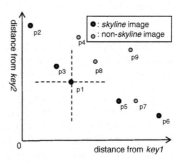

---

[3]Note that the converse is not true.

# 6 Conclusion

This paper proposed MK-CBIR for giving users easy key selection and presented as its concrete implementation $\Phi_{sky}$, $\Omega_{sum}$, $\Omega_{max}$, and $\Omega_{min}$. Experimental results on retrieval performance are summarized as follows:

1. *Recall* of $\Omega_{max}$ is over 0.3 higher than that of $k$-NN search and the highest among those of MK-CBIR. It is comparatively stable against noisy keys.
2. *Recall* of $\Omega_{min}$ is roughly the same as that of $k$-NN search and the lowest among those of MK-CBIR.
3. *Recall* of $\Omega_{sum}$ and *recall* of $\Phi_{sky}$ are roughly the same and between those of $\Omega_{min}$ and $\Omega_{max}$.

It is concluded that "Two heads (MK-CBIR) are better than one (SK-CBIR)". Our future work includes clarification of usable/available utilization of MK-CBIR and integration of MK-CBIR into RF framework.

# References

1. Borzsonyi, S., Kossmann, D., Stocker, K.: The skyline operator. Proc. Int. Conf. Data Eng. (2001)
2. Papadias, D., Shen, Q., Tao, Y., Mouratidis, K.: Group nearest neighbor queries. Proc. Int. Conf. Data Eng. (2004)
3. Papadias, D., Tao, Y., Mouratidis, K., Hui, C.K.: Aggregate nearest neighbor queries in spatial databases. ACM Trans. Database Syst. **30**(2), 529–576 (2005)
4. Yiu, M.L., Mamoulis, M., Papadias, D.: Aggregate nearest neighbor queries in road networks. IEEE Trans. Knowl. Data Eng. **17**(6), 820–833 (2005)
5. Sato, H., Narita, R.: Approximate search algorithm for aggregate k-nearest neighbor queries on remote spatial databases. Int. J. Knowl. Web Intell. **4**(1), 3–19 (2013)
6. Sato, H., Narita, R.: Approximate processing for aggregate range queries on remote spatial databases. Int. J. Knowl. Web Intell. **4**(4), 314–335 (2014)
7. Salton, G., Buckley, C.: Improving retrieval performance by relevance feedback. J. Am. Soc. Inf. Sci. **41**(4), 288–297 (1990)
8. Zhou, X.S., Huang, T.S.: Relevance feedback in image retrieval: a comprehensive review. ACM Multimedia Syst. J. **8**(6), 536–544 (2003)
9. Kulkarni, P.A., Shahane, N.M.: A survey on relevance feedback mechanisms used for CBIR systems. Int. J. Emerg Technol. Adv. Eng. **4**(11), 239–244 (2014)
10. Samukawa, M.: Image Retrieval by Estimating and Visualizing Ambiguous Intention from a Few Key Images, Departmental Bulletin of Graduate School of Science and Engineering, Chuo University 36 (2006), (in Japanese)
11. Stollnitz, E.J., Derose, T.D., Salesin, D.H.: Wavelets for computer graphics: a primer, Part 1. IEEE Comput. Graphics Appl. **15**(3), 76–84 (1995)
12. Berg, M.D., Kreveld, M.V., Overmars, M., Schwarzkopf, O.: Computational Geometry: Algorithms and Applications, 3rd edn. Springer, Santa Clara (2008)
13. Zhang, S., Mamoulis, N., Cheung, D.W.: Scalable skyline computation using object-based space partitioning. In: Proceedings of ACM SIGMOD, pp. 483–494 (2009)

# A Simple Medium Access Scheme Based on Spread Spectrum Suitable for Wireless Ad Hoc Networks

Shiori Watanabe and Yukihiro Kamiya

**Abstract** We propose a simple medium access scheme for wireless ad hoc network (WAHN). The proposed method is based on the spread spectrum (SS) technique to take advantage of the robustness against fading and differentiation of signals. However, conventionally, SS needed a mechanism to allocate a unique spreading code to each node. Since WHAN does not include base stations, this problem must be solved. In this paper, we solve this problem by let each node share a unique code and use it after expanding the chip duration. Although this proposed scheme decreases the data rate if the node expands widely the chip duration, it is better than CSMA. The performance is verified through computer simulations.

**Keywords** Wireless ad hoc network · Digital signal processing · Spread spectrum · Medium access

## 1 Introduction

Wireless ad hoc networks (WAHNs) achieve flexible networking due to the absence of base stations. It means that WAHN is robust against natural disasters such as earthquakes under which base stations are often destructed or shutdown. However WAHN suffers from low efficiency due to the lack of the base stations. This is because the base stations function as network coordinators so that each nodes in the network can send packets avoiding the collision with those sent by the other nodes. Therefore we need a mechanism to let the nodes avoid collisions autonomously, or to let receiving nodes extract a desired signal even though the collision happens.

Code division multiple access (CDMA) is a well-known multiple access scheme widely employed by modern mobile communication systems [1]. Although CDMA is highly efficient, it requires base stations for code assignment, *i.e.*, base stations assign a unique spreading code to each node in order to realize multiple access.

S. Watanabe · Y. Kamiya (✉)
Aichi Prefectural University, Nagakute, Japan
e-mail: kamiya@ist.aichi-pu.ac.jp

© Springer International Publishing Switzerland 2015
E. Damiani et al. (eds.), *Intelligent Interactive Multimedia Systems and Services*,
Smart Innovation, Systems and Technologies 40,
DOI 10.1007/978-3-319-19830-9_29

So it is obvious that CDMA is not applicable to WAHN as it is. However CDMA is still attractive because of the fact that it is based on spread spectrum (SS). So it can take advantage of the RAKE combining that realizes high-speed data transfer through multipath fading channels. Thus the problem is how to realize CDMA without base stations.

It should be remembered that wireless local area networks (LANs) do not allow concurrent communications in the network because of the difficulty of the unique code allocation by simple base stations [2]. This matter is pointed out in [3]. Therefore some of related works assume that the unique code allocation is done by unspecified methods [5], or by a time-consuming method [6].

Carrier sense multiple access (CSMA) might be a solution to this matter. It is defined in IEEE802.11 for LAN [7]. CSMA simplifies the task of base stations by forcing the nodes to follow a simple rule as follows: Suppose that a node wants to send a packet. Then the node does not start sending immediately, but the node listens to the channel. If the node detects the power of signals sent by other nodes, it will wait for a while. After that, the node listens to the channel again and start sending if the node does not detect the signal power. According to this simple rule, the collision of packets can be avoided even if the base station does not exist. However the efficiency is low since only one node can send packets while the other nodes are waiting.

Naturally we can be interested in a new way which is applicable to WHAN, combining the advantages of CDMA and CSMA. It must be a highly-efficient access scheme which does not need the base stations. In addition, it must be robust against fading.

An overview of the new idea is as follows: The proposed system allows each node to share an identical spreading code, instead of realizing the unique code assignment by base stations. Each node uses the spreading code after expanding the chip duration with a unique expansion rate. By this chip duration expansion, the proposed system differentiates the spreading codes of the nodes. The way to choose the chip expansion rate is explained in the following chapters.

In this paper, we propose such a new scheme for WAHN. Through computer simulations, it will be clarified that this proposed method improves the performance of WAHN even though it is based on a simple idea.

Following sections are organized as follows: Sect. 3 provides in-depth explanation of the proposed method based on the mathematical formulations given in Sect. 2. Computer simulations are provided in Sect. 4 for the performance verifications. Finally Sect. 5 concludes this paper.

## 2 Signal Modelling and System Overview

Figure 1 illustrates an image of ad hoc communications where a receiving node is interfered with surrounding nodes. The problem is how to extract a desired signal polluted by interfering signals.

Figure 2 illustrates a block diagram of the $M$ transmitters and a receiver based on the spread spectrum (SS) techniques. These are mathematically formulated as follows:

A binary data sequence of $m(= 1, \cdots, M)$-th transmitter is defined by $\vec{d}_m$ of size $(1 \times D)$ as the following,

$$\vec{d}_m = \left[ d_0^{(m)} \ d_1^{(m)} \ \cdots \ d_D^{(m)} \right] \tag{1}$$

where $d_0, \cdots, d_D \in (0, 1)$. This data sequence is modulated and we obtain a symbol vector $\vec{g}_m$ of size $(1 \times G)$ as follows:

$$\vec{g}_m = \left[ g_0^{(m)} \ g_1^{(m)} \ \cdots \ g_G^{(m)} \right]. \tag{2}$$

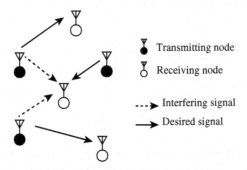

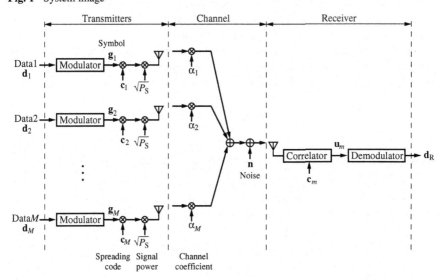

**Fig. 1** System image

**Fig. 2** System image

At the $m$-th transmitter, the symbol vector is multiplied by a spreading code $\vec{c}_m$ of size $(1 \times C_m)$ defined as:

$$\vec{c}_m = \begin{bmatrix} c_0^{(m)} & c_1^{(m)} & \cdots & c_{C_m}^{(m)} \end{bmatrix} \tag{3}$$

where $c_0, \cdots, c_{C_m} \in (-1, 1)$.

By using this spreading code, the $m$-th transmitter generates an SS signal expressed by $\vec{s}_{SS}^{(m)}$ of size $(1 \times C_m G)$ as follows:

$$\vec{s}_{SS}^{(m)} = \sqrt{P_S} \left\{ \text{vec} \left( (\vec{c}_m)^T \vec{g}_m \right) \right\}^T \tag{4}$$

where $P_S$ denotes the signal power. The superscript $\cdot^T$ denotes the transpose of a vector or a matrix. The function $\text{vec}(\cdot)$ is to pile on the columns of a matrix. For example, let us suppose a matrix as follows:

$$\vec{A} = \begin{bmatrix} a_1 & a_3 \\ a_2 & a_4 \end{bmatrix}. \tag{5}$$

So $\text{vec}(\vec{A})$ is as follows:

$$\text{vec}(\vec{A}) = \begin{bmatrix} a_1 & a_2 & a_3 & a_4 \end{bmatrix}^T. \tag{6}$$

The SS signal $\vec{s}_{SS}^{(m)}$ will be band-limited and digital-to-analog converted so that it is radiated from an antenna after the multiplication with a carrier.

Then the receiver receives the sum of SS signals sent by $M$ transmitters with random time offsets. Before going to the formulation of the received signal at the receiver, the timing offset of the signal is formulated.

Since the $M$ transmitters send the signal asynchronously, the receiver receives the signal with unknown random time offsets. The signal sent by the $m$-th transmitter received with the unknown time offset $z_m$ is expressed as:

$$\widetilde{\vec{s}_{SS}^{(m)}} = \begin{bmatrix} \vec{O}_m & \vec{s}_{SS}^{(m)} \end{bmatrix} \tag{7}$$

where $\vec{O}_m$ expresses the time offset as a vector of size $(1 \times z_m)$ in which all entities are 0. Therefore the size of $\widetilde{\vec{s}_{SS}^{(m)}}$ is $\left( 1 \times (C_m G + \hat{z}) \right)$ where $\hat{z} = \max z_m$.

Now we can think about the received signal at the receiver as the sum of the signals sent by the $M$ transmitters expressed by a vector $\vec{r}$ of size $\left( 1 \times (C_m G + \hat{z}) \right)$ as

$$\vec{r} = \sum_{m=1}^{M} \alpha_m \widetilde{\vec{s}_{SS}^{(m)}} + \sqrt{\frac{P_N}{2}} \vec{n} \tag{8}$$

where $\alpha_m$ is the channel coefficient of the channel in which the signal from the $m$-th transmitter went through to the receiver. The vector $\vec{n}$ of size $\left(1 \times (C_mG + \hat{z})\right)$ contains samples of the unit-power complex additive white Gaussian noise (AWGN) and $P_N$ denotes the noise power.

The received signal vector $\vec{r}$ is fed into the correlator in which weight coefficients are set as $\vec{c}_m$. The correlator output is formulated by $\vec{u}_m$ as

$$\vec{u}_m = \begin{bmatrix} \vec{u}_0 \ \vec{u}_1 \ \cdots \ \vec{u}_{U-1} \end{bmatrix} \tag{9}$$

where $U - 1 = (C_mG + \hat{z}) - C_m$. In addition,

$$u_k = \vec{c}_m(\hat{r}_k)^{\mathrm{T}}, \quad \hat{r}_k = \begin{bmatrix} r_k \ r_{k+1} \ \cdots \ r_{k+C_m-1} \end{bmatrix} \tag{10}$$

where $k = 0, \cdots, U - 1$.

Finally, the demodulator recovers the data sequence from $\vec{u}_m$.

# 3 Proposed Method

## 3.1 An Overview of the Proposed Method

As mentioned above, our goal is to achieve a highly efficient medium access scheme like CDMA taking advantage of the robustness against fading, but not dependent on the base stations, like CSMA. The core of the idea is to let the nodes share an identical spreading code. Since the network suffers from collisions if the nodes use the spreading code as it is, each node expands the chip duration of the spreading code to make it possible to be unique.

The image of the chip duration expansion is illustrated in Fig. 3. All nodes share the sequence $\vec{c}_1$. Then $\vec{c}_2$ is generated based on $\vec{c}_1$ by expanding the width two-times wider than that of $\vec{c}_1$. Likewise $\vec{c}_{10}$ is ten-times wider than $\vec{c}_1$. Each node chooses one of those expanded sequence to multiply with the symbol sequence. Now let $\vec{c}_\gamma$ denote the spreading code in which a chip is $\gamma$-times wider than $\vec{c}_1$. Let us call $\gamma$ as the chip expansion rate.

## 3.2 Medium Access Scheme

Although the idea to differentiate the spreading codes is simple, the following two questions remain:

**Fig. 3** Images of the chip
duration expansion

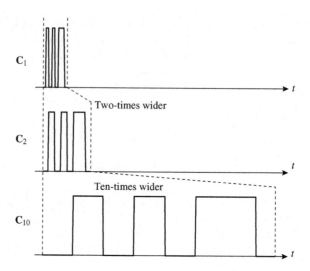

1. How a transmitting node decides the chip-expansion rate $\gamma$?
2. How a receiving node identify $\gamma$ which is used in the transmitting node? This is
   because, based on the spread spectrum technique, the receiving node must use
   the identical spreading code that is used in the transmitting node.

   This problem is expected to be solved by a CSMA-like medium access scheme.
Firstly, the procedure for transmitting nodes is described as follows:

**Step 1** Suppose that Node1 is trying to send a packet. Then Node1 initiates the carrier sensing, *i.e.*, Node1 receives the signals and stores the samples of the
signals in memories.

**Step 2** Node1 feeds the samples to a correlator in which $\vec{c}_1$ is set as a weight vector.

**Step 3** Node1 examines the correlator output and detect periodical peaks of the output which is generated as the auto-correlation between the received signal
and the weight vector of the correlator.

**Step 4** Node1 repeats Steps 2 and 3 by replacing the weight vector of the correlator
with other possible spreading code such as $\vec{c}_2, \vec{c}_3, \cdots \vec{c}_M$.

**Step 5** Node1 chooses $\vec{c}_\gamma$, $(\gamma = 1, \cdots, M)$ if no periodical peaks of the output is
observed. Now Node1 sends a SS signal spread by the selected $\vec{c}_\gamma$.

Next, the procedure for receiving nodes is described as follows:

**Step 1** Suppose that Node2 is idle. Then Node2 monitors the channel receiving the
signals and stores the samples of the signals in memories.

**Step 2** Node2 feeds the samples to a correlator in which $\vec{c}_1$ is set as a weight vector.

**Step 3** Node1 examines the correlator output and detect periodical peaks of the output which is generated as the auto-correlation between the received signal
and the weight vector of the correlator.

**Step 4** Node2 repeats Steps 2 and 3 by replacing the weight vector of the correlator
with other possible spreading code such as $\vec{c}_2, \vec{c}_3, \cdots \vec{c}_M$.

**Step 5**  Node2 demodulates the correlator output if the periodical peaks of the output is observed.

**Step 6**  If the data extracted by the demodulation indicates that the data is for Node2, it starts the necessary procedures to communicate with the transmitter.

## 3.3 An Expected Drawback Caused by the Proposed Method

A drawback by the proposed method is that the data rate is low if the transmitting node selects a large value of the chip expansion rate $\gamma$. To cope with this drawback, we need to think about some other aspects of the medium access control such as scheduling and QoS. At least, we can say that it is better, even if the data late is low, than the node can send nothing. In this paper, we do not focus on this matter remaining it as one of further considerations.

In addition, the proposed method forces the node to handle many spreading codes. Although the spreading codes are generated by a single code expanding the duration, the processing time under a real-time system is of our concern. Therefore we also put the measurement of the processing time in the further considerations.

## 4  Computer Simulations

In this section, we conduct computer simulations for performance verifications. Table 1 summarizes simulation conditions. Figure 4 illustrates the simulated system.

**Table 1**  Simulation conditions

| Signal | |
|---|---|
| Data length ($D$)[bit] | 10000 |
| Spreading code length ($L$)[chip] | 127 (an M-sequence) |
| Sampling | 1 [sample/chip] |
| SNR | -11dB before despreading |
| **Node** | |
| The number of surrounding nodes ($N$) | 2 |

In this simulation, we focus on a node labeled as "Center node in Fig. 4, surrounded by $M$ nodes. Suppose that the center node is receiving a signal to communicate with one of the surrounding nodes. At the same time, the center node suffers from interferences from the other surrounding nodes. In this simulation, it is assumed that the nodes employ different value of $\gamma$, the chip expansion rate, without overlapping. In addition, we assume that the signal power is set identical among the

**Fig. 4** Simulated system image

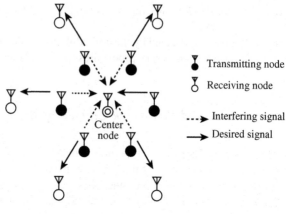

**Fig. 5** Simulation result

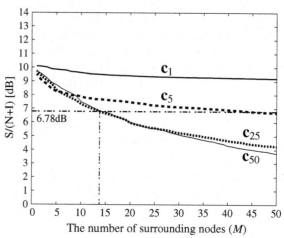

transmitting nodes so as to set SNR, the signal-to-noise power ratio, as -11dB prior to the dispreading at the receiver.

Figure 5 shows S/(N+I), the signal-to-noise-plus-interference power ratio, versus the number of surrounding nodes. This figure compares curves of which the center node employs the spreading code such as $\vec{c}_1, \vec{c}_5, \vec{c}_{25}$ and $\vec{c}_{50}$.

In addition, we decide that the communication is successful if S/(N+I) is more than 6.78dB. This S/(N+I) achieves BER of $10^{-3}$ if BPSK is employed [1].

According to Fig. 4, it is observed that the nodes with $\vec{c}_1, \vec{c}_5, \vec{c}_{25}$ and $\vec{c}_{50}$ can successfully communicate with their pair-wised node till the total number of the surrounding nodes is 14.

Recall that the proposed scheme is to eliminate the base stations and to improve the efficiency of CSMA. So remember that CSMA allows one pair of nodes to communicate while it prohibits other nodes sending signals. The proposed scheme improves the efficiency, *i.e.*, it means that the number of the simultaneous communications is much larger than that is achieved by CSMA.

# 5 Conclusions

In this paper, we proposed a simple scheme to realize SS-based and base station-less medium access suitable for wireless ad hoc networks. The core of the idea is very simple; each node employs expanded spreading code to let the receiver differentiate the signals. We verified the performance by S/(N+I) through computer simulations.

We need further considerations to cope with the drawbacks explained in Sect. 3.3.

# References

1. Proakis, J.G.: Digital Communications, McGraw-Hill (2008)
2. Lal, S., Sousa, E.S.: Distributed resource allocation for DS-CDMA-based multimedia ad hoc wireless LANs. IEEE J. Sel. Areas Commun. **17**(5) (1999)
3. Andrews, J.G., Drexel, S.W., Haenggi, M.: Ad Hoc networks: to spread or not to spread. IEEE Commun. Mag. (2007)
4. Zhang, L., Soong, B.-H.: Multi-code multi-packet transmission (MCMPT) in wireless CDMA Ad Hoc networks under Rayleigh fading channels. IEEE Commun. Lett. **9**(11) (2005)
5. Sankaran, C., Ephremides, A.: The use of multiuser detectors for multicasting in wireless Ad Hoc CDMA networks. IEEE Trans. Inf. Theor. **48**(11) (2002)
6. Qu, Q., Milstein, L.B., Vaman, D.R.: Cross-layer dstributed joint power control and scheduling for delay-constrained applications over CDMA-based wireless Ad-Hoc networks. IEEE Trans. Commun. 58(2) (2010)
7. Prasad, A., Prasad, N.: '802.11 WLANs and IP Networking: Security, QoS, and mobility Artech House Universal Personal Communications (2005)

# Implementation of Tree-Based Data Collection Scheme for Arduino-Compatible Board

Katsuhiro Naito, Kento Nakanishi, Kazuo Mori and Hideo Kobayashi

**Abstract** This paper develops a new Arduino-compatible board for sensor networks and proposes a simple tree-based data collection scheme for low-power microcontrollers. The developed board has a special circuit for a stabilized power source for external sensors and wireless modules, and implements a charge controller from a solar cell to Li-Po battery for autonomous long-lived operation. The implemented tree-based data collection scheme consists of a routing function and a media access control function for low-power microcontrollers. We can apply it as a small size program which is executed in a small size of the execution area by reducing routing information, and it can construct a tree-based route from a sink node to each node on a hop by hop basis. The developed media access control function employs a special time-division frame structure to leverage a sleep operation with low-power consumption for long-lived operation, and to reduce packet collisions between adjacent nodes. The experimental evaluations show that the proposed data collection scheme can operate on the developed board with a low-power microcontroller even if the microcontroller has only 8 [KB] Static Random Access Memory (SRAM). Additionally, the implemented tree-based data collection scheme can construct an adequate route from nodes to a sink node, and can realize a reliable data collection.

**Keywords** Sensor networks · Tree-based routing · Media access control · Arduino-compatible board

K. Naito (✉)
Department of Information Science, Aichi Institute of Technology, 1247 Yachigusa, Yakusa, Toyota, Aichi 470-0392, Japan
e-mail: naito@pluslab.org

K. Nakanishi · K. Mori · H. Kobayashi
Department of Electrical and Electronic Engineering, Mie University,
1577 Kurimamachiya, Tsu, Mie 514-8507, Japan
e-mail: kmori@elec.mie-u.ac.jp

H. Kobayashi
e-mail: koba@elec.mie-u.ac.jp

© Springer International Publishing Switzerland 2015
E. Damiani et al. (eds.), *Intelligent Interactive Multimedia Systems and Services*,
Smart Innovation, Systems and Technologies 40,
DOI 10.1007/978-3-319-19830-9_30

# 1 Introduction

Sensor networks have been focused as a new type of sensing methods by using a multi-hop communication technology [1, 2]. Hence, a lot of routing protocols for sensor networks have been proposed [3–6]. Applications of sensor networks are generally classified into two types; a periodic measurement application and an event-driven measurement application. The first one is to measure the environment by some sensors periodically. The second one is to start some actions by detecting some events. This paper assumes a periodic measurement application for a field sensing with low-power microcontrollers.

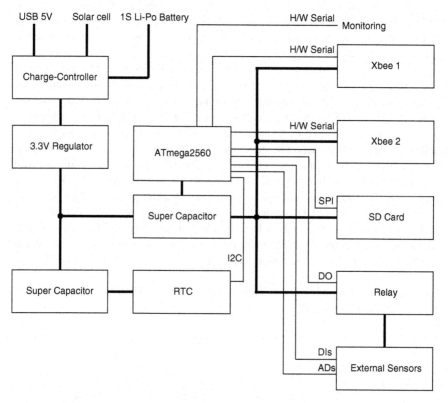

**Fig. 1** System model of Arduino-compatible board

Sensor networks consist of many sensor nodes for measurement and reporting, and a sink node for data collection. Multi-hop communication is utilized to convey measurement information from the sensor nodes to the sink node. Additionally, almost all sensor nodes operate by a battery. Therefore, low-energy operation is an important function for realizing long-lived networks [7, 8]. Employing a low-power microcontroller and using a sleep operation of a hardware are conventional ways to reduce consumed power. However, employing low-power microcontrollers causes

**Fig. 2** Arduino-compatible board

serious limitation of system resource e.g. 8KB SRAM in Atmel ATmega2560 microcontroller. Additionally, timing synchronization for communication between nodes is important to perform a sleep operation effectively. Therefore, special media access control mechanisms have been proposed [9–11]. An interference reduction is another issue in sensor networks because multi-hop networks generally suffer from a self interference between nodes. Therefore, various researches to mitigate any interference effect have been proposed [12–16].

The authors have studied about cross-layer mechanisms supporting a media access control (MAC) and a routing for low-power microcontrollers in sensor networks. The proposed MAC supports a special frame structure to increase a chance of a sleep operation and to decrease packet collisions. The proposed routing protocol exchanges neighbor information between nodes, and constructs a tree-based route from a sink node to nodes. The early implementation employed Sun SPOT devices supporting Squawk Java Virtual Machine [17, 18]. The processor board has one ARM architecture 32 bit CPU with ARM920T core operating at 180 MHz. It had 512 KB RAM, 4 MB flash memories and a 2.4 GHz IEEE 802.15.4 chip. However, we found that the

**Table 1** Specifications of Arduino-compatible board

| Development environment | Arduino software |
|---|---|
| Microcontroller | ATmega2560 (8 [MHz]) |
| Memory | 8 [KB] SRAM, 4 [KB] EEPROM, and 256 KB Flash Memory |
| Regulator | 3.3 [V], 800 [mA] MAX |
| Charge controller | 1 Solar cell, 1 Li-Po battery |
| Digital I/O Pins | 54 (2 for Xbee monitor, 2 for Xbee sleep and 1 for relay) |
| Analog Input Pins | 16 |
| Hardware Serial | 4 (2 for Xbee socket) |
| RTC | Maxim DS3231 |

ARM based CPU is too powerful to realize long-lived operations from the experimental evaluation.

This paper presents a new circuit design of a microcontroller board for sensor networks to realize a stable operation and accurate measurements when some external sensors and wireless modules are connected to the board. Additionally, it also proposes an implementation design of a tree-based data collection scheme for low-power microcontrollers such as Atmel 8-bit AVR RISC-based microcontrollers that implements a few kilobytes internal SRAM.

The developed board has a special circuit for a stabilized power source for external sensors and wireless modules because these peripheral devices require large current and may suffer from a voltage depression that causes an unexpectedly restart and a measurement error. Moreover, it also implements a charge controller from a solar cell to Li-Po battery for an autonomous long-lived operation, and a SD card access for logging measurement values because these functions are generally required in practical sensor networks. The developed board is designed based on Arduino Board to employ Arduino software for developing a software. As a result, we can employ it to develop a practical sensor network system easily.

The implemented tree-based data collection scheme consists of a routing function and a media access control function for low-power microcontrollers. It achieves a small size of execution program by reducing routing information because low-power microcontrollers generally do not have enough memory space to execute a large size of program. The developed routing function can construct a tree-based route from a sink node to each node on a hop by hop basis. The developed media access control function employs a special time-division frame structure to leverage a sleep operation with low-power consumption for long-lived operation, and to reduce packet collisions between adjacent nodes.

In the experimental evaluations, we employs the developed board and the tree-based data collection scheme to realize a sensor network testbed. The basic evaluation shows that 8 [KB] SRAM in ATmega 2560 microcontroller is enough size to implement the mechanisms to realize a multi-hop sensor network. Additionally, the implemented tree-based data collection scheme can construct an adequate route from nodes to a sink node, and can realize a reliable data collection.

## 2  Arduino-Compatible Board

Various kinds of Arduino-compatible boards have been released recently. However, almost all boards do not assume external sensors for accurate measurement and wireless communication functions. Additionally, some boards supporting wireless modules do not consider required current source capacity to activate peripheral modules. As a result, some boards may behave in an erratic way due to inrush current when peripheral equipments such as wireless modules, sensors etc. are activated. Moreover, boards for sensor networks generally require low-power operation because each board is requested to operate autonomously with a battery. However, almost all

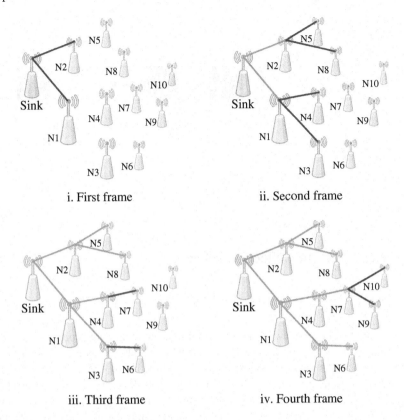

i. First frame

ii. Second frame

iii. Third frame

iv. Fourth frame

**Fig. 3** Overview of routing process

boards do not support a charge control function from a solar cell. Therefore, few boards are appropriate for a sensor network application even if many boards have been released.

Figure 1 shows the system model of the developed Arduino-compatible board and Fig. 2 shows the overview of the board. The developed board has a special circuit consisting of a sufficient large output current voltage regulator and a capacitor for external sensors and wireless modules to avoid a voltage depression that causes unstable operation of a board and erroneous measurement. Additionally, it also implements a charge controller from a solar cell to Li-Po battery to realize an autonomous long-lived operation, a SD card access to record measurement values, a relay switch to reduce power consumption of external sensors, and a real time clock (RTC) chip for accurate time acquisition. It employs an Atmel ATmega 2560 microcontroller to achieve a compatibility with Arduino software to realize an easy development of a software. The detail parameters are presented in Table 1.

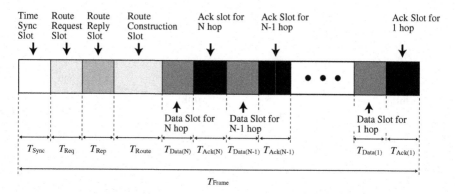

**Fig. 4** Frame structure

# 3 Tree-Based Data Collection Scheme

## 3.1 Overview of Tree-Based Routing

Figure 3 shows the overview of the proposed tree-based routing process. The proposed routing protocol can construct a route between nodes by using only neighbor information because the assumed Arduino-compatible board does not have enough memory space to store much routing information for all nodes. Therefore, a route construction process is performed on a hop by hop basis. The first step is the route construction between the sink node and 1-hop nodes (i.e., N1 and N2). The second step is the route construction between 1-hop nodes and 2-hop nodes. In a similar fashion, the proposed routing protocol can build a tree-based route from the sink node to the edge nodes.

## 3.2 Frame Structure

The proposed mechanism for the media access control employs the special frame format in Fig. 4 to reduce the collision probability and to increase a chance for a sleep operation. The frame format consists of some time slots for desired purposes: synchronizing a local clock, requesting for a new route, replying to the route request, requesting a route construction, data transmission, and acknowledgement transmission. Each sensor node attempts to transmit a packet with the Carrier Sense Multiple Access/Collision Avoidance(CSMA/CA) mechanism in each slot. The proposed format also employs sub-slots for data transmission and acknowledgement transmission to reduce self-interference effect between different hop communication. The purposes of each slot are presented as follows.

- TSYNC (Time SYNChronization) slot
  General Arduino-boards do not have a RTC function. Therefore, a time synchronization mechanism is required to synchronize start timing of the frame. A sink node transmits TSYNC packets periodically. Then, neighbor nodes around the sink node synchronize the start time according to the TSYNC packets. The neighbor nodes that receive the TSYNC packet also transmit a TSYNC packet for their neighbor nodes to disseminate the start timing of the frame.
- RREQ (Route REQuest) slot
  A node, that does not have any route to a sink node, requests for a new route to its neighbor nodes in the RREQ slot. Therefore, the neighbor nodes can recognize that the node tries to find a new route when they receive the RREQ packet.
- RREP (Route REPly) slot
  The nodes that receive the RREQ packets reply a RREP packet in the RREP slot. A RREP packet includes node's hop count information. Therefore, the node that receives these RREP packets can recognize the nearest node to the sink node.
- RCON (Route CONstruction) slot
  The RCON slot is meant for an actual route construction process between nodes. The node without a route requests a route construction by transmitting a RCREQ (Route Construction REQuest) packet to its upstream node. Then, the node that receives this RCREQ packet replies a RCREP(Route Construction REPly) packet to its downstream node. Finally, the node that receives the RCREP packet replies a RCACK(Route Construction ACKnowledgement) to verify the completion of the route construction process.
- Data slot
  Each node transmits observed environmental information during the data slot. As we employ multi-hop communication, the data packets are forwarded from downstream nodes to the sink node through some nodes. Additionally, the data slot is subdivided into some sub-slots for the hop count from the sink node because some packets are interfered with each other. Nodes with a same hop count share their sub-slot in CSMA/CA.
- ACK(ACKnowledgement) slot
  Nodes reply an ACKnowledgement packet when they receive a data packet from their downstream nodes. The AKC slot is also divided into some sub-slots according to the same reason for the data slot.

## 3.3 Routing Procedure

Figure 5 shows the example of packet transmission in the route construction process with the node location in Fig. 3. In the initial state, the sink node is the only node that has a route. In the first frame, all nodes transmit a RREQ packet to find a route. The sink node replies to nodes N1 and N2 in the RREP slot. Then, both nodes attempt to construct a route to the sink node in the RCON slot. In the second frame, all nodes except N1 and N2 also transmit a RREQ message. The node N1 replies to the nodes

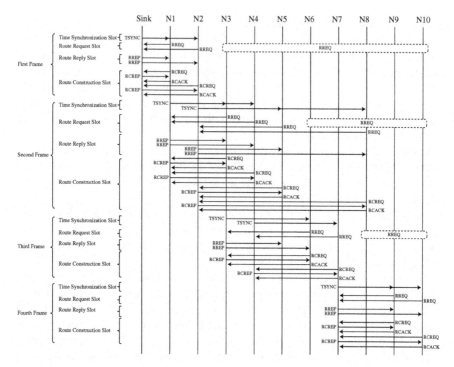

**Fig. 5** Signaling for routing

N3 and N4, and the node N2 replies to the nodes N5 and N8. The nodes N3, N4, N5, and N8 try to construct a route in a same manner in the first slot. Finally, the tree-based route has been constructed after the four frame cycles.

### 3.4 Data Transmission

Figure 6 shows the example of the packet transmission in the data transmission process with the node location in Fig. 3. The proposed protocol employs an acknowledgement operation in each hop count. Therefore, nodes reply an acknowledgement packet when they receive a data packet from their downstream nodes. The proposed data transmission process is started at the edge nodes with the maximum hop count (i.e., N9 and N10) because the forwarding process of data packets should be completed within a frame period.

## 4 Experimental Results

We have implemented the proposed data collection scheme on the developed Arduino-compatible board for the sensor network testbed. We employed Xbee S1 as a wireless communication module. In the evaluation, each node reports observed

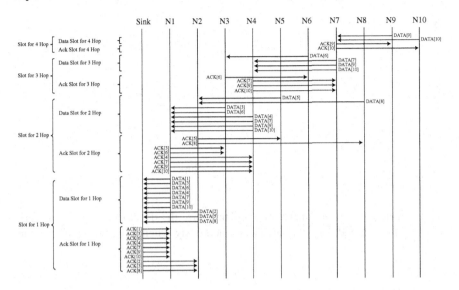

**Fig. 6** Data and ACK transmission

**Fig. 7** Location pattern 1

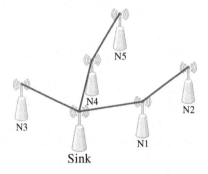

information to the sink node every one minute interval. The system has operated for 24 h to evaluate the packet arrival ratio. We assumed two types of node location in Figs. 7 and 8. We can confirm the routing procedures when a node has some candidate upstream nodes in the location of Fig. 7, and can validate the multi-hop relaying function in the location of Fig. 8. The specific frame parameters are detailed in Table 2.

From the experimental results, we can find that the average packet arrival ratio is 97.0 % in Fig. 7, and 98.7 % in Fig. 8. The main reason of the packet loss is an interference from WiFi signals because we have set the transmission power of Xbee to the minimum transmission level to reduce transmission range.

The developed board supports a sleep operation of ATMega 2560 chip and Xbee module. The consumption current during operation status is about 60 [mA] and that

**Fig. 8** Location pattern 2

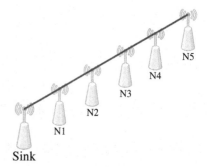

Sink

during sleep status is about 0.55 [mA]. Therefore, the developed board can work with a solar cell and a Li-Po battery autonomously when the board activates the sleep mode.

**Table 2** Frame format parameters

| Node | Arduino-compatible board + XBee S1 (IEEE 802.15.4) |
| --- | --- |
| Number of nodes | Sink:1, Nodes:5 |
| Measurement Interval | 1 [min] |
| Frame period | 60 [sec] |
| TSYNC Slot | 10 [sec] |
| RREQ Slot | 1 [sec] |
| RREP Slot | 3.5 [sec] |
| RCON Slot | 5.5 [sec] |
| Data Sub-Slot | 2 [sec] |
| ACK Sub-Slot | 2 [sec] |
| Maximum hop count | 10 [hop] |
| Data packet size | 68 [Byte] |

# 5 Conclusion

This paper has proposed a new type of Arduino-compatible board for sensor networks, and has proposed a tree-based data collection scheme for low-power microcontrollers. The developed microcontroller board realizes a stable operation even if peripheral devices are attached to the board. The proposed protocol can work with low-power microcontrollers by reducing an amount of exchanging routing information, and can bring about a sleep operation of the board by employing the TDMA frame structure. From the experimental evaluations, we have found that the proposed protocol can work well on the developed Arduino-compatible board, and that the developed board can support low-power operation.

**Acknowledgments** This work is supported in part by the Grant-in-Aid for Scientific Research (C)(26330103), Japan Society for the Promotion of Science (JSPS) and the Integration research for agriculture and interdisciplinary fields, Ministry of Agriculture, Forestry and Fisheries, Japan.

# References

1. Wang, X., Qian, H.: Constructing a 6LoWPAN wireless sensor network based on a cluster tree. IEEE Trans. Veh. Technol. **61**(3), 1398–1405 (2012)
2. Huang, Y.-K., Pang, A.-C., Hsiu, P.-C., Zhuang, W.: Distributed throughput optimization for ZigBee cluster-tree networks. IEEE Trans. Parallel Distrib. Syst. **23**(3), 513–520 (2012)
3. Incel, O.D., Ghosh, A., Krishnamachari, B., Chintalapudi, K.: Fast data collection in tree-based wireless sensor networks. IEEE Trans. Mob. Comput. **11**(1), 86–99 (2012)
4. Delaney, D.T., Higgs, R., O'Hare, G.M.P.: A stable routing framework for tree-based routing structures in WSNs. IEEE Sens. J. **14**(10) (Oct. 2014)
5. Han, Z., Wu, J., Zhang, J., Liu, L., Tian, K.: A General Self-Organized Tree-Based Energy-Balance Routing Protocol for Wireless Sensor Network. IEEE Trans. Nucl. Sci. **61**(2), 732–740 (Apr 2014)
6. Chen, C.-P., Mukhopadhyay, S.C., Chuang, C.-L., Liu, M.-Y., Jiang, J.-A.: Efficient coverage and connectivity preservation with load balance for wireless sensor networks. IEEE Sens. J. **15**(1), 48–62 (Jan 2015)
7. Jurdak, R., Baldi, P., Lopes, C.V.: Adaptive low power listening for wireless sensor networks. IEEE Trans. Mob. Comput. **6**(8), 988–1004 (2007)
8. Guha, S., Basu, P., Chau, C.-K., Gibbens, R.: Green wave sleep scheduling: optimizing latency and throughput in duty cycling wireless networks. IEEE J. Sel. Areas Commun. **29**(8), 1595–1604 (Sep 2011)
9. Shanti, C., Sahoo, A.: DGRAM: a delay guaranteed routing and MAC protocol for wireless sensor networks. IEEE Trans. Mob. Comput. **9**(10), 1407–1423 (2010)
10. Huang, P., Xiao, L., Soltani, S., Mutka, M.W., Ning, X.: The evolution of MAC protocols in wireless sensor networks: a survey. IEEE Commun. Surv. Tutorials **15**(1), 101–120 (2013)
11. Khanafer, M., Guennoun, M., Mouftah, H.T.: A survey of beacon-enabled IEEE 802.15.4 MAC protocols in wireless sensor networks. IEEE Commun. Surv. Tutorials **16**(2), 856–876 (2014)
12. Teo, J.-Y., Ha, Y., Tham, C.-K.: Interference-minimized multipath routing with congestion control in wireless sensor network for high-rate streaming. IEEE Trans. Mob. Comput. **7**(9), 1124–1137 (2008)
13. Yen, H.-H., Lin, C.-L.: Integrated channel assignment and data aggregation routing problem in wireless sensor networks. IET Commun. **3**(5), 784–793 (2009)
14. Ghosh, A., Incel, O.D., Kumar, V.S.A., Krishnamachari, B.: Multichannel scheduling and spanning trees: throughputDelay tradeoff for fast data collection in sensor networks. IEEE/ACM Trans. Networking **19**(6), 1731–1744 (2011)
15. Young, M., Boutaba, R.: Overcoming adversaries in sensor networks: a survey of theoretical models and algorithmic approaches for tolerating malicious interference. IEEE Commun. Surv. Tutorials **13**(4), 617–641 (2011)
16. Chiwewe, T.M., Hancke, G.P.: A distributed topology control technique for low interference and energy efficiency in wireless sensor networks. IEEE rans. Industr. Inf. **8**(1), 11–19 (2012)
17. Naito, K., Ehara, M., Mori, K., Kobayashi, H.: Implementation of field sensor networks with SunSPOT devices. IPSJ The Fifth International Conference on Mobile Computing and Ubiquitous Networking (ICMU 2010) (Apr 2010)
18. Iwasaki, Y., Naito, K., Mori, K., Kobayashi, H.: Implementation of energy saving mechanisms for sensor networks with SunSPOT devices. IPSJ The Sixth International Conference on Mobile Computing and Ubiquitous Networking (ICMU 2012) (May 2012)

# Analysis of Driving Behaviors at Roundabout Intersections by Using Driving Simulator

Naoto Mukai and Misako Hayashi

**Abstract** Japanese road traffic act was revised in November, 2011. In the revised traffic act, the configuration of roundabout intersection and its traffic regulation are defined. The roundabout intersection is already widely used in Europe, but Japanese people are less familiar with it. According to the report of MLIT (Ministry of Land, Infrastructure, Transport and Tourism), there are only about 140 roundabout intersections in Japan, but most of them are narrow and small compared to Europe. Tentative examinations about the effect of the roundabout intersection were carried out in a few cities (e.g., Karuizawa-Cho and Iida-shi). The results of the examination indicate that the intersection environment can be improved by introducing the roundabout intersections. It seems that the roundabout intersection will become more popular in Japan. Therefore, we aim at analyzing the driving behaviors at roundabout intersections to model a virtual driver for traffic simulation. The traffic simulation based on the model will assist novice drivers to avoid accidents in roundabout intersections. In this paper, we construct a virtual driving course based on an existing roundabout intersection at Ichinomiya-shi, Aichi by using a driving simulator "UC-win/Road". Moreover, we analyze the log data such as accelerator and brake pedals from some viewpoints.

**Keywords** Shared space · Roundabout intersection · Virtual driving course · Driving simulator · and UC-win/Road

N. Mukai (✉) · M. Hayashi
Culture-Information Studies, School of Culture-Information Studies,
Sugiyama Jogakuen University, 17-3, Hoshigaoka-motomachi, Chikusa-ku,
Nagoya, Aichi 464-8662, Japan
e-mail: nmukai@sugiyama-u.ac.jp

M. Hayashi
e-mail: hma11da097@st.sugiyama-u.ac.jp

© Springer International Publishing Switzerland 2015
E. Damiani et al. (eds.), *Intelligent Interactive Multimedia Systems and Services*,
Smart Innovation, Systems and Technologies 40,
DOI 10.1007/978-3-319-19830-9_31

347

# 1 Introduction

The concept of "Shared space" is proposed by Hans Monderman, a Dutch traffic engineer. It is a way-out traffic concept which is widely-used in Europe [2, 8, 10]. A city on the concept minimizes the use of traffic lights and traffic signs, and enhances safety awareness of both drivers and pedestrians. The shared space was already introduced as a trial in Kyoto in 2011. The results of the shared space in Kyoto was reported by Miyagawa et al. in the 6th Japanese Conference on Mobility Management (http://www.jcomm.or.jp/). The report indicated that the shared space makes walking easier without the change of amount of traffic. However, the shared space is still unfamiliar for Japanese people. It is important to offer the experience of shared space for a lot of people to introduce into our real-life.

In order to achieve our purpose, we focused on an intersection called by "Roundabout" as a first step. The roundabout is a traffic-lights-free intersection which is frequently used in the shared space. In the roundabout intersection, there is a center circle, and vehicles must drive round the circle in one direction way. The roundabout intersection can be introduced into Japan by the revision of Japanese road traffic act in November, 2011. The report by MLIT(Ministry of Land, Infrastructure, Transport and Tourism) showed that there are only about 140 roundabout intersections in Japan (http://www.mlit.go.jp/road/ir/ir-council/roundabout/). Moreover, most of them are narrow and small compared to other roundabout intersections in Europe. Thus, we still have various problems to start the operation of roundabout intersections for real in Japan. Hasegawa et al. evaluated the comparison between roundabout and signalized intersections [6, 7]. These papers indicated that their proposal method called ADS(Advanced Demand Signal) scheme is always effective, but the roundabout is only effective for off-peak roads. They focused on the effect of roundabout intersections, but do not consider the support for novice drivers. We thought that the support of novice drivers at roundabout intersections is needful to ensure driver's safety.

Thus, we constructed a virtual city based on an actual roundabout intersection at Ichinomiya-shi, Aichi by using "UC-win/Road". UC-win/Road is a driving simulator software developed by Forum8 Co., Ltd. (http://www.forum8.com). The software is widely used for various purposes such as fuel economy [9], electric vehicle [3], and so on [1, 4, 5]. The software can record the detailed driver's log data such as the angle of accelerator and brake pedals. We analyze the driving behaviors based on the log data to model a virtual driver for traffic simulations. Moreover, we developed a navigation function (e.g., go straight, turn right, round a circle, and so on) as a plugin for the software. The plugin displays navigation instructions by text and image on the screen of users. In this paper, our objective is to bring out the characteristic behaviors of drivers at roundabout intersections, but our final goal is to develop a traffic simulator of shared space for training novice drivers.

The outline of this paper is as follows. Section 2 surveys the present condition of roundabout intersections in Japan by reference to the report of MLIT. The virtual driving way we constructed by using UC-win/Road is shown in Sect. 3, and the navigation plugin for the virtual driving way is shown in Sect. 4. Section 5 analyzes the user's driving log data obtained from preliminary experiments using some students in our university. Finally, the summary of this paper is described in Sect. 6.

## 2 Roundabout Intersection

According to the report of MLIT, there are three types of roundabout intersections: "Mini Roundabout", "One-lane Roundabout", and "Multi-lane Roundabout" as shown in Table 1. Most of roundabout intersections in Japan belong to "Mini roundabout" or "One-lane Roundabout". For example, we pick up a roundabout intersection at Ichinomiya-shi, Aichi as shown in Fig. 1 (captured from "Google Earth"). This roundabout intersection is famous as a landmark of Ichinomiya-shi, and a characteristic statue is built in the center circle. The perimeter of the intersection is about 40 m and the number of lanes is one, thus this roundabout intersection belongs to "One-lane Roundabout". In the roundabout intersection, vehicles drive round the circle in a clockwise direction, and vehicles in the circular road have a priority to other incoming vehicles. An advantage of roundabout intersection compared to signalized intersection is to improve the safety in the intersection. In addition, a traffic is not disturbed if a disaster occurs. The Japanese government actively considers introduction of roundabout intersections, thus roundabout intersections will become more popular in the future. However, roundabout intersections are unfamiliar for Japanese people, thus we thought that the experience of driving in virtual driving way is useful for unexperienced drivers in order to reduce traffic accidents.

## 3 Virtual Driving Way at Ichinomiya-Shi, Aichi

We visited the roundabout intersection in Ichinomiya-shi, Aichi and took some pictures of the periphery by digital camera as shown in Fig. 2. The left picture shows a driver view on the approach road, and the right picture also shows a driver view near the center circle. In the left picture, there is a traffic sign for a roundabout intersec-

**Table 1** Types of roundabout intersections

| Type | Perimeter | Speed | Principal use |
|------|-----------|-------|---------------|
| Mini roundabout | 13–27 m | 25–30 km/h | Downtown, residential area, and so on |
| One-lane roundabout | 27–55 m | 30–40 km/h | Downtown, residential area, Highway interchange, and so on |
| Multi-lane roundabout | 46–91 m | 40–50 km/h | Arterial road |

tion, and a speed limit for this road is 40 km/h. We can see a trapezoidal statue on the center circle in the both pictures.

On the basis of the pictures, we developed a virtual driving way by using UC-win/Road. First, we design a road network correspond to the roundabout intersection, and the difference in height and gradient of roads are adjusted according to the geographical data of GSI (Geospatial Information Authority of Japan). Next we define a travel rule in the roundabout intersection. The circular road is one-lane, and vehicles can only turn left. Moreover, we put lines for stop and traffic signs for roundabout intersections on the incoming roads.

Figure 3 is screenshots of a virtual driving way we developed. The upper figure shows an overhead view, and the lower figure shows a driver's view. There are two intersections on the main street which crosses the center of the figure. The left intersection is a signalized intersection, and the right intersection is a roundabout intersection. In the lower figure, you can see the trapezoidal statue and traffic sign of roundabout intersections. Moreover, you can watch the movie of the experience of driving on this virtual driving way on the web (http://cpwc.forum8.co.jp/works_2014/CPWC2014_07-J-0010.html). This movie was created by staff of Forum8 Co., Ltd. which is a developer of UC-win/Road.

## 4 Navigation Plugin

It is important to provide a navigation facility for vehicle drivers for safe driving. Instruction for driving at signalized intersections is already realized by various existing navigation systems. On the other hand, instruction for driving at roundabout intersections is at a stage of development. Thus, we uniquely developed a navigation facility as a plugin for UC-win/Road. Our plugin displays instructions by text and image on the screen of users. Figure 4 shows the examples of instructions: (a) is an instruction for stopping here in front of the circular roads, (b) is an instruction for turning left in the circular roads. These instructions are displayed on the screen if the state of vehicle satisfies the corresponding trigger condition. Each trigger condition

**Fig. 1** Roundabout intersection at Ichinomiya-shi, Aichi

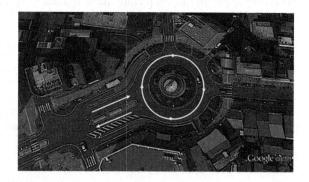

(a) On the Approach Road          (b) Near the Center Circle

**Fig. 2** Pictures of roundabout intersection at Ichinomiya-shi, Aichi

**Fig. 3** Virtual driving way

(a) Overhead View

(b) Driver's View

consists of four parameters: X, Y, DX, and DY. X and Y are the position of vehicle, and DX and DY are the direction of vehicle. This plugin won "Environmental Design and IT Award" in The Second Cloud Programing World Cup (http://vdwc.forum8. co.jp/studentBIM1-cpwc.htm).

## 5  Analysis of User's Driving Log

We performed preliminary experiments to obtain driver's log data by using UC-win/Road. We adopted 9 students in our University as drivers for the virtual driving

**Fig. 4** Instructions for the
roundabout intersection

(a) An Instruction for Stopping Here

(b) An Instruction for Turning Left

way. All of students have their driver's licenses for over a year, but almost students
feel like bad drivers themselves. The trend of the students is summarized in Table 2.
They did not know about roundabout intersections, thus we explain about the driving
rule of roundabout intersections to them in advance. The driving route of them is
shown in Fig. 5. In the route, a driver crosses both a signalized intersection and a
roundabout intersection. Instructions for driving by our plugin are displayed at 8
spots as shown in Table 3. The spot numbers correspond to the numbers in Fig. 5.
Each student drives the course two times: the first time is without the plugin, and the
second time is with the plugin.

**Table 2** Trend of students

| Frequency | Number of Students |
|---|---|
| Everyday | 1 |
| Once a month | 2 |
| Very Little | 6 |

| Confidence | Number of Students |
|---|---|
| Yes | 1 |
| No | 8 |

We focus on three driver's log data: accelerator pedal, brake pedal, and steering.
Here, we pick up one novice driver of the students. Figure 6 shows the ratio of accelerator pedal, brake pedal, and steering while driving. Table 4 shows the statistics
for the log data: "AVG" is average, "SD" is standard deviation, and "RMT" is root
mean square. The value range of accelerator and brake pedals is 0 to 1, and high
value represents pressing down the pedal strongly, and the value range of steering

**Fig. 5** Instruction spots on the virtual driving way

**Table 3** Instruction for driving

| Spot number | Instruction text |
|---|---|
| 1 | Go straight |
| 2 | Stop here |
| 3 | Turn left at third branch |
| 4 | Exit here |
| 5 | Turn left |
| 6 | Turn left |
| 7 | Keep straight |
| 8 | Turn left |

is −1 to 1, the positive value represents steering to the right, and the negative value represents steering to the left. We can find that our plugin can reduce the SD and RMS of ratios of brake and steering. It implies that the driver can put pedal and turn smoothly. On the other hand, the ratio of accelerator pedal only increases. It implies that the concern of drivers can be relieved by the plugin, and this change leads to the reduction of driving time from start to goal.

We can see that some other drivers indicate a similar tendency. According to the feedback after the driving, most of drivers felt that the driving in the roundabout intersection is difficult, but 7 students also felt that the fear of driving can be relieved by this experience. Moreover, all of drivers answered "I'm easy to drive when using the plugin". These results indicate that our virtual driving course can compensate for lack of driving experience at roundabout intersections, and our plugin can support of smooth driving.

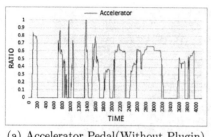

(a) Accelerator Pedal(Without Plugin)

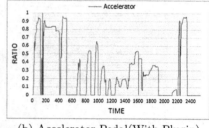

(b) Accelerator Pedal(With Plugin)

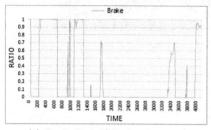

(c) BrakePedal(WithoutPlugin)

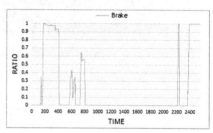

(d) Brake Pedal(With Plugin)

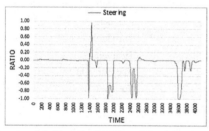

(e) Steering(WithoutPlugin)

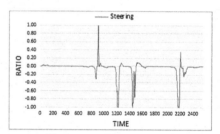

(f) Steering(With Plugin)

**Fig. 6**  Driver's log data

**Table 4**  Statistics for log data

| Log | Without plugin | | | With plugin | | |
|-----|--------|-------|----------|--------|-------|----------|
|     | Accel. | Brake | Steering | Accel. | Brake | Steering |
| AVG | 0.287 | 0.245 | −0.084 | 0.301 | 0.195 | −0.041 |
| SD  | 0.289 | 0.403 | 0.250 | 0.327 | 0.370 | 0.188 |
| RMS | 0.407 | 0.472 | 0.263 | 0.444 | 0.418 | 0.192 |

## 6 Conclusions

In this paper, we focused on a roundabout intersection which is received all of attention in recent years. The revision of road traffic act will support the spread of the roundabout intersection in Japan. However, the awareness of the roundabout intersection is still low for Japanese people. The training and support for driving in the intersection are needed in order to ensure the safety and efficacy, thus we developed a virtual driving course and navigation plugin for UC-win/Road. The virtual driving course is based on an actual roundabout intersection at Ichinomiya-shi, Aichi. We analyzed driver's log data obtained from experiments using the virtual driving course. The result indicated that drivers can drive smoothly by using navigation, and this experience can relieve the fear of driving in roundabout intersections. We achieved some positive results as above, but there are still some problems remaining. One is to model behaviors of novice drivers in roundabout intersections. We think that the model will help drivers to learn driving technique in large-scale traffic simulation. Our final goal is to develop a traffic simulator of shared space for training unexperienced drivers.

**Acknowledgments** This work was supported by Grant-in-Aid for Young Scientists (B).

# References

1. Abdelhameed, W.A.: Micro-simulation function to display textual data in virtual reality. Int. J. Archit. Comput. **10**(2), 205–218 (2012)
2. Elfferding, S.: The legal framework for shared space in germany: characteristics and challenges of sharing public space in towns. IATSS Rev. **35**(2), 112–122 (Aug 2010)
3. Kok, D., Knowles, M., Morris, A.: Building a driving simulator as an electric vehicle hardware development tool. In: Proceedings of Driving Simulation Conference Europe (2012)
4. Lorentzen, T., Fukuda, T., Kobayashi, Y., Ito, Y.: Vr simulation for sustainable transportation planning: public paticipation in sakai city's lrt design. In: Proceedings of ITS AP Bangkok (2009)
5. Lorentzen, T., Kobayashi, Y., Ito, Y.: Virtual reality for consensus building: case studies. Smart Graphics **5531**, 295–298 (2009)
6. Mirokuji, S., Aso, T., Hasegawa, T.: Comparisons criteria on roundabout and signalized intersections. IEICE Tech. Rep. **110**(327), 15–20 (Nov 2010)
7. Mirokuji, S., Aso, T., Hasegawa, T.: Performance comparisons between roundabouts and signalized intersections. IEICE Tech. Rep. **109**(459), 113–118 (Mar 2010)
8. Nishikawa, K., Yamamoto, S.: Research on shared space, part 1: concept of shared space. In: Proceedings of Japanese Society For The Science of Design, vol. 58, p. 66 (2011)
9. Scott, H., Knowles, M., Morris, A., Kok, D.: The role of a driving simulator in driver training to improve fuel economy. In: Proceedings of Driving Simulation Conference Europe (2012)
10. Yamamoto, S., Kiyoshi, N.: Research on shared space, part 2: field study of shared space. In: Proceedings of Japanese Society For The Science of Design, vol. 58, p. 67 (2011)

# Is Experience of Novel Reading Useful to Compose Technical Papers?

Toyohide Watanabe and Koichi Asakura

**Abstract** In this paper, we discuss the effect of novel reading experience on composing technical papers and investigate how the novel reading experiences in which many students have been as their interesting activities should be applicable to the paper composition work. Our goals were to extract the common features between novels and technical papers in comparison with their different viewpoints, and then to find out the creative ability of paper writing from intelligent work in the novel reading process.

**Keywords** Paper writing · Novel reading · Time-dependency · Cause-result relationship · Logical writing · Story structure

## 1 Introduction

The technical papers have to be organized under a predefined writing style so as to make the positions clearly understandable. The ordinary and basic writing styles are based on the 4-notions principle "ki-sho-ten-ketsu": "Introduction", "Development", "Turn" and "Conclusion", which are regarded as the most effective and successful story representation means.

These notions take their own roles in sequence:

1. Introduction: to show initially a topic as a subject with background, motivations and objectives;

T. Watanabe (✉)
Nagoya Industrial Science Research Institute, Nagoya, Japan
e-mail: watanabe@nagoya-u.jp

K. Asakura
Department of Information Systems School of Informatics, Daido University, Nagoya, Japan
e-mail: asakura@daido-it.ac.jp

© Springer International Publishing Switzerland 2015                                    357
E. Damiani et al. (eds.), *Intelligent Interactive Multimedia Systems and Services*,
Smart Innovation, Systems and Technologies 40,
DOI 10.1007/978-3-319-19830-9_32

2. Development: to explain newly-occurring topics in detail with other related topics or under corresponding observable phenomena;
3. Turn: to validate indirectly-related topics with a view to making the features of the subjective topic clear or a view to characterizing the subjective topic for the writing purpose;
4. Conclusion: to evaluate the subjective topic remarkably or to discuss the possibility and/or effects in the future.

At least, the technical papers must be organized under such an outline: of course, many practical papers are specified explanatorily with other description parts. This writing style does not always different from the structure of novels: it is often said that the scenario writer should focus on the story as a fundamental sequence among related events first and then pay attention to the interactions between actors, based on the events [1–3].

In this paper, we discuss how to use the experiences, which have been accumulated for a long period of time as knowledge and/or know-how in our novel reading or novel understanding, when we must write down technical papers at first such as graduation theses and research papers. Also, we analyze the similarities and differences between novels and technical papers, comparatively. Thus, our research point is to investigate how to apply the experiences of novel reading, which were constantly continued since our elementary school age, to the first chance of paper writing effectually, and to find out how to teach the know-how or skill to compose papers on the basis of a lot of novel reading experience. Concerning this research point, the focusing idea is a viewpoint that logical story structure should be followed strictly even if individual events or explanation terms are directly or indirectly discussed.

## 2 Framework of Writing Style in Technical Papers

Figure 1 is a framework of the technical paper configuration in computer engineering and information science. Of course, many variations exist even if the framework in Fig. 1 were used by some researchers in practice so that the papers could be well explained under good writing style. The work proposed in the background and motivation must be in the original goal and planned with the idea in the practical explanation process, first. Of course, in order to make the originality clear in comparison with the other currently-proposed/published researches, the research objectives and research perspectives have to be discussed as a paper positioning means anywhere. Second, the work must explain the detailed procedural methods and global relationships between these methods so that everyone can use the methods in each research application. Third, the work must show the practical effects or the experimental results with respect to an analytical evaluation and perspective verification. Also, it is necessary for the work to make the applicable limitation of the described methods or the possibility of improved method discussible. Finally, the work must be concluded remarkably for the research objectives, and it is better to summarize the expansive issues and their attainability in the next step. In accordance with this framework, we illustrated the main organization of this article in Fig. 2. This article differs a little bit from ordinary papers about method research or system development, which mainly focus on algorithms, experimental evaluations or system prototyping with procedures.

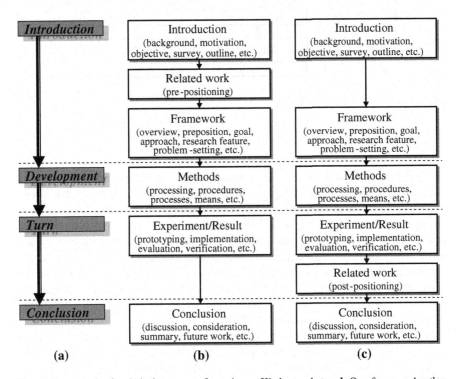

(a)                          (b)                          (c)

**Fig. 1** Framework of technical paper configuration. **a** Ki-sho-ten-ketsu. **b** One frame. **c** Another frame

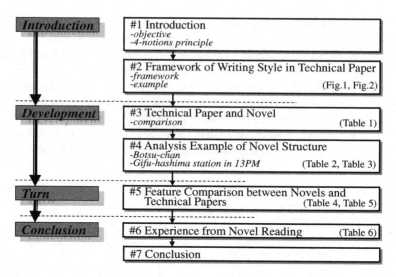

**Fig. 2** Story structure in my paper

## 3 Technical Paper and Novel

The distinction between technical papers and novels may depend on whether or not there are actors as arranged in Table 1. Although various types of novels have been published under categories such as "mystery," "adventure," "history" and so on in all types, actors or pseudo actors (or virtual entities) such as pseudo-humans, androids and so on, take their assigned roles in the story, and their interactions are successively based on time-dependent events, and generate the story dramatically. The novel is defined as a collection of events; and also each event represents the behaviors of individual actors, and the continuous behaviors of actors construct the story with individual scenes. On the other hand, we do not require such actors in the technical papers. Thus, whether or not the actors exist in the descriptions may distinguish novels from technical papers at least: so, the interactions among actors compose individual events. It is important for us to focus on the fact that the interactions could be organized on the basis of the concept of "time." The story in many cases describes the behaviors of the main actors, the final goal of the chief-actor, and/or the situation between co-related actors under a sequence of time-dependent interactions, in which the author intends with his motivation and creativeness. Of course, we can neither observe that the technical papers are represented with the actors nor find out even that the representation is time-dependent.

Namely, in novels, the writing style or story organization is to describe the actions of actors or the interactive relationships among actors along a time axis, while in the technical paper the writing style or paper construction is to specify the logical relationship among individual explanations under the cause-result relationships.

**Table 1** Structural features

|  | Technical papers | Novels |
|---|---|---|
| Writing principle | ki-sho-ten-ketsu (introduction, development, turn, conclusion), IDC (introduction, development, conclusion), IBC (introduction, body, conclusion) | ki-sho-ten-ketsu |
| Descriptive unit | Explanation, concept, etc. | Plot, passage, event |
| Story control | Cause-result relationship among explanations | Time-dependency among events |
| Description entities | Fact, concept, explanation | Actor, action, interaction |
| Writing target | Phenomena/situation explanation | Interaction among actors, situation explanation |
| Writing view | Why, What, How | 5W1H (Who, What, When, Where, Why, How) |

# 4 Analysis Example of Novel Structure

Here, we concretely analyze a novel with respect to its writing structure: "Botchan (坊ちゃん)," written by Soseki Natsume [4]. This novel is comparatively short in length, and is very popular in Japan. Table 2 shows the analyzed schema in a point of story structure. This novel consists of 11 chapters in total, and these chapters are divided into the 4-notions principle as a basic constructive framework for understanding or composing technical papers. The first column is 4-notions, and also the second column shows chapters and contains corresponded chapters owing to the constructive relationship interpreted through co-related events. The third column is the topic for each chapter. In this case, each topic has its own role for the other topics. The fifth column shows the keywords which first occur and can be clue words/phrases to help understand the story flow. Chapters 1–3 are in the "Introduction" because they are the historical/personal characteristics of the main actor Botchan in Chap. 1, and Botchan's understanding of school work as the teacher in Chaps. 2 and 3. Chapters 4–6 describe incidents and their related behaviors under Botchan's understanding in Chaps. 2 and 3. Chapters 7–10 write down successively from the new topic in Chap. 7, which was not addressed in Chaps. 4–6 for "Development", and carry out new scenes with the new topics in Chap. 7. Finally, Chap. 11 concludes the goal for incidents and related events expanded until Chap. 10. Also, this novel combines the content in Chap. 1 with the ending part in Chap. 11. As you can see, it is clear so that the novel of "Botchan" is well structured under the 4-notions principle.

**Table 2**  Analysis of "Botchan"

|            | chap. | topic | chapter role | keywords |
|------------|-------|-------|--------------|----------|
| Introduction | 1 | historical records of Botsu-chan | background | |
| | 2-1 | understanding of night duty | propos-1.1 for 4 | ice-water of 1 sen and 5 rins |
| | 2-2 | ice-water from Yama-arashi | propos-1.2 for 6-1 | |
| | 3 | watch of pupil's daily behaviors | propos-2 for 4 | |
| Development | 4 | insect incident in night duty | issue-1 for 6-2 | spring water in night duty |
| | 5 | Aka-shatsu's advice | issue-2 for 6-1 | Aka-shatsu's trick ↑ |
| | 6-1 | quarrel against Yama-arashi | issue-3 for 9-1 | mental amusement ↑ ¦ |
| | 6-2 | staff-meeting about insect incident | continuation | ¦ ¦ |
| Turn | 7-1 | relationship between Uranari and Madonna | explan-1.1 for 8 | ¦ ¦ |
| | 7-2 | relationship between Aka-shatsu and Madonna | explan-1.2 for 8 | ¦ ¦ |
| | 8 | Uranari's movement | expan-1 for 9-2 | ¦ ¦ |
| | 9-1 | amicable settlement of Yama-arashi | expan-2.1 for 10 | ¦ ¦ |
| | 9-2 | good-bye party of Uranari | expan-2.2 | ¦ ¦ |
| Conclusion | 10 | celebration and quarrel | expan-3 for 11-1 | Aka-shatsu's trap ↓ ↓ |
| | 11-1 | director's disposal for quarrel | con-1.1 for 11-2 con-1.2 for 11-3 | |
| | 11-2 | different disposals for Yama-arashi and Botsu-chan | con-1.3 | |
| | 11-3 | return of Tokyo after retirement | | wait at grave |

In Table 2, some dependent topics are indicated by the chapter role in the fourth column: for example, the relationship between Uranari and Madonna as explanation 1.1 in Sect. 7.1 and the relationship between Aka-shatsu and Madonna as explanation 1.2 in Sect. 7.2 derive together the movement of Uranari as expansion 1 in Chap. 8. This movement of Uranari derives the good-bye party for Uranari as expansion 2.2 in Sect. 9.2. Also, in the column of keywords, arrows are used to indicate the effective range of clue actors/keywords. For example, the keyword "ice-water of 1 sen and 5 rins ("sen" and "rin" are old monetary units that are no longer used in Japan.)" for the topic "ice-water from Yama-arashi" in Sect. 2.2 corresponds to the topic of "amicable settlement of Yama-arashi" in Sect. 9.1. Therefore, Botchan is always unfamiliar to Yama-arashi in the scene from Chap. 3 to Chap. 8.

Moreover, we arrange another genre: the detective novel. The detective novel we used is "Gifu-hashima station at 13 PM (岐阜羽島駅 25 時)," written by Kyotaro Nishimura [5]. The total number of pages is 160 with 2 columns of 17 lines and 24 characters per page. The analysis illustrated in Table 3 differs a little bit from the previous table in Table 2. In Table 3, the fourth column indicates the keywords, like the fifth column in Table 2. The fifth column contains clue events, which are discovered in this topic, as constructs with the logical relationships for the successive chapters. Here, we divided "Turn" into 5 successive sub-Turns in order to change the story flow drastically because these notions are able to manipulate

**Table 3** Analysis of "Gifu-hashima station at 13 PM"

| | Chapter | Topic | Keywords | Clue event |
|---|---|---|---|---|
| Introduction *origin of incident* | 1 | Yasuoka accident | Death of 2 men of property | Dr. Asakura's book |
| | 2 | Tsutsui accident | | |
| | 3 | Dr. Asakura's book | Eternal youth and immortal | Gifu-hashima |
| | 4 | Common properties between Yasuoka accident and Tsutsui accident | | A man of property, young wife, old husband |
| Development *Appearance of Dr. Asakura* | 5 | Gifu-hashima and Nagoya Dr. Asakura's lecture and consultation | | Dr. Asakura's lecture |
| | 6 | | | |
| | 7 | Hospital entering | | |
| Turn-1 *Arrangement of hospital* | 8 | Amano in hospital arrangment of hospital | Request of treatment | Treatment of contribution |
| | 9 | | | |
| | 10 | Operation for Amano | 3 operation rooms | 3 operations |
| Turn-2 *Activity of institute* | 11 | Eternal youth and immortal club | Prohabited research | Reiko's re-visit of institute testament |
| | 12 | Reiko | | |
| Turn-3 *Place of institute* | 13 | Investigation of institute | Sea-side institute | 1 operation room in Hashima grocery gift |
| | 14 | Reiko's re-visit | | |
| Turn-4 *Real circumstance ofplant* | 15 | Investigation of sea-side plant re-investigation of sea-side plant | Research of clone Narita's safety | Invitation from Dr. Asakura emancipation after 1 week |
| | 16 | | | |
| Turn-5 *Narita's disappearance* | 17 | Research praise of Totsukawa information collection until August 4 accident in August 5 fires in Irako cape and Gifu-hashima | Birth of clone | Birth of grandson emancipation of Narita at 13PM |
| | 18 | | | |
| | 19 | | | |
| | 20 | | | |
| Conclusion *End* | 21 | Dr. Asakura's escape investigation of hospital rescue for Abe | | |
| | 22 | | | |
| | 23 | | | |

Note: propos: proposition, explan: explanation, expan: expansion, con: conclusion

independent descriptions. Namely, these notions are successively derived from "Development."

## 5 Feature Comparison Between Novels and Technical Papers

The structure of the novels can be mapped into that of the technical papers without any hard interpretations as analyzed in Sect. 4. This is because even if they are different mutually in their described objectives, they should have their own goals and stories: the technical papers must write down the proposals of authors' insistences as their goals, and then validate the proposal with some practical experiments or logical deductions, after explaining related investigations or using some practical methods, and discuss the proposals and the related-works/products with evaluations. This framework is also applicable to the novel writing process. Of course, the details of the writing styles are different. The most important principle in the technical paper is to construct a logical description on the basis of the facts and verifiable explanations. The principle in the novel is to compose an intentional representation derived from the author's writing purpose [6, 7]. Namely, the story structures between the technical papers and the novels, analyzed in the examples, are arranged in Table 4.

Their different views are observed between the writing features of the novels and those of the technical papers. These features mainly depend on the composition principles as knowledge resources associated inherently with their distinguished utilization materials. However, we cannot conclude that since they are different on many writing points, constructive views, descriptive means, etc., they are independent at all so that the experience which has been grown up in the novel reading does not take any applicable roles in composing technical papers. At least, the description viewpoints between technical papers and novels are more or less different with respect to the author's insistence or specification means. The main differences between novels and technical papers are as follows:

- Whether are there actors or not?: in novels, actors or pseudo actors are necessary, but they do not exist in technical papers;
- Whether are there events or not?: in novels, events are basic description units, but events are not clearly defined in technical papers;
- Whether are individual descriptions specified with time-dependent relationships or not?: in novels, time-dependent relationships among individual events are always specified in a whole ordered form or a semi-ordered one, but the time dependency is not clearly observed among individual explanations in technical papers;
- Whether is author's personality representative or not?: in novels, the author's personality is necessary, but the author's own personality is not to be represented in the description of technical papers;

– Whether is the description creative or fact-based?: in novels, the description may be rather creative, but the description is always done using fact-based explanations or verifiably in practical cases in technical papers.

**Table 4** Composition between technical paper and novels

| | Technical paper | Novels | |
| | My paper | Botsu-chan | Gifu-hashima station in 13PM |
|---|---|---|---|
| Ki (introduction) | 1: Introduction 2: Framework of writing style in technical papers 3: Novel structure | 1: History of Botsu-chan 2-1: Undersatnding of night duty 2-2: Ice-water from Yama-arash 3: Watch of pupil's daily behaviors | 1: Origin of incident |
| Sho (development) | 4: Technical papers and novels 5: Analysis example of novel structure | 4: Insect incident in night duty 5: Advise from Aka-shatsu 6-1: Quarrel against Yama-arashi 6-2: Staff meeting in insect incident | 2: Appearance of Dr. Asakura |
| Ten (turn) | 6: Structural features between novels and technical papers | 7-1: Relationship between Uranari and Madonna 7-2: Relationship between Aka-satsu and Uranari 8: Movement of Uranari 9-1: Amicable settlement of Yama-arashi 9-2: Good-bye party for Uranari | 3: Arrangement of hospital 4: Activity of institute 5: Place of institute 6: Real circumstance of institute 7: Narita's disappearance |
| Ketsu (conclusion) | 7: Experience from novel reading 8: Conclusion | 10: Celebration and quarrel 11-1: Director's disposal for quarrel 11-2: Different disposals between Yama-arashi and Botsu-chan 11-3: Return to Tokyo after retirement | 8: Abe's confession |

**Table 5** Meaningful features

| | Technical papers | Novels |
|---|---|---|
| Emotionality | X | O |
| Redundancy | X | O |
| Logicality | O | Δ (O and/or x) |
| Un-expectedness | X | O |
| Leap | X | O |
| Conversation style | X | O |
| Writing means | Explicit | Explicit or implicit |
| Description point | Cause-result relationship | Time-dependency |

Table 5 arranges the descriptive features from the viewpoint of the representation or the author's motivation for novel composition. These features explicitly distinguish novels from technical papers. These distinguished features discriminate novels and technical papers exclusively as shown in Table 5.

# 6  Experience with Novel Reading

Although the novel is taken from the technical paper in focusing on the description means as shown in Table 6, the structure of novels is not always different from that in technical papers with respect to the story structure. However, we can look upon them as the same view for the story structure on the basis of the 4-notions principle. If we can pay attention to the story structure in novels and read the story contents along with the successive flow of time-dependent events, we can make sure of the successive flow with the cause-result relationships, and this carefully focused experience can lead us to improve our composition ability with technical papers. Table 6 arranges the detailed story structures for two novels, using some detailed chapter concepts in Fig. 1. Concerning Table 6, we can find out strict composition

**Table 6** Story structure

| 4-notions principle | Framework of technical paper | Novels | | | |
|---|---|---|---|---|---|
| | | Botsu-chan | | Gufu-hashima station in 13PM | |
| Introduction | Introduction *background, motivation, objective...* | 1 | Historical records of Botsu- chan | 1 | Yasuoka accident |
| | | | | 2 | Tsutsui accident |
| | Related work | | | | |
| | Framework *approach, goal, Problem-setting,...* | 2 | -Understanding of night duty - ice-water from Yama-arashi | 3 | Dr. Asakura's book |
| | | 3 | Watch of pupil's daily behaviors | 4 | Common properties between Yasuoka accident and Tsutsui accident |

(continued)

**Table 6** (continued)

| 4-notions principle | Framework of technical paper | Novels | | | |
|---|---|---|---|---|---|
| | | Botsu-chan | | Gufu-hashima station in 13PM | |
| Development | Methods *processing, process, means,...* | 4 | Insect incident in night duty | 5 | Gifu-hashima and Nagoya |
| | | 5 | Aka-shatsu's advice | 6 | Dr. Asakura lecture and consultation |
| | | 6 | -Quarrel against Yama-arashi - staff-meeting about insect incident | 7 | Hospital entering |
| Turn | Experiment/Result *prototyping, evaluation, verification,...* | 7 | -Relationship between Uranari and Madonna - relationship between Aka-shatsu and Madonna | 8,9,10 | Arrangement of hospital (...) |
| | | | | 11,12 | Activity of institute (...) [y] |
| | | 8 | Uranari's movement | | |
| | | 9 | -Amicable settlement of Yama- arashi - good-bye party of Uranari | 13,14 | Place of institute (...) |
| | | | | 15,16 | Real circumstance of institute (...) |
| | | | | 17,18, 19,20 | Narita's dis-appearance (...) |
| Conclusion | Conclusion *discussion, consideration, summary, future-work,...* | 10 | Celebration and quarrel | 21 | Dr. Asakura's escape |
| | | 11 | -Director's disposal for quarrel - different disposals for Yama- arashi and Botsu-chan -return of Tokyo after retirement | 22 | Investigation of hospital |
| | | | | 23 | Rescue for Abe |

procedures with respect to the relationships between clue words/events in the chapters/sections. If we pay careful attention to the cause-result relationships along with individual clue words/events in each chapter/section, it is not always difficult for us to make use of our experience in reading novels or in writing technical papers in regard to structure design and story specification. Thus, the experience which has been acquired naturally since our first reading novel year by year is effective and successive as know-how or skill for the composition process of technical papers.

# 7 Conclusion

In this paper, we first described the story structure of technical papers, analyzed the story structures of novels, and compared the analyzed results with those in technical papers. Next, we addressed how to apply experience with novel reading to the composition work of technical papers. We used a lot of time in reading many novels since elementary school or prep school. However, such long-term experience is not always used or cannot be successfully utilized as know-how or skills in composing technical papers against the volume of exhausted periods or endeavors when we must write down our graduation theses at first. Thus, it is necessary to read the novels analytically in the story structure so as to make effective use of the accumulated experience in the logical writing of technical papers.

# References

1. Chi, P.-Y., Lieberman, L.: Raconteur: from Intent to Stories. In: 15th International Conference on Intelligent User Interfaces, pp. 301–304 (2010)
2. Si, M., Marsella, S. C., Riedl, M. O.: interactive drama authoring with plot and chapter: An Intelligent System that Fosters Creativity. In: AAAI 2008 Spring Symposium on Creative Intelligent Systems (2008)
3. Watanabe, T., Arasawa, R.: Computer-supported novel composition, based on externalization. In: 18th Annual Conference on Knowledge-based and Intelligent Information & Engineering Systems, KES2014, pp. 1662–1671 (2014)
4. Natsume, S.: Botchan. Chikuma-shobo, p. 148 [in Japanese]
5. Nishimura, K.: Gifu-hashima Station at 13 PM, Shincho-sha, p. 166 [in Japanese]
6. Zhang, J.: The nature of external representation in problem solving. Cogn. Sci. 21(2), 179–217 (1997)
7. Shirouzu, H., Miyake, N., Masukawa, H.: Cognitively active externalization for situated reflection. Cogn. Sci. 26(4), 469–501 (2002)

# A Movement Algorithm for Evacuee Agents in Disaster Simulators: Towards the Development of Evacuation Guidance Systems Based on Ant Colony Systems Using MANET

Koichi Asakura and Toyohide Watanabe

**Abstract** Our research goal is to develop evacuation guidance map systems for disaster situations such as major earthquakes. Our guidance system provides a safe-road map in disaster areas that is constructed based on ant colony systems. We describe a movement algorithm for evacuee agents utilizing an evacuation guidance system in a disaster simulator. In the simulators, evacuee agents move to safe shelters in accordance with information on the guidance system. We evaluated the effectiveness of the evacuation guidance systems using the movement algorithm. Experimental results show that the evacuee agents utilizing the guidance system can reach the shelters faster.

**Keywords** Mobile agents · Geographic information systems · Mobile ad-hoc network · Disaster simulators

## 1 Introduction

Researchers have paid attention to mobile ad-hoc network (MANET) technologies for communication systems handling emergency situations such as major earthquakes [1–4]. In such situations, the communication infrastructure, such as the mobile phone network, may break down or malfunction, thereby making them useless for gathering information. In a MANET, mobile devices can construct a communication network and share information with no communication infrastructures

K. Asakura (✉)
Department of Information Systems, School of Informatics, Daido University, 10-3,
Takiharu-cho, Minami-ku, 457-8530 Nagoya, Japan
e-mail: asakura@daido-it.ac.jp

T. Watanabe
Nagoya Industrial Science Research Institute, 1-13, Yotsuya-dori, Chikusa-ku,
464-0819 Nagoya, Japan
e-mail: watanabe@nagoya-u.jp

© Springer International Publishing Switzerland 2015                                369
E. Damiani et al. (eds.), *Intelligent Interactive Multimedia Systems and Services*,
Smart Innovation, Systems and Technologies 40,
DOI 10.1007/978-3-319-19830-9_33

[5, 6]. In other words, the communication network can be maintained autonomously by only mobile devices.

Our research goal is to develop a map information system for evacuation guidance in disaster situations. In disaster areas, information on safe roads is very important for evacuees. However, evacuees have a very difficult time trying to acquire such information in real time. This is because disaster situations can change drastically, and roads may become impassable. In order to solve this problem, we have already proposed a map construction system based on ant colony systems [7, 8]. In this system, information on safe roads is acquired by evacuees with mobile devices. This information is collected by considering the recentness of information based on ant colony systems. The system constructs up-to-the-minute safe-road maps in disaster areas.

In order to evaluate effectiveness of the evacuation guidance system, this paper proposes a movement algorithm for evacuee agents in disaster simulators. This movement algorithm enables evacuee agents to move to shelters in accordance with information on safe roads, and this information is provided by the guidance system. The effectiveness of the evacuation guidance system can be evaluated with disaster simulators.

The rest of this paper is organized as follows. Section 2 describes an evacuation guidance system for disaster areas. Section 3 describes a movement algorithm of evacuee agents utilizing the guidance system. Section 4 details our simulation experiments. Finally, Sect. 5 concludes this paper and explains our future work.

## 2 Evacuation Guidance System

This section describes the evacuation guidance system. The system provides evacuees with a safe-road map enabling rapid evacuation to shelters. It is based on ant colony systems in order to take the recentness of information into account. First, we describe ant colony systems in Sect. 2.1. Next, we describe how to apply ant colony systems to the evacuation guidance system in Sect. 2.2.

### 2.1 Ant Colony Systems

In order to find food, ants move around their colony. While moving, they lay pheromones down on the ground. When other ants find trails of pheromones, they follow the path and drop more pheromones. This behavior enables ants to generate paths from their colony to the food. Because pheromones are a volatile liquid, they evaporates over time. Thus, more pheromones are left on the shorter paths, enabling ants to find the shortest path.

The ant colony system is a computational method proposed for solving combinatorial optimization problems [9–12]. It is a distributed algorithm in which several agents cooperate to find good solutions. In the ant colony system, agents take cooperative action based on the aforementioned pheromones.

## 2.2 Construction of Safe-Road Maps

We have already proposed a method for constructing safe-road maps by using the concept of pheromones in ant colony systems [7, 8]. This method represents the recentness and correctness of information on safe roads by pheromones. On the basis of the concept of pheromones in ant colony systems, roads that are taken by many evacuees are considered to have more pheromones. Also, roads that have been recently used for evacuation are considered to have more pheromones. Thus, by measuring the quantity of these pheromones, we can select safer roads to shelters for evacuation.

Figure 1 shows a comparative example of the constructed safe road maps with and without ant colony systems. In these figures, the shelter is denoted as a rectangle, roads taken by evacuees are represented by thick lines, and an evacuee is denoted as a bullet. Figure 1(a) shows information on the road network without any colony systems. In this map, all taken roads are treated equivalently as safe roads for evacuation. Figure 1(b) shows a safe-road map based on ant colony systems. In this figure, the taken roads are shown in grayscale ramps. This grayscale shows the quantity of pheromones, namely the reliability of the information. Here, we consider that road segment $r_1$ becomes impassable while the evacuee moves to the shelter. In Fig. 1(a) without ant colony systems, the evacuee may take road $r_1$ because the evacuee tends to use the shortest path to the shelter. Thus, the evacuee cannot reach the shelter and has to walk back to the path. However, in Fig. 1(b) with ant colony systems, the pheromone on road $r_1$ decreases in comparison with the surrounding roads. This shows that road $r_1$ has not been taken recently for some reason. Thus, the evacuee may use road $r_2$ to get to the shelter safely. This example demonstrates that more reliable safe-road maps can be constructed using ant colony systems.

## 3 Movement Algorithm for Evacuee Agents

As described in the previous section, an evacuation guidance system with ant colony systems provides safe-road maps to evacuees. In order to demonstrate the effectiveness of the system, we have to pay attention to the behavior of evacuees. Namely, we have to clarify the movement of evacuees with the evacuation guidance system. In this section, we describe a movement algorithm for evacuee agents utilizing the guidance system in disaster simulators. In this paper, we focus on evacuees who know the locations of shelters and the shortest path from the current position to the shelters. We denote such evacuees as local evacuees in this paper.

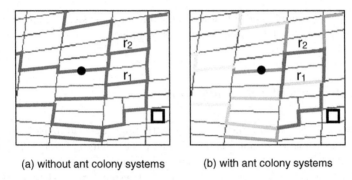

(a) without ant colony systems          (b) with ant colony systems

**Fig. 1**  Comparative example of safe-road maps

## 3.1 Movement of Evacuees

In general, local evacuees take the shortest path to shelters during evacuation. If a road along the shortest path is impassable, they find an alternative path that does not include the impassable road. This process is repeated until they reach the shelter.

If local evacuees can use safe-road maps provided by the evacuation guidance system, they will likely calculate the path that consists almost entirely of roads with a large quantity of pheromones and not just use the shortest path. Thus, the movement of local evacuees utilizing the guidance system can be summarized as follows:

– If a local evacuee is located on a road with pheromones, the evacuee constructs a path that consists of roads with pheromones.
– If a local evacuee is not located on a road with pheromones, the evacuee takes the shortest path from the current location to the nearest road with pheromones.

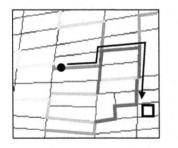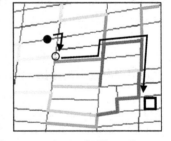

(a) agent on the road with pheromones    (b) agent on the road without pheromones

**Fig. 2**  Moving trajectories of evacuee agents

Figure 2 shows examples for movement of local evacuees. Figure 2(a) shows the moving trajectory of the agent on the road segment with pheromones. The agent can move to the shelter smoothly by tracking the pheromones. Figure 2(b) shows the moving trajectory of the agent who is not on the road segment with pheromones. The agent first moves to the nearest road segment with pheromones, denoted as the circle in the figure. Then, the agent can track the pheromones to the shelter.

## 3.2 Movement Algorithm

Figure 3 shows a movement algorithm for evacuee agents called *AgentMovement*. In this algorithm, a destination and a path to the destination are calculated first by the algorithm *PathConstruction* described later. Then, agents move to the destination using the path. Once the agents reach the destination or once they reach an impassable road, a new path is calculated. This process is continued until the agents reach the shelter.

Figure 4 shows the algorithm *PathConstruction*. In this algorithm, information on safe-road maps is denoted as a set of tuples:

$$SafeRoads = \{< r_i, p_i >, \cdots, < r_n, p_n >\}, \tag{1}$$

where $r_i$ denotes a road and $p_i (0 \le p_i \le 1)$ denotes the quantity of pheromones for $r_i$. If the agent is located on a road in *SafeRoads*, the destination is set for the shelter, and the shortest path to the shelter is constructed from *SafeRoads*. Otherwise, the destination is set for the nearest road in *SafeRoads*, and the shortest path to the nearest road is constructed. The threshold value $\alpha$ is used for selecting roads in *SafeRoads*. This value represents the robustness of constructed path. Evacuation paths constructed with a high number of $\alpha$ are considered as robust paths because the algorithm do not select roads with a low quantity of pheromones.

```
ALGORITHM AgentMovement
begin
  loop
     path ← PathConstruction().
     while path ≠ empty
        next ← pop the first entry of path.
        if the agent cannot reach next because of impassable road then
           break
        endif
        move to next.
     endwhile
  until the agent reaches the shelter
end
```

**Fig. 3** Movement algorithm for evacuee agents

```
ALGORITHM PathConstruction
Input
    SafeRoads: a set of the tuple of roads and pheromones <rᵢ, pᵢ>.
    pos: the current position of the agent.
    α: threshold value.
Output
    path: a sequence of intersections to the destination.
begin
    UsableRoads ← { rᵢ | <ri, pi> ∈ SafeRoads ∧ pᵢ > α }.
    if pos is on the road segment in UsableRoads then
        path ← the shortest path to the shelter in UsableRoads.
    else
        destination ← the nearest road segment in UsableRoads.
        path ← the shortest path to destination.
    endif
end
```

**Fig. 4** Path construction algorithm

# 4 Experiments

In order to demonstrate the effectiveness of the evacuation guidance system, we conducted simulation experiments using evacuee agents with our movement algorithm.

## 4.1 Simulation Scenario

The experiments featured a virtual disaster area 2.0 km wide and 1.5 km high. Figure 5 shows the simulation area. One shelter was placed in the area, as shown in the figure. In order to evaluate the effectiveness of the guidance system, we provided the following three scenarios for evacuees:

1. 500 evacuee agents that did not utilize the guidance systems were deployed.
2. 500 evacuee agents that utilized the guidance systems were deployed. The threshold value $\alpha$ for this group was set to 0.9. This means that the guidance system constructs evacuation paths utilizing road segments with a high quantity of pheromones.
3. 500 evacuee agents that utilized the guidance systems were deployed. The threshold value $\alpha$ was set to 0.6 for these agents.

Furthermore, in order to evaluate robustness of the guidance system's ability to respond to changes in the situation, we generated two scenarios for the disaster areas:

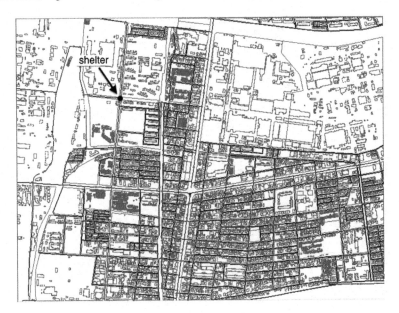

**Fig. 5** Simulation area

**Large-scale disaster:** 5 % of road segments in the area were impassable at the beginning. Also, 1 % of road segments became impassable every 30 min during the simulation.

**Small-scale disaster:** 3 % of road segments were impassable at the beginning, and 0.5 % of road segments became impassable every 30 min.

In both scenarios, evacuees were randomly deployed on roads, and 0.2 persons/minute appeared in accordance with a Poisson distribution. Impassable roads were also chosen randomly. In the experiments, we observed the average and maximum evacuation time to the shelter.

Figure 6 shows an example of a safe-road map constructed by the evacuation guidance system. We observed that an alternative long route was constructed because of impassable roads.

## 4.2 Experimental Results

Table 1 shows the experimental results in a small-scale disaster situation. From this table, we found that the average time for each scenario was almost the same. Thus, most evacuee agents could move to the shelter using the shortest path in the small-scale disaster scenario because the number of impassable roads is not very high. However, differences in the maximum evacuation time were evident. The evacuee agents that did not utilize the guidance systems needed a lot of time for the evac-

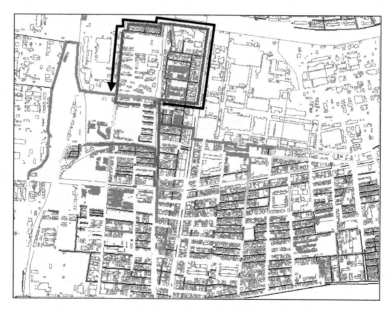

**Fig. 6** Example of alternative route in evacuation guidance system

**Table 1** Elapsed time for evacuation in small-scale disaster scenario

| | No guidance | With guidance | |
| | | $(\alpha = 0.9)$ | $(\alpha = 0.6)$ |
|---|---|---|---|
| Average time | 247.2 | 246.5 | 250.4 |
| Maximum time | 783.6 | 493.5 | 580.0 |

[sec]

uation. Such agents could not acquire information on impassable roads during the evacuation, which causes their evacuation time to become longer than that in other scenarios.

Table 2 shows the experimental results in a large-scale disaster situation. This scenario had obvious differences in evacuation time. Evacuee agents that did not utilize the guidance system had far worse average time and maximum time. Furthermore, the evacuation time of agents utilizing the guidance systems also differed in accordance with the threshold value $\alpha$; agents with $\alpha = 0.6$ got to the shelter faster than agents with $\alpha = 0.9$. In the large-scale disaster scenario where the condition of the road segments changed drastically, evacuees who were too picky in choosing road segments could not reach the shelter fast.

These experimental results demonstrated that the evacuation guidance system is very effective for shortening evacuation time, and that the appropriate value of the threshold depends on the changing conditions of the disaster areas.

**Table 2** Elapsed time for evacuation in large-scale disaster scenario

[sec]

| | No guidance | With guidance | |
|---|---|---|---|
| | | ($\alpha = 0.9$) | ($\alpha = 0.6$) |
| Average time | 343.9 | 314.8 | 269.2 |
| Maximum time | 1270.2 | 703.6 | 570.6 |

# 5 Conclusion

This paper presented a movement algorithm for evacuee agents with evacuation guidance systems. This algorithm enabled us to evaluate the effectiveness of the guidance system in a disaster simulator. The experimental results demonstrated that the evacuation guidance system can shorten evacuation time.

For future work, we will design a system architecture for the evacuation guidance system. Finally, we will implement the evacuation guidance system with MANET technologies. By using our algorithm, we have to consider the system architecture in detail.

**Acknowledgments** The authors thank Mr. Yuki Nakagawa for his implementation work for the simulation systems when he was a student of Daido University.

# References

1. Midkiff, S.F., Bostian, C.W.: Rapidly-deployable broadband wireless networks for disaster and emergency response. In: 1st IEEE Workshop on Disaster Recovery Networks (2002)
2. Meissner, A., Luckenbach, T., Risse, T., Kirste, T., Kirchner, H.: Design challenges for an integrated disaster management communication and information system. In: 1st IEEE Workshop on Disaster Recovery Networks (2002)
3. Pomportes, S., Tomasik, J., V'eque, V.: Ad hoc network in a disaster area: a composite mobility model and its evaluation. In: International Conference on Advanced Technologies for Communications, pp. 17–22 (2010)
4. Raffelsberger, C., Hellwagner, H.: Evaluation of MANET routing protocols in a realistic emergency response scenario. In: 10th International Workshop on Intelligent Solutions in Embedded Systems, pp. 88–92 (2012)
5. Toh, C.K.: Ad Hoc Mobile Wireless Networks: Protocols and Systems. Prentice Hall, New Jersey (2001)
6. Murthy, C., Manoj, B.: Ad Hoc wireless networks: architectures and protocols. Prentice Hall Communications Engineering and Emerging Technologies Series. Prentice Hall PTR, New Jersey (2004)
7. Asakura, K., Fukaya, K., Watanabe, T.: A map construction system for cisaster areas based on ant colony systems. In: 17th International Conference on Knowledge-based and Intelligent Information and Engineering Systems, KES2013 (2013)
8. Asakura, K., Watanabe, T.: Construction of navigational maps for evacuees in disaster areas dased on ant colony systems. Int. J. Knowl. Web Intell **4**(4), 300–313 (2013)

9. Dorigo, M., Maniezzo, V., Colorni, A.: Optimization by a colony of cooperating Agents. IEEE Trans. Syst. Man Cybern Part BC ybern **26**(1), 29–41 (1996)
10. Dorigo, M., Gambardella, L.M.: Ant colonies for the traveling salesman problem. BioSystems. **43**(2), 73–81 (1997)
11. Dorigo, M., Gambardella, L.M.: Ant colony system: a cooperative learning approach to the araveling salesman problem. IEEE Trans. Evol. Compu **1**(1), 53–56 (1997)
12. Dorigo, M.: Stützle, T: Ant Colony Optimization. Bradford Company, Scituate (2004)

# Author Index

© Springer International Publishing Switzerland 2015                              379
E. Damiani et al. (eds.), *Intelligent Interactive Multimedia Systems and Services*,
Smart Innovation, Systems and Technologies 40,
DOI 10.1007/978-3-319-19830-9

Printed in the United States
By Bookmasters